Advances in Petroleum Science and Technology

Advances in Petroleum Science and Technology

Editor: Natalie Mitchell

R CALLISTO
REFERENCE
www.callistoreference.com

Callisto Reference,
118-35 Queens Blvd., Suite 400,
Forest Hills, NY 11375, USA

Visit us on the World Wide Web at:
www.callistoreference.com

ISBN: 978-1-64116-352-1 (Hardback)

Cataloging-in-Publication Data

Advances in petroleum science and technology / edited by Natalie Mitchell.
    p. cm.
Includes bibliographical references and index.
ISBN 978-1-64116-352-1
1. Petroleum. 2. Petroleum engineering. 3. Petroleum--Refining. 4. Petroleum products. I. Mitchell, Natalie.
TN870 .A28 2020
665.5--dc23

Table of Contents

Preface

Petroleum is a substance that occurs naturally in a yellowish-black form. It is found in a liquid form in the geological formations underneath the Earth's surface. It includes all solid, liquid and gaseous hydrocarbons as well as crude oil. Under surface temperature and pressure, pentane and other heavier hydrocarbons exist in solid and liquid forms. Under the same conditions the lighter hydrocarbons such as methane, propane, ethane and butane exist in gaseous forms. Petroleum is a fossil fuel that is obtained from the ancient fossilized organic materials such as zooplankton and algae. Petroleum formation is the result of the hydrocarbon pyrolysis at high temperature and pressure in a number of endothermic reactions. Alkenes, asphalt, lubricants, aromatic petrochemicals and paraffin wax are a few of the useful products derived from the mixture of hydrocarbon and non-hydrocarbons. The excessive use of petroleum as a fuel is a threat to the environment. It is a major cause of global warming and ocean acidification. This book contains some path-breaking studies in the field of petroleum science and technology. Most of the topics introduced herein cover new techniques and the applications of this field. It will help the readers in keeping pace with the rapid changes in this area.

After months of intensive research and writing, this book is the end result of all who devoted their time and efforts in the initiation and progress of this book. It will surely be a source of reference in enhancing the required knowledge of the new developments in the area. During the course of developing this book, certain measures such as accuracy, authenticity and research focused analytical studies were given preference in order to produce a comprehensive book in the area of study.

This book would not have been possible without the efforts of the authors and the publisher. I extend my sincere thanks to them. Secondly, I express my gratitude to my family and well-wishers. And most importantly, I thank my students for constantly expressing their willingness and curiosity in enhancing their knowledge in the field, which encourages me to take up further research projects for the advancement of the area.

Editor

A Numerical Simulation Study on Wellbore Temperature Field of Water Injection in Highly Deviated Wells

Liu Jun[1, 2]*, Wang Wei[1], Zhang Jianzhong[3], and Xiao Guohua[3]

[1]School of Civil Engineering and Architecture, Southwest Petroleum University, Chengdu 610500, China
[2] Modern Design and Simulation Lab for Oil and Gas Equipment, Southwest Petroleum University, Chengdu 610500, China
[3]Drilling Technology Research Institute of Jidong Oil Field Company, Hebei, 063000, China

ABSTRACT

According to the temperature distribution of water injection well-bore in highly deviated wells under different conditions and unstable temperature field heat conduction principles, a true three-dimensional model was established to analyze the law of variation on temperature of highly deviated wells during the water injection process, and to analyze the factors that influence the water injection well-bore in highly deviated wells; this can improve the computational accuracy of the transformation of water injection string made by the temperature, and help us predicate the transformation of water injection string more accurately during the production process. The results show that the temperature distribution of highly deviated wells are greatly affected by water injection inlet temperature or injection of water inlet temperature, water injection rate, and water injection time. With the temperature field model of highly deviated wells made in this paper, the construction quality and oil production can be highly improved by reasonably controlling the temperature. All the findings which have been obtained form a theoretical basis and the experimental research for the development of mechanical analysis simulation software have been used to calculate temperature distribution and variation of water injection string in highly-deviated wells and have a great theoretical significance and practical value.

Keywords: Highly Deviated Well, Water Injection, Temperature Profile, Numerical Calculation, Experimental Research

INTRODUCTION

Currently, oil exploration and development in China are facing high pressure and high temperature environment, and simultaneously many development blocks are far from artificial islands and Jacket well location, so tilt angle and displacement are growing in wells.

This type of oil is buried deeply, and well tilt angle and wellbore temperature change a lot. When string movement is restricted, a large temperature difference will result in stress and the deformation of large string changes, and then the safety of pipe string is affected [1]. Meanwhile, the formation fracture

*Corresponding author
Liu Jun
Email: 99803392@qq.com

pressure of reservoir can be easily affected by temperature [2], and the changing of wellbore temperature can easily lead to wellbore instability [3, 4]. Besides, the continuous injection of cold water into the formation makes temperature drop around bottom formation, and then produces the oil within the reservoir; thus the water viscosity changes, which will affect the oil displacement effect of water [5, 6]. Therefore, it is very significant that wellbore temperature field of water injection in highly deviated wells is accurately calculated. The true three-dimensional model is established to analyze the highly deviated well water injection process temperature variation and the influential factors of wellbore temperature field by unstable temperature field heat conduction principles herein so that the value of accurate temperature changes of the water injection in different working conditions could be calculated and the influential factors of wellbore temperature distribution could be analyzed. This research also provides valid data for the further analyses of temperature effect on the mechanical behavior caused by the temperature of the water injection string. Moreover, this research serves as a theoretical guidance for rational regulation of wellbore temperature.

THEORETICAL MODEL
Fundamental Assumptions

According to the actual situation of highly deviated wells injection wellbore temperature field, the following assumptions are put forward based on the numerical simulation models:

- Wellbore continuously and rigidly contact with pipe string, and the rigidity of pipe string is considered.
- The axis of pipe string is in accordance with the borehole axis.

- The hole trajectory of highly deviated wells is described by the Frenet formula [7].
- The computational element of the pipe string is a circular arc in the space on the inclined plane.
- Before water injection, the thermals status of the pipe string, fluid in the wellbore, and the ground are equal and in equilibrium situation. Simultaneously, the initial temperature distribution obeys a stable geothermal gradient change.
- After water injection, heat transfer is considered in the direction of wellbore radius and along the vertical depth; besides, during water injection, putting the original fluid in the wellbore extrusion of mass transfer causes heat conduction to be taken into account.
- When the wellbore temperature fields are discretized in the spatial domain, the properties of matters in each micro body domain such as heat transfer coefficient and specific heat are relatively stable.
- Water injection rate, water injection temperature, all the year round average temperature under the ground, and geothermal gradient are constant.
- Away from the wellbore axis enough in the distance of $r \geq R_{max}$, formation temperature is original ground temperature.
- Under the surface of the earth $Z=Z_m$, temperature T_m does not change over time; under the surface of the earth $Z>Z_m$, original formation temperature satisfies the linear relationship of $T_{dz}=T_m+\alpha z$, where α is the formation temperature gradient.
- By considering a variety of conditions in oil production operations, this important tip is found out that a process of injecting liquid from string and liquid flowing out from the wellbore annulus must be considered for the next process when the wellbore temperature field is calculated.

Establish Model

On the basis of the above assumptions, in order to the first law of thermodynamics, and the basic principles of heat transfer, the liquid in the tubing string, the string wall, liquid in the annulus, and stratigraphic unit were taken as the control volume in order to derive these control bodies energy balance equation and the mathematical model of temperature field. As shown in Figure 1, any infinitesimal bodies were taken as the control volume in the wellbore-stratigraphic system, and it was assumed that the radial distance is dr, and the borehole distance along the axial direction is dz.

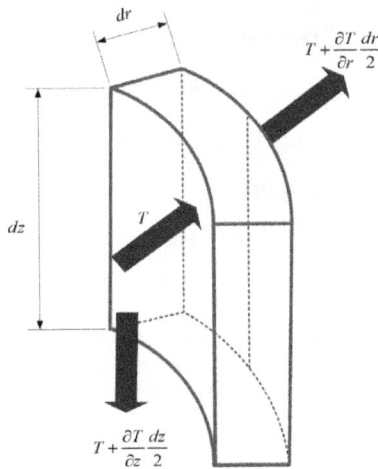

Figure 1: A schematic diagram of micro element body.

The micro unit center temperature is set as T, which, by Fourier heat conduction law, can be learned; at the place of z-dz/2 the temperature is $T-(\partial T/\partial z)(dz/2)$; at the place of r+dz/2 the temperature is $T+(\partial T/\partial z)(dz/2)$; at the place of r-dz/2 the temperature is $T-(\partial T/\partial r)(dz/2)$; at the place of r+dr/2, the temperature is $T+(\partial T/\partial r)(dz/2)$. For example, taking a unit of the pipe string is to establish a thermodynamic equilibrium equation. The thickness of a micro unit in the column is dz, and the radial distance is equal to the string

inside radius r_{ti}. According to the assumptions, if the changes in fluid potential and fluid density were not considered, the heat of the micro element body carried over by the injected fluid from the top of wellbore per unit time may be written as follows:

$$Q_1 = c_f \rho_f \pi r_{ti}^2 v_f \left(T_f - \frac{\partial T_f}{\partial z} \frac{dz}{2} \right) \quad (1)$$

The heat of the micro element body carried over by the injected fluid from the bottom of wellbore per unit time may be written as given by:

$$Q_2 = c_f \rho_f \pi r_{ti}^2 v_f \left(T_f + \frac{\partial T_f}{\partial z} \frac{dz}{2} \right) \quad (2)$$

Respecting the convective heat transfer between fluid and pipe wall, and according to the Newton's law of convective heat transfer formula, the heat of the micro element body carried over from pipe wall per unit time can be given by

$$Q_3 = \alpha_{tf} \cdot 2\pi r_{ti} dz \left(T_t - \frac{\partial T_t}{\partial r} \frac{r_{to} - r_{ti}}{2} - T_f - \frac{\partial T_f}{\partial r} \frac{r_{ti}}{2} \right) \quad (3)$$

The heat of the micro element body carried over by friction per unit time reads:

$$Q_4 = \frac{1}{4} \pi \lambda_f \rho_f dz r_{ti} v_f^{\,3} \quad (4)$$

$$Q_5 = c_f \rho_f \pi r_{ti}^2 dz \frac{\partial T_f}{\partial t} \quad (5)$$

Heat energy variation per unit time in a micro element body can be defined as follows:

$$Q_1 - Q_2 + Q_3 + Q_4 = Q_5 \quad (6)$$

Using the heat balance principles, Equation 6 can be rewritten as follows:

$$c_f \rho_f \pi r_{ti}^2 v \left(T_f - \frac{\partial T_f}{\partial z} \frac{dz}{2} \right) -$$

$$c_f \rho_f \pi r_{ti}^2 v_f \left(T_f + \frac{\partial T_f}{\partial z} \frac{dz}{2} \right) + \alpha_{tf} \cdot 2\pi r_{ti} dz$$

$$\cdot \left(T_t - \frac{\partial T_t}{\partial r} \frac{r_{to} - r_{ti}}{2} - T_f - \frac{\partial T_f}{\partial r} \frac{r_{ti}}{2} \right)$$

$$+ \frac{1}{4} \pi \lambda_f \rho_f dz r_{ti} v^3 = c_f \rho_f \pi r_{ti}^2 dz \frac{\partial T_f}{\partial t} \quad (7)$$

The thermal equilibrium given in Equation 7 can be transformed into the following form:

$$-c_f \rho_f r_{ti} v_f \frac{\partial T_f}{\partial z} + \frac{2}{r_{ti}} \alpha_{tf} (T_t - T_f) + \frac{1}{4 r_{ti}} \lambda_f \rho_f v_f^3 = c_f \rho_f \frac{\partial T_f}{\partial t} \tag{8}$$

On the basis of the same principle, the heat balance equation of the other units can be established; the true three dimensional mathematical model of temperature field for wellbore-stratigraphic system can be defined as given below.

$$\begin{cases} -c_f \rho_f r_{ti} v_f \frac{\partial T_f}{\partial z} + \frac{2}{r_{ti}} \alpha_{tf} (T_t - T_f) + \frac{1}{4 r_{ti}} \lambda_f \rho_f v_f^3 = c_f \rho_f \frac{\partial T_f}{\partial t} \\ k_t \frac{\partial^2 T_t}{\partial z^2} - 2\alpha_{tf} \frac{r_{ti}}{r_{to}^2 - r_{ti}^2}(T_t - T_f) + 4 k_t \frac{r_{to}}{(r_{to}^2 - r_{ti}^2)(r_{to} - r_{ti})}(T_a - T_t) = c_t \rho_t \frac{\partial T_t}{\partial t} \\ k_a \frac{\partial^2 T_a}{\partial z^2} - 4 k_t \frac{r_{ci}}{(r_{ci}^2 - r_{to}^2)(r_{to} - r_{ti})}(T_a - T_t) + 4 k_a \frac{r_{ci}}{(r_{ci}^2 - r_{to}^2)(r_{ci} - r_{to})}(T_c - T_a) = c_a \rho_a \frac{\partial T_a}{\partial t} \\ k_c \frac{\partial^2 T_c}{\partial z^2} - 4 k_a \frac{r_{ci}}{(r_{co}^2 - r_{ci}^2)(r_{ci} - r_{to})}(T_c - T_a) + 4 k_c \frac{r_{co}}{(r_{co}^2 - r_{ci}^2)(r_{co} - r_{ci})}(T_e - T_c) = c_c \rho_c \frac{\partial T_c}{\partial t} \\ k_e \frac{\partial^2 T_e}{\partial z^2} - 4 k_c \frac{r_{co}}{(r_{ce}^2 - r_{co}^2)(r_{co} - r_{ci})}(T_e - T_c) + 4 k_e \frac{r_{ce}}{(r_{ce}^2 - r_{co}^2)(r_{ce} - r_{co})}(T_w - T_e) = c_e \rho_e \frac{\partial T_e}{\partial t} \\ \frac{\partial^2 T_w}{\partial z^2} + \frac{\partial^2 T_w}{\partial r^2} + \frac{1}{r}\frac{\partial T_w}{\partial r} = \frac{c_w \rho_w}{k_w} \frac{\partial T_w}{\partial t} \end{cases} \tag{9}$$

where, C_f (J/(kg°C)) is the specific heat of the fluid in the tubing string; ∂t_f (W/(m²°C)) is the coefficient of convective heat transfer between fluid in the string and pipe wall; λ_f is the friction coefficient of the fluid in the string, which is related to fluid Reynolds number and flow pattern; ρ_f (kg/m³) represents fluid density in the string; v_f (m/s) stands for the speed of the fluid motion. k_t, k_a, k_c, k_e, and k_w are respectively the heat conductivity coefficient of injected fluid, cylinder wall, within the annulus fluid, formation rock, injection, and the original liquid mixing unit; Similarly, c_t, c_a, c_c, c_e, and c_w are respectively their specific heat capacity in J/(kg°C); ρ_t, ρ_a, ρ_c, ρ_e, and ρ_w are respectively their the density in kg/m³; r_{to}, r_{co}, r_{ci}, r_{ti}, and r_{ce} (m) are respectively the radius of the outer wall of pipe string, the radius of the outer wall of the casing, the string inner radius, the radius of the inner wall of the casing, and the cement ring diameter.

Boundary Conditions

When the wellbore temperature field model is established, boundary conditions are established as follows:

T_m is the temperature in somewhere under local annual average surface, and in general, it is a fixed value at the depth of 20 meters. Since the wellbore is axisymmetric, in the center of the hole diameter $(\partial T/\partial r)_{r=0}=0$. Besides, since infinity away from the center line of the hole diameter, $T_{r=Rmax}=T_{dz}$. The adiabatic boundary for bottom hole formation is $(\partial T/\partial z)_{z=l}=0$. The temperature of the fluid injected into from wellhead tubing or the annulus is constant; back to the exit of the boundary for the liquid temperature, $(\partial T/\partial z)_{z=0}=0$ [8].

Initial Conditions

According to the research condition in this paper, the initial conditions are established as follows:

When $t=0$, the temperature of wellbore and formation is the known formation temperature distribution. Since the recovery operation is carried out continuously during each process, the next working procedure of the initial temperature condition is one end of the previous process temperature conditions.

Working Condition

When water injection process temperature field was calculated, the heat conduction caused by mass transfer during fluid injection from wellhead and the original fluid in the wellbore, which is squeezed out, must be taken into account. U (m³/min) is the volume of injected fluid flow from wellhead. Since the injected fluid is liquid, the volume is always constant and equal to U. The velocity of fluid at the cross-section A_z of the fluid flow path along longitudinal Z can be written as:

$$u_z = U / A_Z \tag{10}$$

There are an L_m number of A_{zi} in the well depth of L (i=1, 2, ..., L_N), and then the total time t_L of injected fluid flow to the bottom of a well is worked out by:

$$t_L = \sum L_i A_{zi} / U \ (i = 1 \cdots L_N) \tag{11}$$

Several typical working conditions are considered as follows:

Condition one: when time is $t \leq t_k$, there are a number of interfaces of $Z=Z_{ab}$ in the flow path. At the well depth of $Z<Z_{ab}$, the fluid is the injected fluid, meanwhile when $Z>Z_{ab}$, the original fluid is in the wellbore. Z_{ab} can be worked out by t, t_k, and μ_z. Therefore, a unit with two media mixing is appeared on space domain and satisfies the following heat balance equation:

$$(c_o \rho_0 U T_{12} - c_1 \rho_1 U T_{34}) - \lambda_0 (\partial T - \partial n)_{So} - \lambda_1 (\partial T - \partial n)_{S1} = V_0 c_0 \rho_o (\partial T / \partial t)_{vo} + V_1 c_1 \rho_1 (\partial T / \partial t)_{v1} \tag{12}$$

Units below the well depth of $Z<Z_{ab}$ and $Z>Z_{ab}$ are the same fluid, which satisfies the following heat balance equation:

$$c_i \rho_i U (T_{12} - T_{34}) - \lambda_i (\partial T / \partial n)_S = V_i c_i \rho_i (\partial T / \partial t)_V \tag{13}$$

Condition two: when time is $t>t_k$, there are no mixed units, and then the heat can be calculated by Equation 1.

Condition three: when the fluid in the oil tube and oil casing is static and stagnant, there are several dielectric interfaces in the flowing path and more than two kinds of medium mixing unit in spatial domain as shown in Figure 2. The actual distribution and heat transfer near the interface of two kinds of different medium within the cell are extremely complex, but all can be solved by the heat balance Equation 2 and Equation 3.

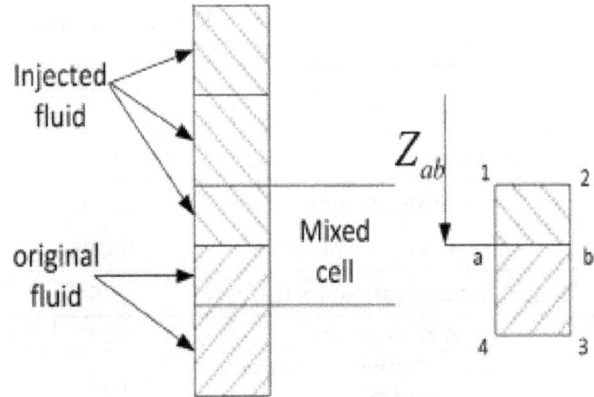

Figure 2: Fluid mixing units.

Solution Method for Model

The spatial domain is discrete by high precision of isoparametric elements adopted in this paper, and an optional implicit finite difference time domain of variable step is used in loop iteration, which obviously improves the accuracy and effectiveness of the calculation of the whole temperature field [9,10]. Next, the established temperature field model is solved by FORTRAN language programming, which is used to analyze the effect of different factors on the wellbore temperature distribution.

Factors Influencing the Temperature of Highly Deviated Well Water Injection Wellbore

As an example, the parameters of a high angle well showed in Tables 1 and 2 are used to analyze the law of wellbore temperature field distribution with the various influential factors [11]. The controllable factors in the process of water injection included a water injection rate, water inlet temperature, water injection time, etc. In addition, wellbore temperature will also be affected by environmental factors and the structure of the borehole, etc.

Table 1: Parameters of highly deviated well.

Parameter Name	Value	Parameter Name	Value
Drill pipe inner diameter (m)	0.031	Specific heat of string (KJ/kg°C)	0.11
Drill pipe outer diameter (m)	0.0365	Thermal conductivity of string (W/m²°C)	0.64
Inner casing diameter (m)	0.06068	Cement sheath diameter (m)	0.3
Outer casing diameter (m)	0.06985	Specific heat of cement sheath (KJ /kg°C)	0.21
Specific heat of formation (KJ/kg°C)	0.25	Specific heat of water (KJ/kg°C)	1.0
Well depth (m)	2955	Water injection rate (m/s)	0.2
Thermal conductivity of formation (W/m²°C)	0.02	Thermal conductivity of cement sheath (W/m²°C)	0.016

Table 2: Parameters of wellbore trajectory.

Well depth (m)	Deviation angle (°)	Azimuth (°)	Well depth (m)	Deviation Angle (°)	Azimuth (°)
0	0	82.08	1673.88	4.85	301.27
231.47	0.69	40.48	1962.03	0.85	328.45
521.99	1.26	246.32	2260.76	2.41	23.14
810.03	0.83	277.98	2548.93	2.32	160.57
1097.78	11.28	301.58	2819.09	0.72	143.98
1385.81	15.24	298.55	2955.7	0.13	269.33

Impact of Injection Water Inlet Temperature on the Temperature Distribution of Highly Deviated Wellbores

The inlet temperature of water is a very important factor in oil field development. Under the conditions of a depth of 2955.7 m, a water injection rate of 0.15 m³/min, and a water injection time of 120 min, adjusting different water inlet temperature and the chart about wellbore temperature change law of the highly deviated well are displayed in Figure 3. When the inlet temperature of water varies from 15 to 40°C, the bottom hole temperature increases by about 10°C. That is, improving the wellhead injection water temperature is equivalent to increasing more energy entering the wellbore by the pump, which increases the temperature of bottom hole. The amplitude of the increase in bottom hole temperature is associated with an increased range of

inlet temperature of the water injection; however, the temperature distribution of wellbore is not obviously affected by the water inlet temperature.

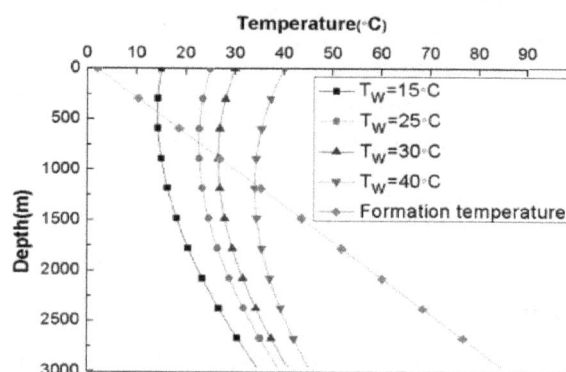

Figure 3: Wellbore temperature profiles and water temperature curve.

Impact of Water Injection Rate on the Temperature Distribution of Highly Deviated Wellbores

Water injection rate plays an important role in oil field development [12]; under the conditions of a water injection time of 120 min and a water injection initial temperature of 15°C, water injection rate is respectively 0.05, 0.08, 0.10, and 0.12 m³/min; the wellbore temperature profile curves are presented in Figure 4. Figure 4 shows that when the water injection rate is low, injection water and rock stratum around the wellbore have sufficient time exchanging heat; therefore, the wellbore temperature is more close to the ambient temperature than the higher water injection rate. With increasing the water injection rate, the convective heat transfers between the injected water and environment time is accordingly reduced. Eventually, the temperature of bottom wellbore is decreased.

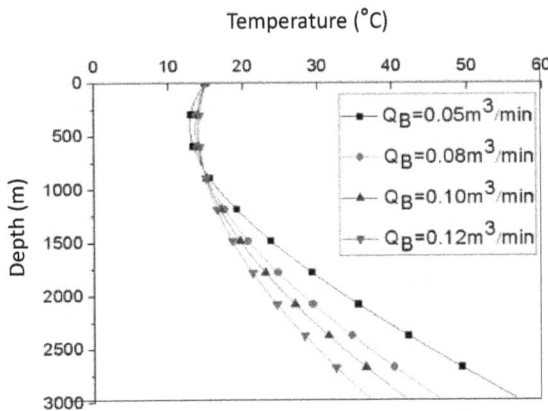

Figure 4: Wellbore temperature profiles and water injection rate curves.

Impact of Water Injection Time on the Temperature Distribution of Highly Deviated Wellbores

Wellbore temperature is higher in a shorter injection time as shown in Figure 5. With increasing injection time, the temperature of formation around the wellbore is reduced. Also, when heat exchange between wellbore and formation tends to be stable, wellbore temperature is relatively low.

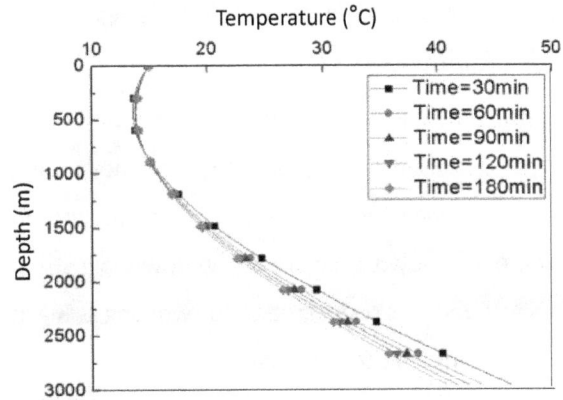

Figure 5: Wellbore temperature profiles and water injection time curves.

Impact of Geothermal Gradient on the Temperature Distribution of Highly Deviated Wellbores

Under the conditions of a water injection time of 120 min, a water injection initial temperature of 15°C, and a water injection rate of 0.1 m³/min, wellbore temperature profiles and geothermal gradient curves are presented in Figure 6. The picture shows that with increasing geothermal gradient, wellbore temperature increased significantly. As the well becomes deeper, the temperature of the bottom hole becomes higher and higher. Hence, the precise prediction of the geothermal gradient willhave a great influence in order to predict wellbore temperature profile [13].

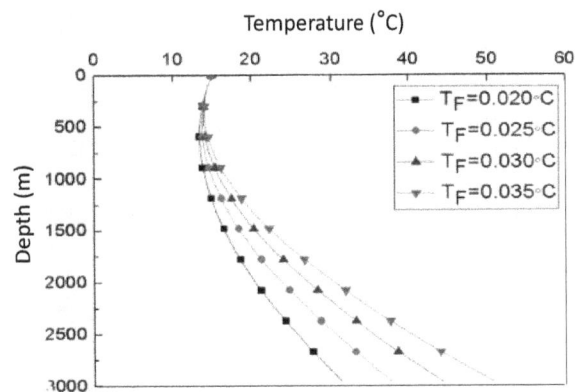

Figure 6: Wellbore temperature profiles and geothermal gradient curves.

The Experiment to Measure the Temperature Field Distribution

The stress measurement system of the downhole stringis composed of two parts: the downhole string stress test device and the computer data acquisition software. As shown in Figure 7, the downhole string stress test device is connected to the working string in the pipe string and to the measures.

The tester structure diagram is shown in Figure 8. The measuring device is built in a single-chip controlled manner by setting the sampling and external pressure, temperature, axial load, and torque data, and the data collected by the single-chip are stored in the memory. When the operation is finished, the data in the test device is transferred through dedicated communication lines to computer data software.

As shown in Figure 9, the temperatures of different depths and injection water times were measured by the tester and used in Jidong oil field. At the well depth of 30 m, the temperature of the wellbore is about 16°C. It is low and does not change over the injection time. At a depth of 2854 m, close to bottom hole, the temperature is about 82°C, and reduces obviously over the injection water time. Near the wellhead, the temperature of water is close to the temperature of the stratum, and the heat transfer is not obvious. When the well depth is larger, the temperature difference of water and stratum heavily causes heat transfer, and then the corresponding wellbore temperature is decreased; finally thermal equilibrium is established.

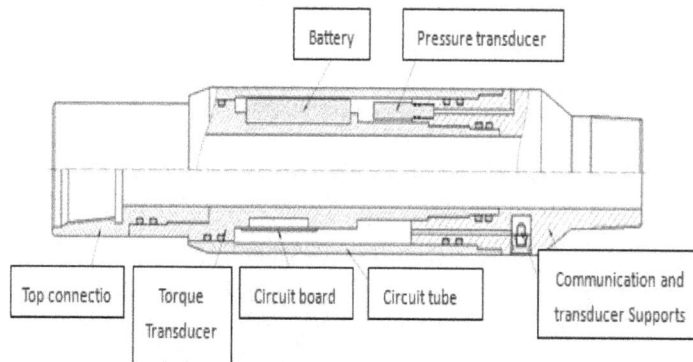

Figure 7: Tester structure diagram.

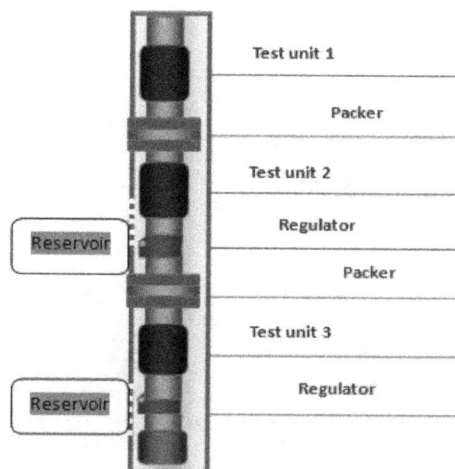

Figure 8: A schematic diagram of test process.

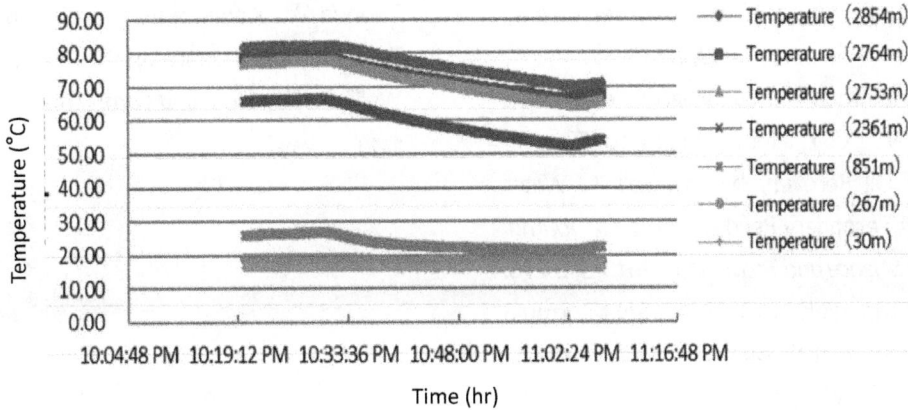

Figure 9: The temperature of different depths and injection water times.

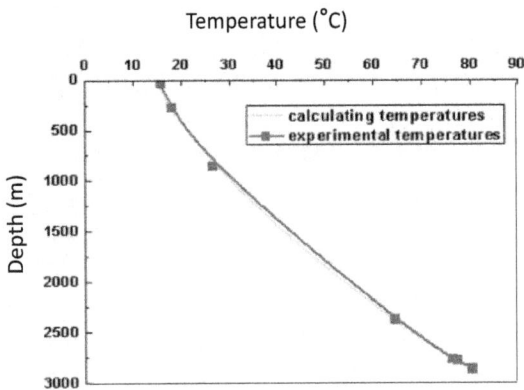

Figure 10: Comparison of the experimental and theoretical results.

CONCLUSIONS

The true three-dimensional model which has been proposed in the current paper can accurately calculate the distribution of temperature in different water injection conditions; besides, this model can provide technical support to prevent wellbore temperature [14] from becoming too high and thus producing a larger temperature stress and avoiding borehole wall collapse [15] in the process of water injection. Figure 10 shows that the results of the experiments and in good agreement with the theoretical calculations. Additionally, water inlet temperature does not obviously affect the wellbore temperature distribution except for water injection rate, water injection time, and geothermal gradient. It is necessary that the geothermal gradient be accurately measured and then water injection rate and water injection time, in descending order, be reasonably controlled. The results are consistent with those reported in the literature [14]. Finally, the results have a great theoretical significance and practical value.

REFERENCES

1. Baoxia Q., Ceji Q., Chongchong S., and Jian S., "Analysis and Design Research for Layered Water Injection String Mechanics," *Petroleum Engineering Technology in Chinese*, **2015**, *1*, 57-59.

2. Fan Y., Baodong Ch., Wenquan J., Yu Sh. et al., "Application Research of P-R State Equation in Thermodynamic Properties Calculation of Natural Gas," *Contemporary Chemical Industry*, **2013**, *42*, 649-650.

3. Xiangqing H., Xiangjun L., and Pingya L., "Influence of Temperature Perturbation on Borehole Wall Stabilization and Oil Field Development," *Natural Gas Industry*, **2003**, *23*, 39-41.

4. Yin S., Towler B. F., Dusseault M. B., and Rothenburg L., "Fully Coupled THMC Modeling of Wellbore Stability with Thermal and Solute Convection Considered," *Transport in Porous Media*, **2010**, *84*, 773-798.

5. Xuefeng J. and Weige J, "Temperature's Impact on High Pour-Point Oil Reservoir Development Effect,"

Journal of Gui Zhou University of Technology (Natural Science Edition), **2009**, *37*, 9-11.

6. Shaddel S., Hemmati M., Zamanian E., and Nejati Moharrami N., "Core Flood Studies to Evaluate Efficiency of Oil Recovery by Low Salinity Water Flooding as A Secondary Recovery Process," *Journal of Petroleum Science and Technology*, **2013**, *3*, 65-76.

7. Xingbo W., Xixiang Ch., Limie L., "Frenet Formula of Arbitrary Parameter," *Journal of Hunan Institute of Technology*, **2005**, *2*, 1-4.

8. Ning W., Bao Jiang S., and Zhi Yuan W., "Boundary Condition Handling for Analytical Calculations of Circulating Mud Temperatures," Proceedings of Thirteenth National Conference on Water Dynamics and the Twenty-Sixth National Symposium on Water Dynamics, **2014**, August.

9. Donea J. and Huerta A., "Finite Element Methods for Flow Problems," **2003**, John Wiley and Sons.

10. Harari, I. and Hauke, G." Semidiscrete Formulations for Transient Transport at Small Time Steps," *International Journal for Numerical Methods in Fluids*, **2007**, *54*, 731-743.

11. Aslanyan I., Barghouti J., and Al-Shammakhy A. J., "Evaluating Injection Performance with High Precision Temperature Logging and Numerical Temperature Modeling," SPE Reservoir Characterization and Simulation Conference and Exhibition, Society of Petroleum Engineers. **2013**, September.

12. Xiaoyan Y., Shenglai Y., Xianghong W., Jie L. et al., "Numerical Simulation Study on the Influence of Water Injection on the Temperature Field of High Pour Point oil Reservoir," *Complex Hydro-Carbon Reservoirs*, **2011**, *4*, 51-53.

13. Shi Y., Song Y., and Liu H., "Numerical Simulation of downhole Temperature Distribution in Producing Oil Wells." *Applied Geophysics*, **2008**, *5*, 340-349.

14. Wu K., Li X., Ruan M., Yu M. et al., "Dynamic Tracking Model for the Reservoir Water Flooding of a Separated Layer Water Injection Based on a Well Temperature Log," *Journal of Petroleum Exploration and Production Technology*, **2015**, *5*, 35-43.

15. Baohua W., Xiaofeng L., and Bingyin W., "Law of Temperature Influence on Wellbore Stability in Hot Well". *Drilling Fluid and Completion Fluid*, **2004**, *21*, 1-11.

Mechanical Properties and Structures of Oil-well Cement Stone Reinforced with Siliceous Whisker

Ming Li[1], Shuang Deng[1], Jianzhou Jin[2], Yujia Yang[1], and Xiaoyang Guo[1]*

[1] State Key Laboratory of Oil and Gas Reservoir Geology and Exploitation (Southwest Petroleum University), Southwest Petroleum University, Chengdu, 610500, China
[2] CNPC Drilling Research Institute, Beijing 102206, PR China

ABSTRACT

To fully utilize the siliceous whiskers and improve the mechanical properties of oil-well cement, the cement composites with different additions of siliceous whiskers were prepared, and analyzed by means of scanning electron microscopy (SEM), mechanical testing, X-ray diffraction (XRD), and infrared spectrum analysis. The results showed that the addition of 2% siliceous whiskers could slightly improve the compressive strength, and markedly increased the flexural and tensile strength as well as toughness; it also decreased the permeability and porosity of oil-well cement composites. Moreover, it was confirmed that the siliceous whisker could not affect the cement hydration process and hydration products. The improvement in mechanical properties and the more compact microstructure could be mainly resulted from enhancement mechanisms such as bridging, crack deflection, and pulling-out in cement matrix.

Keywords: Oil Well Cement, Mechanical Properties, Siliceous Whiskers, Enhancement Mechanism

INTRODUCTION

Usually, cementitious materials have a relatively high compressive strength, which is the most basic property of cement-based composites, but they are more sensitive to tensile stresses than compressive ones. The brittleness indexes such as flexibility, tensile strength, and the impact resistance of cement are relatively low. Therefore, making an enhancement in tensile strength and maintaining the higher toughness simultaneously is a challenge for possible applications of brittle oil-well cements as the basic materials in the petroleum industry [1]. Nowadays, various methods have been used to improve the strength and toughness of cement composites, of which the most effective one is blending with various reinforcing additives, including natural fibers [2], carbon fibers [3], steel fibers [4], carbon nanotubes [5], nanosized particles [6-7], synthetic fibers [8], and the hybrid combination of different types of reinforcing additives [9-11]. Recently, an important reinforcing additive, namely $CaCO_3$ whiskers, has been used as a unique strengthening and toughening enhancer in cement composites [12-13], which is effective to delay the formation and propagation of micro-cracks. It was also found out that $CaCO_3$ whiskers not only can slightly improve the compressive strength, but also can significantly

*Corresponding author
Xiaoyang Guo
Email: guoxiaoyangswpi@126.com

promote the flexibility, load-deflecting curves, and work of fracture [13]. To the best of our knowledge, nevertheless, the abovementioned investigations have not considered the hydration products, structures, and the permeability or porosity of cement composites, which are factors affecting the properties of composites directly [13], and they discussed the mechanical properties alone.

The mechanical properties and microstructures of cement composites depend on the type, dispersity, and content of the reinforcing additives [3,6,15-16]; the adhesion strength between reinforcing additives and cement [4,11,17]; and the hybrid dosages of different reinforcing additives [3,9,18-20]. Siliceous whisker is a new kind of reinforcing material, which has a structure similar to diamond, having a high strength and modulus. Compared with the other reinforcing additives used in oil-well cement completely [1,10,21], siliceous whiskers belong to a sort of whiskers which have a definite slenderness ratio, but barely have defects. Thus, this kind of whiskers has perfect mechanical properties, such as high strength, high Young's modulus, and good resistance to temperature, which allow them to be used as reinforcing materials [22-24]. Generally, the current paper is to assess the most effective addition of siliceous whiskers into oil-well cement and to discuss the mechanical properties and the structures of the composites obtained.

EXPERIMENTAL PROCEDURES
Materials

The siliceous whiskers used in this study were provided by Jiechuang Novel Materials Ltd. (Xuzhou, China), and its main properties are listed in Table 1. Other materials used included class G Portland cement (Sichuan Jiajiang Guijiu Special Cement Co.,

Ltd., China), the chemical composition of which is presented in Table 2. The fluid loss agent G33S and dispersant SXY-2 were commercially supplied. The SEM morphologies of siliceous whiskers are shown in Figures 1. As shown in Figure 1, siliceous whiskers is a kind of whisker with a diameter of 0.2~1.5 µm and a length of 10~100 µm.

Table 1: Properties of siliceous whiskers.

Species	Length (µm)	Diameter (µm)	Tensile Strength (GPa)	Bulk Density (g.cm^{-3})	Elastic Modulus (GPa)
Siliceous whiskers	15~80	0.3~1.4	20.58	3.2	48.02

Table 2: Chemical composition of class G oil-well cement as analyzed by X-ray fluorescence (XRF).

Composition	SiO_2	Al_2O_3	Fe_2O_3	CaO	MgO	K_2O	SO_3	MnO_2
Content (wt.%)	22.7	3.39	4.81	65.5	0.9	0.37	1.21	0.09

(a)

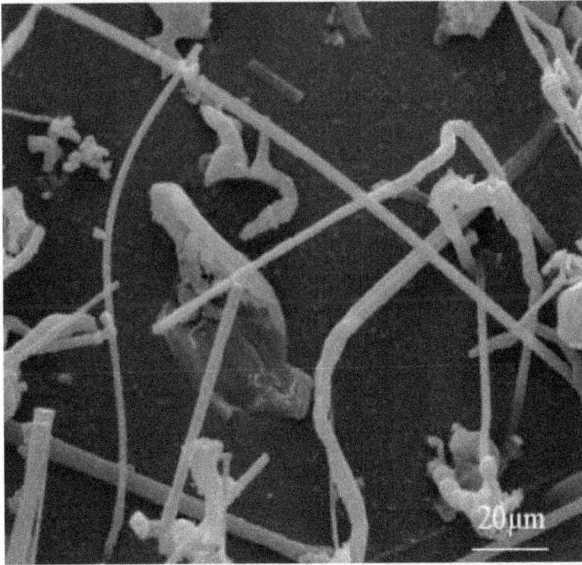

(b)

Figure 1: SEM micrographs of siliceous whiskers.

Specimen Preparation and Experimental Methods

Cement slurries were prepared and cured according to the standards of API recommended practice 10B [25]; also, cement, siliceous whiskers, fluid loss agent G33S, and dispersant SXY-2 were all mixed with the cement powders and were then stirred by a cement paste blender at a water/cement mass ratio (W/C) of 0.4. The detailed mixture proportion of the cement slurries is given in Table 3. The cement slurries were kept in standard curing molds at 100% relative humidity and at a temperature of 90 °C for 3, 7, 14, and 28 days. The cured rectangular specimens (50.8×50.8×50.8 mm³) and cylindrical specimens (φ25.0×25.0 mm³) were used to measure tensile and compressive properties at a crosshead speed of 600 N/s using electronic hydraulic testing machines (Haizhi Technology Co. Ltd., Beijing, China). In addition, the cured rectangular specimens (40.0×40.0×160.0 mm³) were used to measure flexible strength with 3-point-bending using a motorized bending tester

(Jianyi Instrument & Machinery Co. Ltd., Wuxi, China) at a span and a crosshead speed of 100 mm and 0.02 mm/min respectively. Furthermore, the cured rectangular specimens (φ25.0×40.0mm³) kept in standard curing molds at 90 °C at 100% relative humidity for 7 days were used to measure the permeability using a particle volume or core gripper (Yiyong Technology Co. Ltd., Changzhou, China) at a pressure of 2 MPa; the pore size distribution of the cement was tested by mercury injection apparatus (Poremaster 60, Quantachrome Ins, USA). Moreover, the cured cylindrical samples (φ25.0×50.0mm³) were used to measure the strain-stress curve by triaxial rock mechanics testing system (RTR-1000, GCTS Co., USA) according to the standards of GB/T 50266-99 at a constant loading rate of 2 KN/min and at a confining pressure of 20.7 MPa. There were 7 samples tested for each testing, and the standard deviation was expressed as the variation in the results.

Table 3: Composition of the prepared specimens.

Specimens	Cement (g)	Siliceous whiskers (g)	G33S (g)	SXY-2 (g)	Water (g)
V	400	0			
R_1	396	4			
R_2	392	8	4	4	160
R_3	388	12			
R_4	384	16			

In addition, an X-ray diffraction meter (Shimadzu, XRD-7000, Japan) was used to study the diffraction behaviors of siliceous whiskers and the enhanced cement. The scanning electron microscope (SEM, JEOL JSM-6510LV, Japan) was used to observe the In addition, an X-ray diffraction meter (Shimadzu, XRD-7000, Japan) was used to study the diffraction behaviors of siliceous whiskers and the enhanced cement. The scanning electron microscope (SEM, JEOL JSM-6510LV, Japan) was used to observe the microstructure of the siliceous whiskers and the fracture surface of the specimens. Furthermore, the infrared absorption spectrometer (FI IR, Thermo Scientific, Nicolet 6700, America) was used to confirm compositional variation during the hydration process.

RESULTS AND DISCUSSION
Effect of Siliceous whiskers on the Mechanical Properties

Figure 2 shows the mechanical properties of the cement stone; it can be seen that the mechanical properties of the cement composites are improved due to the addition of the siliceous whiskers, and they are increased with an increase in curing time. Compared with sample V (without siliceous whiskers), the compressive, tensile, and flexural strengths of sample R_2 respectively obtain an increment of 10.35%, 132.43%, and 30.39% after 28 days of curing by a 2% addition of siliceous whiskers, which is the most effective addition. The enhancement in the mechanical properties of the cement composites may result from the higher modulus of siliceous whiskers as well as the good adhesion between the whiskers and the cement composites [26].

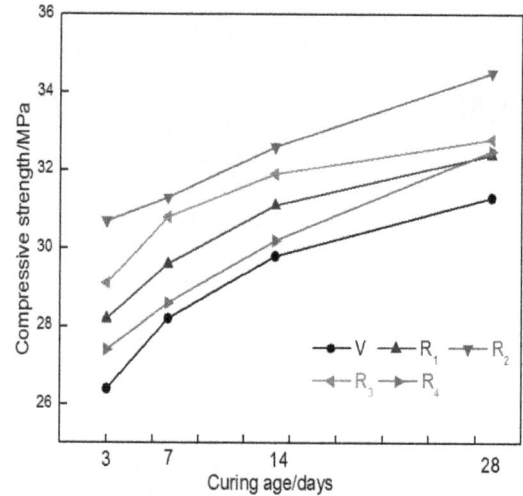

a) Compressive Strength

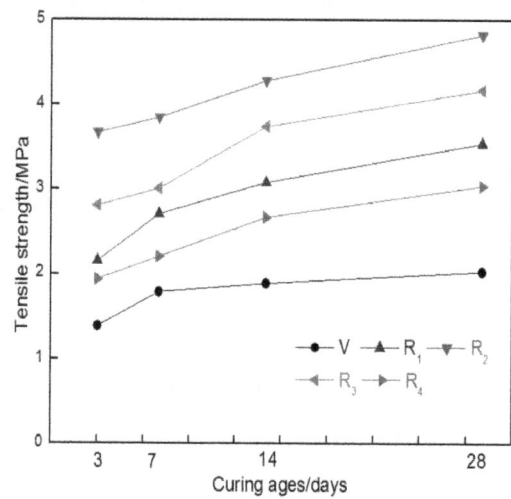

b) Tensile Strength

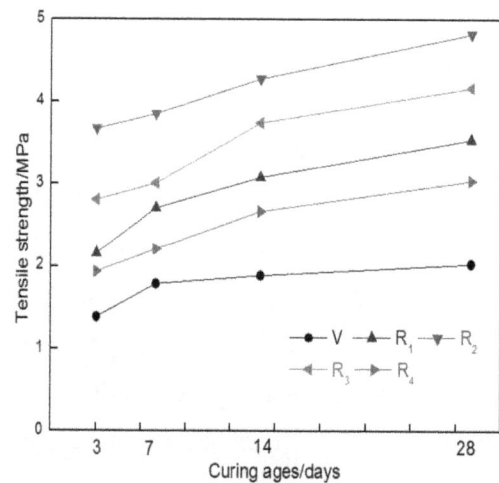

c) Flexural Strength

Figure 2: Effect of adding siliceous whiskers on the cement mechanical properties.

Effect of Siliceous Whiskers on the Permeability and Porosity

Table 4: Pore size distribution curves of specimens V and R_2.

Pore volume /Pore size distribution (%)	>200 nm	50-200 nm	20-50 nm	<20 nm
V	4.372/19.6	15.998/71.6	1.074/4.8	0.91/4.1
R_2	0.002/0.04	4.416/84.9	0.635/12.2	0.151/2.9

It is well known that cement composites are porous materials having a large number of pores and voids. The microstructure of cement composites is the key basic elements of the macro-properties. Generally, the more the pores and voids exist, the lower the mechanical properties of cement composites become. Table 4 shows the pore size distribution of the cement containing whiskers (R_2) and the pure cement (V). It is distinct that the addition of whiskers can improve the pore size distribution of cement composites, and the total pore volume and the pore size of the cement including whisker is significantly lower than the pure cement. The probability of the pore size bigger than 50 nm of the pure cement is 91.2% with a pore volume of 20.37 ml/g, while that of the cement composite containing whiskers decreased to 84.94% with a pore volume of 4.418 mL/g. This indicates that the modified cement contains fewer and smaller pores and voids, which is beneficial to mechanical properties.

Table 5 shows the influence of siliceous whiskers on the permeability and porosity of the cement composites; the permeability and porosity of cement composites with whiskers is decreased nearly by 82.8% and 7.7% respectively compared to the pure cement, representing that siliceous whiskers can effectively enhance the microstructure and make more compact structures; this is also in accordance with the analysis of pore size distribution.

Table 5: Permeability and porosity of specimen V and specimen R_2.

Specimens	Permeability (mD)	Porosity (%)
V	0.2533	83.78
R_2	0.0435	77.30

X-ray Diffraction Analysis of Siliceous Whiskers Enhanced Cement

Figure 3(a) and (b) show the comparison of X-ray diffraction between the virgin cement (V) and the cement containing siliceous whisker (R_2) cured for 7 days and 28 days respectively. The results demonstrated that there was no new kind of hydration in the cement with siliceous whiskers cured for 28 days, while the hydrations were most similar in Figure 3(a) when they cured for only 7 days. The crystal phase of specimen R_2 consists of calcium hydroxide (CH), ettringite (AFt) etc. As a result, siliceous whiskers have no effect on the hydration process and products of cement composites.

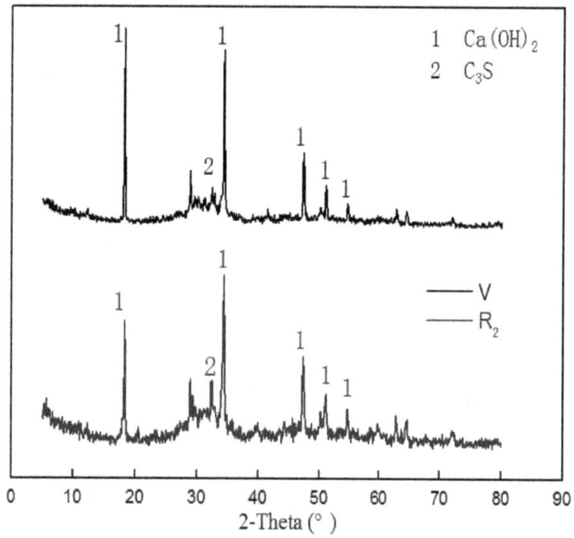

a) 7 days

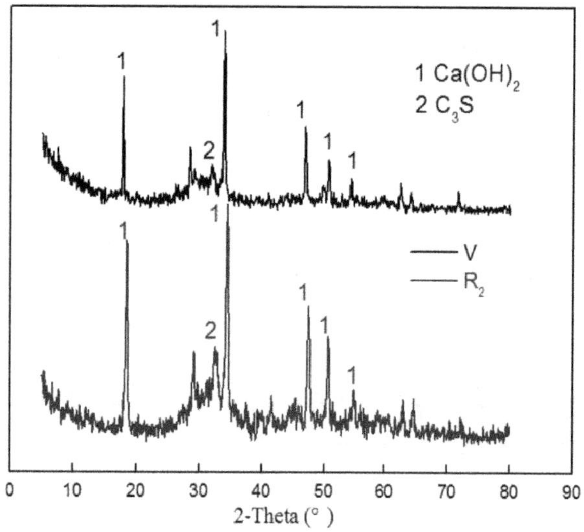

b) 28 days

Figure 3: XRD pattern of specimen V and the specimen R$_2$.

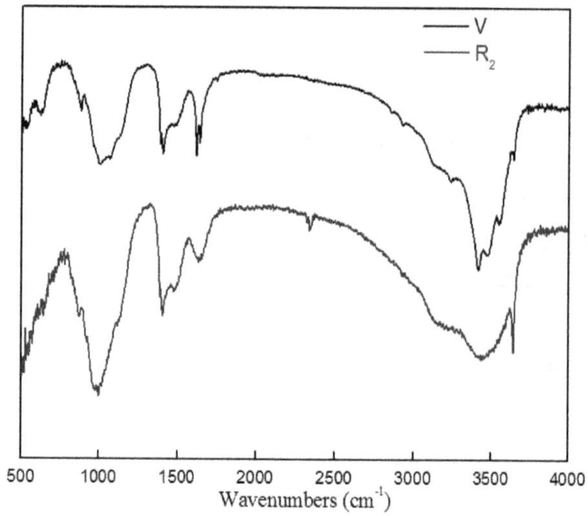

a) 7 days

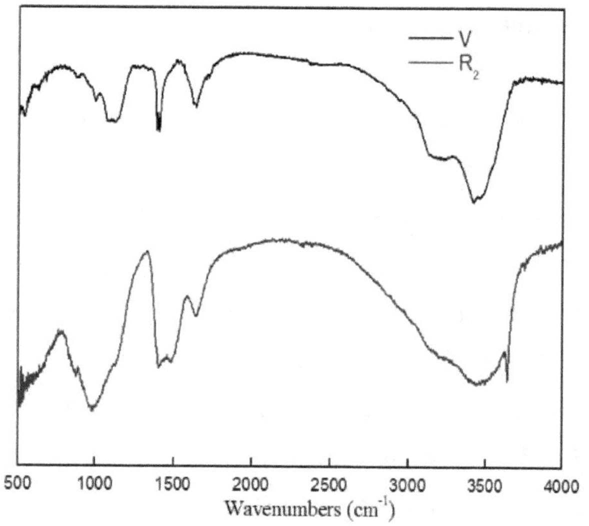

b) 28 days

Figure 4: IR spectra of specimen V and the specimen R$_2$.

Infrared Spectrum Analysis of Siliceous Whiskers Enhanced Cement

Infrared absorption spectrum (IR), also known as molecular vibration spectrum, has been broadly used in the chemical group and bond characterization based on the position of characteristic peaks. Therefore, the infrared spectrum analysis can be used in cement analysis due to the different vibration frequency of anionic groups.

The infrared absorption spectra of the cement with whiskers (R$_2$) and the pure cement (V) curing for 7 and 28 days are displayed in Figure 4. The C-S-H gel is an amorphous crystal, resulting in different asymmetric stretching vibration displacements of

Si-O bond in [SiO4]$^{-4}$ such as the Si-O bond with a displacement range of 900-980 cm^{-1} and the Si-O-Si bond with a displacement range of 980-1000 cm^{-1}; the aggregations will also be V$_3$[SiO$_4$] when displacement reaches 1100 to 1200 cm^{-1}. The strong and narrow sharp peaks at 3640 cm^{-1} corresponds to the characteristic spectral line of OH^{-1} in Ca(OH)$_2$; comparing with the pure cement, the peak of specimen R$_2$ is representing a weaker vibration of OH^{-1}, indicating less Ca(OH)$_2$ in the cement composites. Furthermore, the nearby larger absorption peak at 3640 cm^{-1} is the stretching vibration peak of hydroxyl in the water crystallization. It can be seen that there is no difference in the characteristic peaks representing anionic groups, which is in accordance with the XRD analysis.

Siliceous Whiskers Enhancement Mechanism for the Mechanical Properties of Oil Well Cement

Fractured Surface Morphology

The structure of oil-well cement stone formed is connected with the hydration and its products during the hardening process. The scanning electronic microscope morphology of the composites with siliceous whiskers (R$_2$) and the virgin specimen (V) is shown in Figure 5. It can be seen that a more compact structure is formed due to the special structure of siliceous whiskers. There are fewer micro-cracks, which may be attributed not only to the small sized siliceous whiskers as the fillers of voids in the cement, but also to the lack of any clear gap between the cement matrix and siliceous whiskers.

a) specimen V

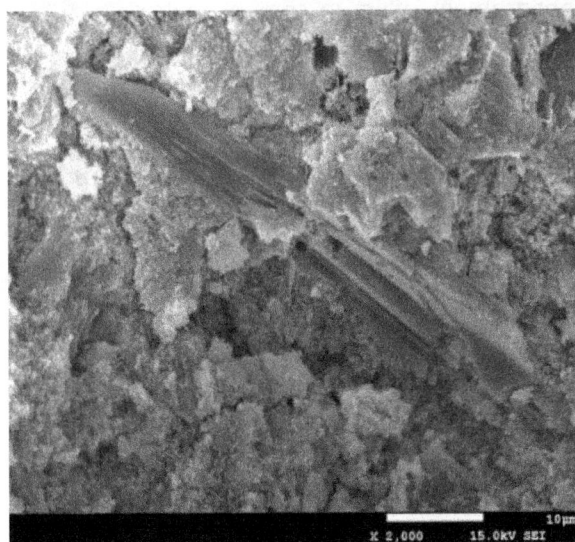

b) specimen R$_2$

Figure 5: SEM micrographs of the fractured surfaces.

Enhancement Mechanism Analysis of Siliceous Whiskers

The reinforcing mechanism of siliceous whiskers on mechanical properties consists of bridging effects, crack deflection, and pull-out effects, which is detected in SEM photography of the cement with siliceous whiskers. Figure 6 shows the SEM photography of the cement with siliceous whiskers. It can be seen that the fracture failure of the cement materials is mostly caused by crack propagation, and the more the energy is consumed during crack propagation, the better the toughness of the cement become. The compact structure of siliceous whiskers cement matrix results in an extra function in cracking resistance and energy dissipation because of the smaller sized and more intensive siliceous whiskers. As shown, the micro-cracks are astricted by the siliceous whiskers which can bridge cracks in micro-cracks, expand cracking path, and increase the absorption and dissipation of energy during fracture, thereby forming an enhancement effect in cement matrix.

a) bridging

b) crack deflection

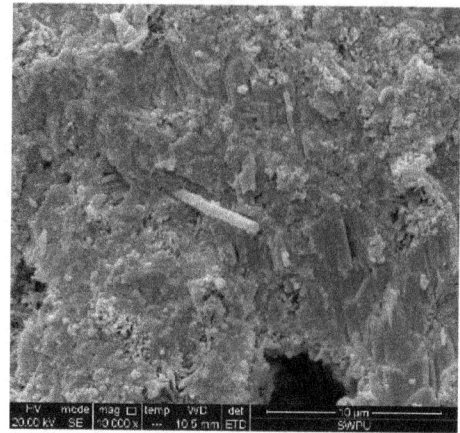

c) pull-out

Figure 6: SEM micrographs of the fractured surfaces of the enhanced cement.

In the initial stage of bearing external load, subjecting to low inner stress, some smaller micro-cracks are formed in the cement matrix. Because of the siliceous whiskers with a smaller size and good mechanical properties, the closure stress can be formed in the crack tip with crack bridging of the whiskers, which results in stress loading on the crack surface instead of stress concentration at the tip of the cracks [28]. Figure 6a shows the micro-cracks in the presence of whiskers; as the load increases, if the micro-cracks gradually grow into larger and even regional cracks, the siliceous whiskers larger in size at this point play a leading role in bridging cracks and effectively limiting the large-scale crack extension until the whiskers are destroyed.

The micro-cracks in the cement cannot develop in the original orientation when approaching the whisker with a smaller angle (<90°) because of the high modulus of whisker; as a result, the propagation direction of cracks would be transformed to spread along the interface between the whiskers and the cement matrix; with an increase in the path of crack propagation (Figure 6b), more loading energy is consumed, resulting in less damage in the microstructure of the cement.

The microscopic appearance of the fracture surface of the cement composites containing whiskers are shown in the Figure 6c. As analyzed above, whiskers can act as bridges to connect the both sides of cracks to decrease the crack extension; also, with an increase in load and bearing duration, some of the siliceous whiskers would be pulled out, which might facilitate better stress transfer from the cement matrix to the stronger whisker. Vast energy is also consumed due to the rubbing action between whiskers and cement. Moreover, the addition of whiskers can significantly improve the microstructure of the cement composites, and more energy will be consumed in crack extension, resulting in an effective enhancement of toughness and mechanical properties of the cement.

CONCLUSIONS

The current paper discussed the effect of siliceous whiskers on the mechanical properties and microstructures of oil well cement composites. The results show that there is a reinforcing effect of siliceous whiskers on mechanical properties; an improvement of 10.35%, 132.43%, and 30.39% was seen in compressive, flexibility, and tensile strengths respectively after 28 days of curing in the presence of 2% siliceous whiskers. Also, the analysis of the results reveal that siliceous whiskers plays a positive effect on permeability, porosity, and pore size distribution, and the microstructure of cement enhanced by whiskers becomes more compact; the permeability and porosity are decreased by 82.8% and 7.7% respectively. Meanwhile, the presence of whiskers can improve the loading damage process in the microstructure

based on the reinforcement mechanisms such as bridging, pulling-out, crack deflecting, and more damage energy can be absorbed and dissipated. The improvement in the mechanical strength and microstructure of oil well cement composites may be significant for the cementing engineering and possible applications of siliceous whiskers.

ACKNOWLEDGMENTS

The authors are grateful for the support by National Science and Technology Major Project (No. 2016ZX05020004-008 and No. 2016ZX05052) and the CNPC Science and Technology Project (2014A-4212).

REFERENCES

1. Choolaei M. M., Rashidi A. M., Arjmand M., Yadegari A., and et al., "The Effect of Nanosilica on the Physical Properties of Oil Well Cement," *Mat. Sci. Eng. A.*, **2012**, *538*, 288-294.

2. Ghavami K., Toledo Filho R. D., and Barbosa N. P., "Behavior of Composite Soil Reinforced with Natural Fibers," *Cem. Concr. Res.*, **1999**, *21*, 39-48

3. Piekarczyk J., Piekarczyk W., and Blazewicz S., "Compression Strength of Concrete Cylinders Reinforced with Carbon Fiber Laminate," *Constr. Build Mater.*, **2011**, *25*, 2365-2369.

4. Wang J. Y., Chia K. S., Liew J. Y. R., and Zhang M. H., "Flexural Performance of Fiber-reinforced Ultra-lightweight Cement Composites with Low Fiber Content," *Cem. Concr. Compos.*, **2013**, *43*, 39-47.

5. Anastasia S., Viktor M., Vyacheslav K., and et al., "Dispersion of Carbon Nanotubes and its Influence on the Mechanical Properties of the Cement Matrix," *Cem Concr Compos.*, **2012**,

34, 1104-1113.

6. Shiri S., Abbasi M. H., Monshi A., and Karimzadeh F., "A Study on Mechanical and Physical Properties of Monocalcium Aluminate Cement Reinforced with Nano-SiO$_2$ Particles," *Compos Part B*, **2014**, *56*, 30-33.

7. Farzadnia N, Abang A., Demirboga R., and Parvaz Anwar M., "Effect of Halloysite Nanoclay on Mechanical Properties, Thermal Behavior and Microstructure of Cement Mortars," *Cem. Concr. Res.*, **2013**, *48*, 97-104.

8. Alamshahi V., Taeb A., Gaffarzadeh R., andRezaee M. A., "Effect of Composition and Length of PP and Polyester Fibres on Mechanical Properties of Cement Based Composites," Constr. Build Mater, 2012, 36, 534-537.9.

9. Dawood E. T. and Ramli M., "High Strength Characteristics of Cement Mortar Reinforced with Hybrid Fibres," *Constr. Build Mater.*, **2011**, *25*, 2240-2247.

10. Pacheco Torgal F. and Jalali S., "Cement Building Materials Reinforced with Vegetable Fibres: A Review," *Constr Build Mater.*, **2011**, *25*, 575-581.

11. Chakraborty S., Kundu S. P., Roy A., Basak R. K., and et al., "Improvement of the Mechanical Properties of Jute Fiber Reinforced Cement Mortar: A Statistical Approach," *Constr. Build Mater.*, **2013**, *38*, 776-784.

12. Cao M. L., Zhang C., and Wei J. Q. "Microscopic Reinforcement for Cement Based Composite Materials," *Constr. Build Mater.*, **2013**, *40*, 14-25.

13. Cao M. L., Zhang C., and Lv H. F.,"Characterization of Mechanical Behavior and Mechanism of Calcium Carbonate Whisker-reinforced Cement Mortar," *Constr Build Mater.*, **2014**, *66*, 89-97.

14. Cao M. L., Zhang C., and Lv H. F., "Mechanical Response and Shrinkage Performance of Cementitious Composites with New Fiber Hybridization," *Constr. Build Mater.*, **2014**, *57*, 45-52.

15. Kim H. K., Nam I. W., and Lee H. K., "Enhanced Effect of Carbon Nanotube on Mechanical and Electrical Properties of Cement Composites by Incorporation of Silica Fume," *Compos Struct.*, **2014**, *107*, 60-69.

16. Kang S. H., Ahn T. H., and Kim D. J. "Effect of Grain Size on the Mechanical Properties and Crack Formation of HPFRCC Containing Deformed Steel Fibers," *Cem. Concr. Res.*, **2012**, *42*, 710-720.

17. Sawsen C., Fouzia K., Mohamed B, Moussa G. "Optimizing the Formulation of Flax Fiber-reinforced Cement Composites," *Constr. Build Mater.*, **2014**, *54*, 659-664.

18. Morsy M. S., Alsayed S. H., and Aqel M., "Hybrid Effect of Carbon Nanotube and Nano-clay on Physico-mechanical Properties of Cement Mortar," *Constr. Build Mater.*, **2011**, *25*, 145-149.

19. Pereira E. B., Fischer G., and Barros J. A. O., "Effect of Hybrid Fiber Reinforcement on the Cracking Process in Fiber Reinforced Cementitious Composites," *Cem. Concr. Compos.*, **2012**, *34*, 1114-1123

20. Larbi A S., Agbossou A., and Hamelin P., "Experimental and Numerical Investigations about Textile-reinforced Concrete and Hybrid Solutions for Repairing and/or Strengthening Reinforced Concrete Beams," *Compos Constr.*,

2013, *99*, 152-162

21. Ulm F. J. and James S., "The Scratch Test for Strength and Fracture Toughness Determination of Oil-well Cements Cured at High Temperature and Pressure," *Cem. Concr. Res.*, **2011**, 41, 942-946.

22. Shi Y. G., Yang J. F., Liu H. L., and et al., "Fabrication and Mechanism of 6H-type Silicon Carbide Whiskers by Physical Vapor Transport Technique," *J. Cryst. Growth.*, **2012**, *349*, 68-74.

23. Han S. and Chung D. D. L., "Strengthening and Stiffening Carbon Fiber Epoxy Composites by Halloysite Nanotubes, Carbon Nanotubes andSilicon Carbide Whiskers," *Appl. Clay Sci.*, **2013**, *83-84*, 375-382.

24. Ramulu M., Paul G., and Patel J., "EDM Surface Effects on the Fatigue Strength of a 15 vol.% SiCp/Al Metal Matrix Composite Material," *Compos Struct.*, **2001**, *54*, 79-86.

25. American Petroleum Institute, API Recommended Practice 10B-2, American Petroleum Institute, Washington, 2nd edn., **2013**.

26. Hossain M. Z. and Awal A., "Flexural Response of Hybrid Carbon Fiber Thin Cement Composites," Constr. Build Mater, 2011, 25, 670-677.

27. Shebl S. S., Seddeq H. S., and Aglan H. A., "Effect of Micro-silica Loading on the Mechanical and Acoustic Properties of Cement Pastes," *Constr. Build Mater.*, **2011**, *25*, 3903-3908.

28. Becher P. F., Hsueh C. H., and Angelini P., "Toughening Behavior in Whisker-reinforced Ceramic Matrix Composites," *Ceram Soc.*, **1988**, *71*(12), 1050-1061.

3

Research on Chaos Characteristic of Crack Evolution in Coal-rock Fracturing

Wanchun Zhao[1], Dan Zhao[1]*, Tingting Wang[2], and Dongfeng Jiang[1]

[1] Department of Petroleum Engineering, Northeast Petroleum University, China
[2] Department of Electrical Engineering and Information, Northeast Petroleum University, China

ABSTRACT

Precisely describing the formation and evolution rules of coal-rock fracturing crack have great value on reservoir fracturing improvement and highly efficient mining of coal bed methane well. In this paper, a non-linear dynamic method is used to study crack damage evolution behavior of coal-rock fracturing. Considering distribution characteristics of natural cracks in coal-rock, and based on damage mechanics, a mathematical model on stress around the tip of coal-rock fracturing crack and crack evolution is developed. Micro-crack amounts, circumferential stress, and axial stress at the crack tip of coal-rock fracturing crack evolution process are used as characteristic indicators to describe crack evolutionary. C-C method is chosen to reconstruct the phase space of coal-rock fracturing crack evolution. Correlation dimension, Lyapunov index, and Kolmogorov entropy are introduced as chaos characteristic quantities of crack evolution system, and the process of coal-rock fracturing crack damage evolution could be calculated and described. Coal-rock mass of Zhangchen mining area in Heilongjiang, China was used as a research object, and the results show that as the radial stress increases, the Kolmogorov entropy and the degree of chaos decrease; also, as the circumferential stress increases, the Kolmogorov entropy increases, and a higher degree of chaos is obtained; increasing the number of micro-cracks evolution raises Kolmogorov entropy up to a critical value, and then the Kolmogorov entropy drops, which means the chaotic degree decreases. The results calculated show that crack formation is a damage evolution process which has chaos characteristics. Finally, we provided a new way for further research on coal-rock fracturing crack evolution regularities.

Keywords: Coal-rock Fracturing, Crack Evolutionary, Chaos Characteristic, Nonlinear Dynamics, Zhangchen Mining Area

INTRODUCTION

Coal bed methane (CBM) is a kind of unconventional natural gas which is created inside the coal-rock during historical evolution of geology. It is also a kind of clean and high-quality energy which can be widely used in industrial fuel, power fuel, chemical fuel, living fuel etc. At present, hydro fracturing is a major technology of CBM production, and more than 90% coal bed methane wells are reconstructed by hydro fracturing.

*Corresponding author
Dan Zhao
Email: zdwork201506@163.com

Coal-rock layer hydro fracturing is a technology that uses high-pressure pump units to inject high viscosity fracturing fluid into well by a delivery capacity higher than accepting fluid ability of reservoirs, which could generate high pressure. When the pressure is over crustal stress around the wall of the well and reach the tensile strength of the rock, layers will generate cracks. Because of the development of the coal-rock crack system, weak structural planes (like bedding plane, cleat face, secondary joint surface etc.) at all levels of coal-rock hydro fracturing crack start breaking, extending and widening, which leads to a very complicated network crack system [1-3]. Crack sizes are different in the crack network system, and crack distributions are confusing and complicated, so description is limited.

Researchers at home and abroad have carried out some research on coal-rock fracturing. Olovyanny et al. [4] set up the mathematical model of coal-rock hydro fracturing, and the initiation fracturing pressure, and the direction of coal-rock hydro fracturing are obtained by calculating; Lekontsev et al. [5] used Berezovskaya coal mine as a research object and performed research on the initiation fracturing and expanding regulations of oriented fracturing crack; Li Tonglin et al. [6] researched into basic theories like basic mechanic properties of coal-rock, crack forming conditions of coal-rock hydro fracturing, crack patterns., cracking angle positions etc. Huang Bingxiang et al. [7] set up the calculation model on cleating begin to fracturing of coal-rock hydro fracturing base on fracture mechanics and found the minimum initiation pressure of coal-rock hydro fracturing and crack extension length; Li Chengcheng and Pan Yishan [8] combine theories with numerical simulation to talk about the influences of different crustal stresses on the pressure of cleating begin to fracturing of coal-rock hydrofracturing and

crack expanding state, set up a mechanical model, and analyze the direction and shape of cracks of coal-rock at different crustal stresses; the results show that the initiation pressure is connected with crustal stress difference, and the more the crustal stress difference becomes, the less the initiation pressure is. Ding Jinli et al. of Tsinghua University [9] studied the necessary conditions of sample hydro fracturing destructions by thin and thick walled cylindrical specimen and soil materials; Chen Mian et al. [10] simulated formation conditions by system of large-sized true triaxial simulation test, and used cement mortars as the experimental materials to study the trends and changes of crack width of hydro fracturing cracks by simulation. Deng Guangzhe et al. [11] used coal samples of Tongchuan mine area to study the controlling parameters of crack expanding behaviors, permeability changes of coal samples, and crack expanding behaviors under complicated stresses.

The development of chaos in recent years provides a theoretical basis for research on rock nonlinear characteristics. Peng Tao and He Manchao [12] studied the chaos characteristics in the process of coal soft rock deformation. Yin Guangzhi et al. [13-15] conducted a series of researches on rock microscopic fracture process and bifurcation and chaos characteristics in micro-crack evolution by nonlinear theories, and they discussed the nonlinear evolution characteristics of rock burst occurrence mechanism. Zhou Hui et al. [16] studied chaos characteristics in the process of mining quake by physical cellular automate model, and set up the nonlinear prediction techniques of mining quake system. Song Weiyuan et al. [17] proposed the methods on the prediction and forecasting of rock burst in a certain amount of time scale, which was based on studying the chaos characteristics of rock burst. Liu Chuanxiao et al. [18] used chaos dynamics to analyze the evolutionary process of the complete stress-strain system from a high chaotic state to a low chaotic state. Li Xibing et al. [19-20] studied the regularities of rock acoustic emission and the situations of chaotic attractors based on a chaos theory, and put forward corresponding prediction methods. Zhu Fangcai et al. [21] used rock-testing machine as a rock mechanics system, and researched chaotic behaviors in rock uniaxial compression process. Jiang Yongdong et al. [22] studied the chaos characteristics in a complete stress-strain system based on the evolutionary process of acoustic emission hits. Ai Chi et al. [23] discovered the chaos characteristics in the research of rock hydrofracturing process, and characteristics of micro-crack evolution. Tao Hui et al. [24] studied the chaos characteristics in the process of rock burst destruction based on micro-earthquake time series. This paper studies the distribution characteristics of natural fracture in coal-rock mass and sets up the basic model of coal-rock fracturing crack evolution by choosing micro-crack amounts, circumferential stress and axial stress at the crack tip of coal-rock fracturing crack evolution process as characteristic indicators to describe crack evolutionary. C-C method is chosen to reconstruct the phase space of coal-rock fracturing crack evolution. Correlation dimension and introducing Lyapunov index and Kolmogorov entropy as chaos characteristic quantities of crack evolution system, and then the process of coal-rock fracturing crack damage evolution could be calculated and described. Coal-rock mass of Zhangchen mining area in Heilongjiang, China is used as a research object, the chaotic characteristic parameters of the three characteristic data are calculated to analyze the chaotic characteristic of crack evolution. The calculated results show that coal-rock fracturing

crack evolution has a chaos characteristic, and crack formation is a damage evolution process, which has chaos characteristics. The current work provides a new way for further research on coal-rock fracturing crack evolution regularities.

THE MODEL OF COAL-ROCK FRACTURING CRACK EVOLUTION

The Mathematical Models of Crack Tip Stress

Describing fracturing micro-crack evolution as a dynamic process in which micro-cracks nucleate, expand, and connect, it deals with crack patterns and expansion rates. The radial damage stress of nonlinear tensor near the crack tip is given by:

$$n_k^{(s)}n_l^{(s)}e_k^{(s)}e_l^{(s)}:\sigma_{ij}^{r\mathrm{dyn}(s)}$$
$$=\frac{K_I^{\mathrm{dyn}(s)}(i)\tilde{l}}{\sqrt{2\pi}(\tilde{l}+a)}\sqrt{\frac{1}{r_1}}$$
$$-n_k^{(s)}n_l^{(s)}e_k^{(s)}e_l^{(s)}:\sigma_{ij}^{h(s)} \tag{1}$$

$$n_k^{(s)}n_l^{(s)}e_k^{(s)}e_l^{(s)}:\sigma_{ij}^{\theta\mathrm{dyn}(s)}$$
$$=\frac{1+\sin\phi}{1-\sin\phi}\left(n_k^{(s)}n_l^{(s)}e_k^{(s)}e_l^{(s)}:\sigma_{ij}^{r\mathrm{dyn}(s)}\right)$$
$$+n_k^{(s)}n_l^{(s)}e_k^{(s)}e_l^{(s)}:\sigma_{ij}^{c} \tag{2}$$

Here $n_k^{(s)}$ and $n_l^{(s)}$ are the groups divalent tensors of crack plane position; $e_k^{(s)}$ and $e_l^{(s)}$ are unit basic vector on the axis direction; $\sigma_{ij}^{r\mathrm{dyn}(s)}$ is radial apparent stress at the crack tip of coal-rock fracturing; $K_I^{\mathrm{dyn}(s)}(i)$ is dynamic stress intensity factor, and r_1 is the distance from crack tip; m is the apparent stress on the vertical direction of the ith crack surface; $\tilde{l}$ is the average characteristic value in forming micro-crack, and α is the length of increasing micro-crack; σ_{ij}^c is the expression forms of residual stress during rock damage.

Hypothesis crack evolution deals with rock mass characteristics around cracks and evolution rate by defining dynamic stress intensity factor $K_I^{\text{dyn}(s)}(i)$ of crack evolution as reads:

$$K_I^{\text{dyn}(s)}(i) = [k_1 K_d^{(s)}(i) + k_2 K_v^{(s)}(i)] K_I \tag{3}$$

where

$$
\left\{
\begin{aligned}
K_d^{(s)}(i) &= \frac{1}{\sqrt{I - \ln \dfrac{A+B}{I - \sum_{s=1}^{N_2} \int_v n_{0f}^{(s)} \otimes n_{0f}^{(s)} dv_0^{f(s)}}}} \\
K_v^{(s)}(i) &= \frac{v_e - v}{\sqrt{v_e^2 - 0.919 \dfrac{\sqrt{2(1-2\mu)}v_e}{\sqrt{(1-\mu)}} \left(1 - \dfrac{\sqrt{1-2\mu}}{\sqrt{2(1-\mu)}}\right)^2 v}} \\
K_I &= (p_{ij}^{(s)}(i) - n_k^{(s)} n_l^{(s)} e_k^{(s)} e_l^{(s)} : \sigma_{ij}^{h(s)}) \sqrt{\pi a}
\end{aligned}
\right. \tag{4}
$$

$$n_i = F_1(\sigma) \cdot (n_0 + m_0) + F_2(\sigma) \cdot (N - (n_0 + m_0)) \cdot \rho_i \tag{5}$$

where, N is the total number of micro cracks in effective growing zone, and ρ_i is the density of micro cracks defined as $\rho_i = (n+m)/N$; F_1 is the survival rate of micro cracks (generally 0.8-0.84), and $F_2(\sigma)$ is the impact capability of fracturing loading stress, zero dimension. As for sandstone fracturing development, we choose:

$$F_2(\sigma) = \frac{1}{(1-D)} \frac{\ln \sigma}{\ln \sigma_c} \tag{6}$$

Substituting Equation 9 into Equation 8 gives:

$$n_i = F_1 \cdot (n_0 + m_0) + \left[\frac{1}{(1-D)} \frac{\ln \sigma}{\ln \sigma_c} \cdot \right]$$
$$(N - n_0 - m_0) \frac{n_0 + m_0}{N} \tag{7}$$

where, σ is hydro fracturing loading stress. Thus, one may obtain:

$$n_i = F_1(n_0 + m_0) + \frac{1}{(1-D)} \frac{\ln(\delta p_f(t) - \sigma_\theta)}{\ln \sigma_c} \cdot$$
$$(N - n_0 - m_0) \frac{n_0 + m_0}{N} \tag{8}$$

where, D is the damaging variable.

Phase Space Reconstruction of Coal-Rock Fracturing Crack Evolution

Crack evolution is a complicated nonlinear dynamic system; its mathematical model is difficult to drive by limited variables. To analyze whether the crack system has chaos characteristics, we need to figure out the following characteristic quantities: correlation dimension, Lyapunov index, Kolmogorov entropy etc. These characteristic quantities need to be calculated in phase space, so we need to reconstruct phase space at first. In this essay, we choose time delay reconstruction method to study the chaos characteristics of cracks, and we use time series {x(t)} to reconstruct the phase space of crack evolution dynamic system.

Hypothesis n-dimension ($n \geq m$) phase space of crack evolution dynamic system is expressed as:

$$x^{(n)} = f_i\left(x, x', \cdots, x^{(n-1)}\right) \tag{9}$$

where, m is embedded dimension, and x^i,s (i=1,2,...,n) are variables that participate in crack evolution dynamic system.

Time-evolution form described in Equation 12 is given by:

$$X(t) = \left(x(t), x'(t), \ldots x^{(n-1)}(t)\right) \tag{10}$$

We can use discrete and difference equations to replace all-order derivatives as given in Equation 13:

$$X(t) = \left(x(t), x(t+\tau), \ldots x(t+(n-1)\tau)\right) \tag{11}$$

Equation 14 is the reconstruct phase space based on time delay reconstruction method, and τ is delay time. M and τ are major parameters in phase space reconstruction system of crack evolution. If τ is too small, two independent coordinate components could not be offered, thereby leading to information redundancy. If τ is too large, two coordinates are completely independent in a statistical sense, and the track of chaos attractor in two directions projection showed no correlation, named "uncorrelation". If embedded dimension m is too small, attractors could be folded and even self-intersected, which leads to reconstructed attractors totally different from the original attractors. If m is too large, noise will be amplified and the calculated amount will be increased, which is undesirable. Therefore, in order to guarantee the quality of reconstructed phase space, we need to use a proper method to get the value of m and τ. C-C method is used to reconstruct phase space.

Using the hypothesis of time series (δ_1, δ_2, ..., δ_i, ..., δ_N) of coal-rock fracturing system, phase space can be reconstructed by C-C method:

$$X_i = \left[\delta_i, \delta_{i+t}, \ldots, \delta_{i+2t}, \ldots, \delta_{i+(m-1)t}\right]^T$$
$$(i = 1,2,\ldots,N-(m-1)t)X_i \in R^m \quad (12)$$

where, t (t=1,2,...,n) is the delay time index.

Correlation integral of time series is defined by:

$$C(m,N,r,t)$$
$$= \frac{2}{M(M-1)}\sum_{j=1}^{M}\sum_{i\neq j+1}^{M}\Theta\left(r-\|X_i-X_j\|\right) \quad (13)$$

where, r>0 and $\Theta(a)=\begin{cases}0 & a<0\\1 & a>0\end{cases}$; C (m, N, r, t) is Probability of distance between two points less than r in phase space; N is the size of data set, and M=N-(m-1)t is phase amount in m-dimension phase space; ||···|| is Euclidean distance. The limit of C(m, N, r, t) is C(m, r, t).

Introducing function $S_1(m, N, r, t)$ defines monitoring statistics as:

$$S_1(m,N,r,t)$$
$$=C(m,N,r,t)-C^m(1,N,r,t) \quad (14)$$

Determining the delay time is to investigate the dependence of $S_1(m, N, r, t)$ to delay time index. In the actual calculation, time series {δi} (i=1,2,...,N) is divided into t equal non-intersect subsequences, i.e.:

$$S_2(m,r,t)$$
$$=\frac{1}{t}\sum_{s=1}^{t}\left[C_s(m,r,t)-C_s^m(1,r,t)\right](m=2,3,\ldots) \quad (15)$$

Numerical experiments indicate that, when $2 \le m \le 5$, $\theta/2 \le r \le 2\theta$, and $N \ge 500$, $S_2(m, N, r, t)$ can represent the correlation of the series, where σ is the mean squared error of the dataset. By setting m=2,3,4,5; r_i=0.5iσ; and i=1,2,3,4, one may define the following average quantities:

$$\bar{S}_2(t) = \frac{1}{16}\sum_{m=2}^{5}\sum_{j=1}^{4}S_2(m,r,t) \quad (16)$$

$$\Delta\bar{S}_2(k) = \frac{1}{4}\sum_{m=2}^{5}\Delta S_2(m,t) \quad (17)$$

$$S_{2cor}(k) = \Delta\bar{S}_2(t) + \left|\bar{S}_2(t)\right| \quad (18)$$

The first zero point of $\bar{S}_2(k)$ or the first minimal value of $\Delta\bar{S}_2(k)$ are selected as the optimum time delay. Together with $\bar{S}_2(k)$ and $\Delta\bar{S}_2(k)$, one may find the global minimal value of $S_{2cor}(k)$ as the length of time window of time series.

Certainty Test of Time Series

The complexity of time series of coal-rock fracturing crack evolution has external random factors and its own internal factors of deterministic dynamics. Therefore, it is necessary to conduct certainty tests of time series of crack evolution.

By setting Y_0 as a fixed vector in Rm, $Y_1,Y_2...Y_i$ represent i neighboring points in phase space, and $Z_1,Z_2,...,Z_i$ are corresponding mapping values. Transfer vector is defined as:

$$V_i = Y_i - Z_i \ (i = 0,1,...,l) \tag{19}$$

Transfer error is given by:

$$E_{error} = \frac{1}{l+1} \sum_{i=0}^{l} \frac{\|Y_i - \langle V \rangle\|^2}{\|\langle V \rangle\|^2} \tag{20}$$

where, <V> is the mathematical expectation of the transfer vector.

Calculation of Correlation Dimension

Reconstructing phase space Rm by time series, one can calculate correlation integral:

$$C(m,r) = \frac{2}{M(M-1)} \sum_{j=1}^{M} \sum_{i \neq j+1}^{M} \Theta(r - \|X_i - X_J\|) \tag{21}$$

Variables in Equation 24 have the same meaning as the one in Equation 16.

As for the fractal structure, because of its self-similarity characteristic, an interval of r has the following scale relation:

$$C(m,r) \propto r^\gamma \tag{22}$$

We can affirm that the index γ in Equation 25 is a dimension, too. Actually, γ is approaching correlation dimension. Correlation dimension is defined by:

$$D_2 = \lim_{t \to 0} \frac{\ln C(m,r)}{\ln r} \tag{23}$$

where, D_2 is correlation dimension.

Calculation of Lyapunov Index

Lyapunov index, is an essentially quantitative index to measure the process of crack evolution of coal-rock fracturing. If the maximum Lyapunov index is positive, the system is chaotic, otherwise the system remains steady. As for the calculation of Lyapunov index, this paper adopts improved small-data method based on Wolf [25]. The main calculating processes are:

(1) Reconstructing phase space R^m by time series of crack system:

$$X_i = \left[\delta_i, \delta_{i+t}, \delta_{i+2t}, ..., \delta_{i+(m-1)t} \right]^T$$
$$(i = 1,2,...,N - (m-1)t) \, X_i \in R^m \tag{24}$$

(2) Searching every X'_j nearest to X_j in phase space and limit them to short separation, that is:

$$d_j(0) = \min \|X_j - X'_j\|, |j - j'| > P \tag{25}$$

where, P is the average period of time series.

(3) For every X'_j in phase space, calculating distance $d_j(t)$ after t discrete time steps:

$$d_j(t) = \min \|X_{i+1} - X'_{i+1}\|$$
$$t = 1,2,\cdots,\min(M-j, M-j') \tag{26}$$

(4) For every t, calculating average $x(t)$ of $\ln d_j(t)$ of all j:

$$x(t) = \frac{1}{q\Delta t} \sum_{j=1}^{q} \ln d_j(t) \tag{27}$$

where, q is the amount of nonzero $d_j(t)$.

(5) Drawing the curve of $x(t)$-t, from the straight slope of which the maximum Lyapunov index λ can be obtained via least square method.

Calculation of Kolmogorov Entropy

Kolmogorov entropy can evaluate chaos extent of the chaos system quantitatively [26]. $K=0$ means system is regular, and $K>0$ means the crack evolution has chaos characteristics. The more complex the crack evolution system is, the larger the K entropy is, and the higher chaos extent becomes.

Considering strange attractor orbits $\{x_1(t), x_2(t),...x_m(t)\}$ of an m-dimension dynamic system, set phase space is divided into im size boxes, and the system condition can be observed in τ time interval. Set $P(i_1,i_2,...im)$ is the joint probability of $X(t=\tau)$ in i1 box ... $X(t=m\tau)$ in im box. Thus, Kolmogorov entropy is given by:

$$K = -\lim_{\Delta t \to 0} \lim_{\varepsilon \to 0} \lim_{m \to \infty} \frac{1}{m\tau} \ln \sum_{i_1 \cdots i_m} P(i_1,i_2...i_m) \log_2 P(i_1,i_2...i_m) \tag{28}$$

For the crack evolution chaos system of coal-rock fracturing K_2 is not zero ($K_2 \neq 0$). Therefore, we can estimate K_1 in virtue of 2-steps Renyi entropy. 2-steps Renyi entropy is defined as:

$$K_2 = -\lim_{\Delta t \to 0} \lim_{\varepsilon \to 0} \lim_{m \to \infty} \frac{1}{m\tau} \ln \sum_{i_1,...,i_m} p^2(i_1,...i_m) \tag{29}$$

Correlation integral C(m,r) has the following relationship:

$$C(m,r) \approx \sum_{i_1,...,i_m} p^2(i_1,...,i_m) \qquad (30)$$

Through setting and analyzing, one may obtain:

$$K_2 = -\lim_{\Delta l \to 0} \lim_{\varepsilon \to 0} \lim_{m \to \infty} \frac{1}{m\tau} \ln C(m,r) \qquad (31)$$

By select a proper delay time τ to reconstruct m-dimension phase space, we obtain:

$$K_2(m,r) = \frac{1}{\tau} \ln \frac{C(m,r)}{C(m+1,r)} \qquad (32)$$

set phase space is divided into im size boxes, and the system condition can be observed in τ time interval. Set P(i_1,i_2,...im) is the joint probability of X(t=τ) in i_1 box ... X(t=mτ) in im box. Thus, Kolmogorov entropy is given by:

$$K_2 = \lim_{\substack{m \to \infty \\ \tau \to 0}} K_2(m,r) \qquad (33)$$

Example Verification of Coal-Rock Crack Evolution

Coal sample for petrology is taken from right six alley, 3# layer, west three area, Zhangchen mining area, Jixi, Heilongjiang. Buried depth is 901-910 m. The sample size is 400×300×180 mm, and the development density of butt cleat ranges 4-5 pieces/10 cm (Figure 1); the density of face cleat ranges 11-12 pieces/10 cm (Figure 2), and the slope angle of cleat is 16°; the length of natural crack ranges 3-6 cm, and crack width is 1.26 mm. The sample size of coal-rock mechanics parameter test is Φ25×50~75mm and Φ50×25~50 mm (Figure 3); the surrounding rock pressure is 10 mPa. The sample size of coal-rock mechanics is 300×300×300 mm (Figure 4). The relationship between the net pressure in crack and acoustic emission is shown in Figure 5.

Figure 1: Cleat feature of coal vertical to face cleat direction in Right 6 Alley.

Figure 2: Cleat feature of coal vertical to butt cleat direction in Right 6 Alley.

Figure 3: The coal sample of compression test (Φ25).

Figure 4: Experiment core for fracturing (300mm×300mm×300mm).

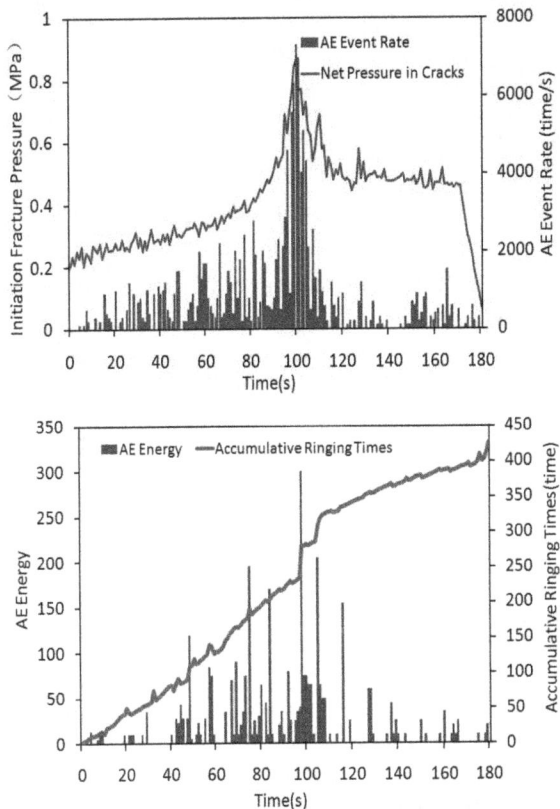

Figure 5: The relationship between the net pressure in crack and acoustic emission.

Reconstruction Parameters

Figures 6-8 display curves of $\bar{S}_2(t)$, $\Delta \bar{S}_2(t)$ and $S_{2cor}(t)$ as a function of delay time index t; it is clear that that the first time $\Delta \bar{S}_2(t)$ reaching the minimum value is earlier than the time $\bar{S}_2(t)$ reaching zero, and the delay time is equal to 3 (t=3). When t=13, $S_{2cor}(t)$ reaches the minimum value, a delay time window of 13 τ_s (τ_w=13τ_s), and the minimum embedding dimension m=5.

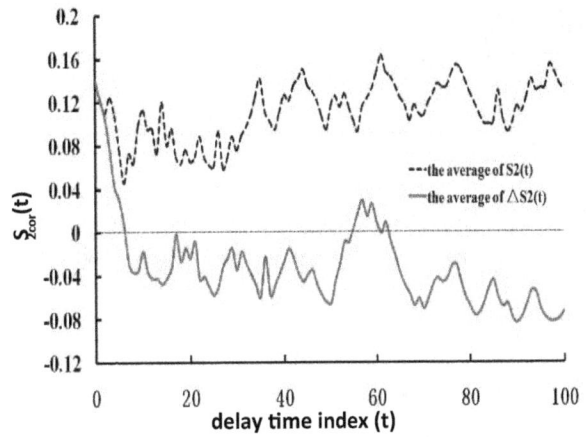

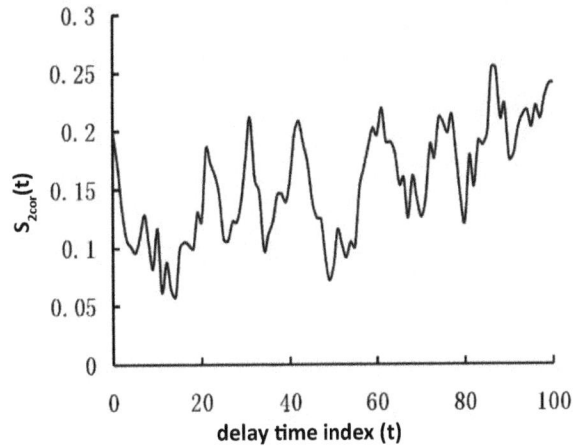

Figure 6: The variation of the number of micro cracks $S_2(t)$, $\Delta \bar{S}_2(t)$ and $S_{2cor}(t)$ with delay time index (t).

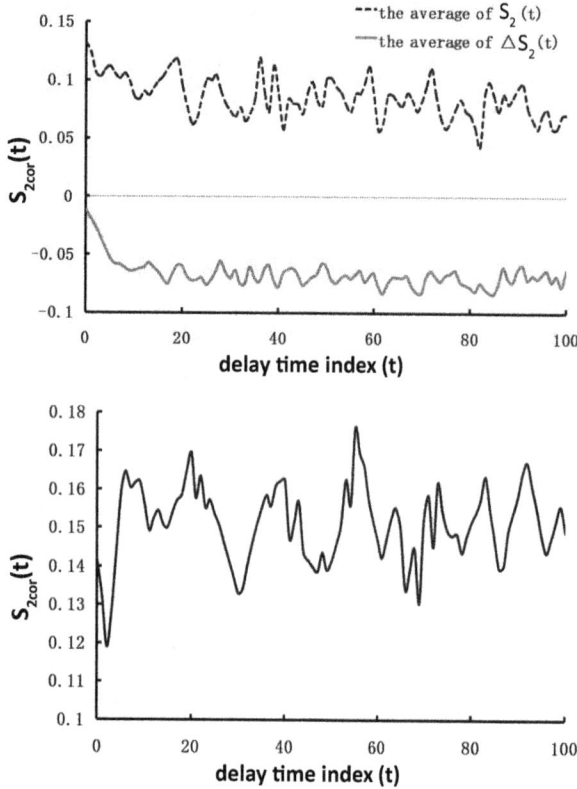

Figure 7: The variation of the circumferential stress $\bar{S}_2(t)$, $\Delta\bar{S}_2(t)$, and $S_{2cor}(t)$ with delay time index (t).

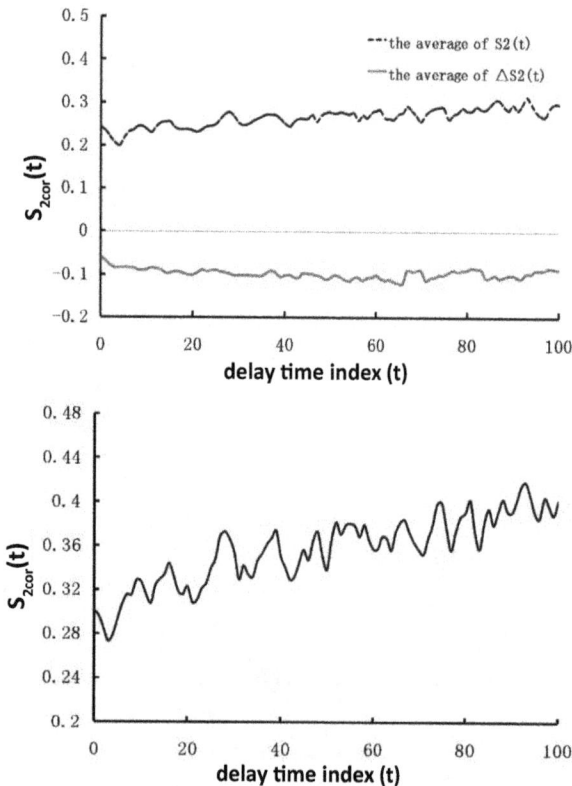

Figure 8: The variation of the radial stress $\bar{S}_2(t)$, $\Delta\bar{S}_2(t)$ and $S_{2cor}(t)$ with delay time index (t).

We then need to reconstruct micro-crack amount series, the phase space of circumferential stress, and the axial stress time series at the crack tip, and choose embedded dimension m=1,2,...12. i(m,r) can be calculated by plotting curve lnC(m,r) versus ln$r$ as follows:

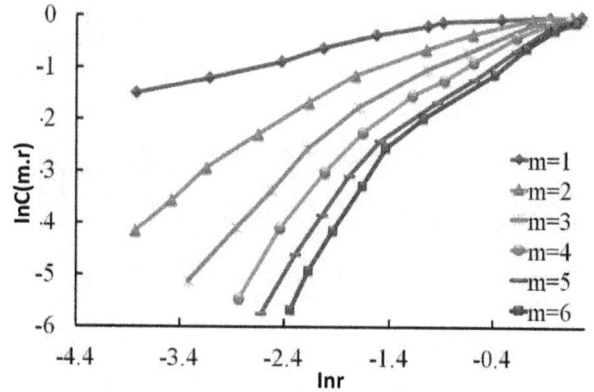

Figure 9: Micro crack amount time series curve lnC(m,r)-ln$r$.

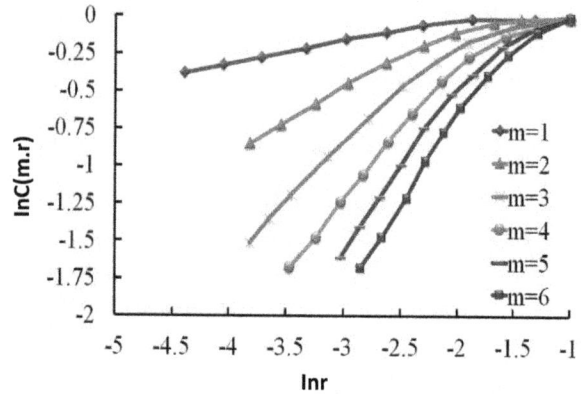

Figure 10: Crack tip circumferential stress time series curve lnC(m,r)-ln$r$.

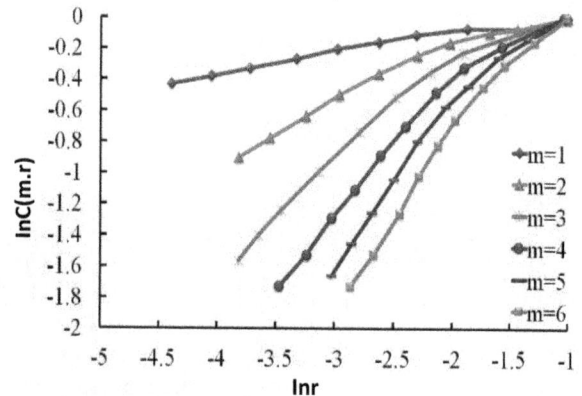

Figure 11: Crack tip axial stress time series curve lnC(m,r)-ln$r$.

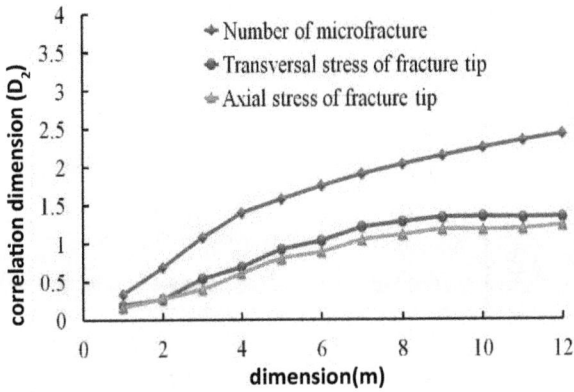

Figure 12: Correlation dimension (D_2) variation with embedded dimension m.

From Figures 9-11, we can see obvious straight line segments in curve lnC (m,r) versus lnr. When m is small, the slopes of straight line segments are small and the distances between lines are large. The slopes increase and distances decrease and gradually move closer with the increase of m. When m≥5, the straight line segments intend to be in parallel with each other and become increasingly intensive. The slopes do not change and can be considered as a fixed value, which is defined as the correlation dimension as a characterization of these series.

Figure 12 shows the variation of correlations dimension (D_2) versus embedded dimension (m). The correlation dimensions of these three series are 2.12, 1.34, and 1.23. Due to their non-integer, we can preliminary concluded that the time series of coal-rock fracturing crack system contains chaotic components.

Calculation of Lyapunov Index

Figures 13-15 show the variation of x(t) versus time (t) evolution for various when embedding dimensions m= 3, 4, 5, 6, and 7. From Figures 13-15 we can infer that curves are all increasing functions and the maximum Lyapunov index is positive, which means adjacent tracks are divergent; Moreover, they illustrate that the system has chaotic characteristics. For those curves close to straight lines after embedding dimension, the

least square fitting method is used to handle the part. We obtained the maximum Lyapunov index of three time series respectively equal to 0.0625, 0.0063, and 0.023.

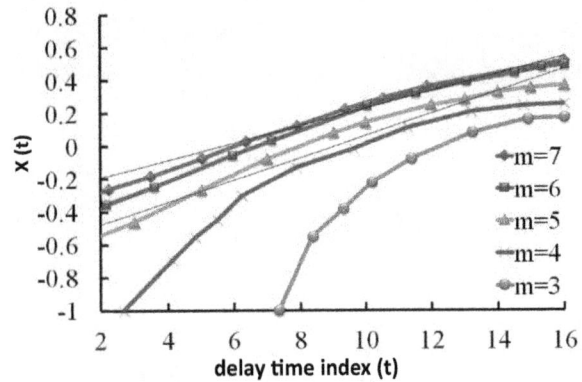

Figure 13: Maximum Lyapunov index of micro crack as a function of delay time.

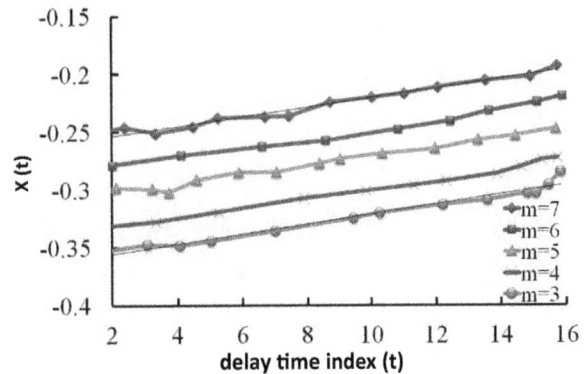

Figure 14: Maximum Lyapunov index of crack tip circumferential stress as a function of delay time.

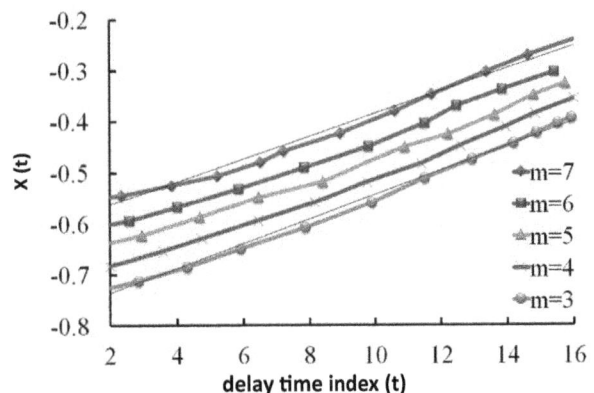

Figure 15: maximum Lyapunov index of crack tip axial stress as a function of delay time.

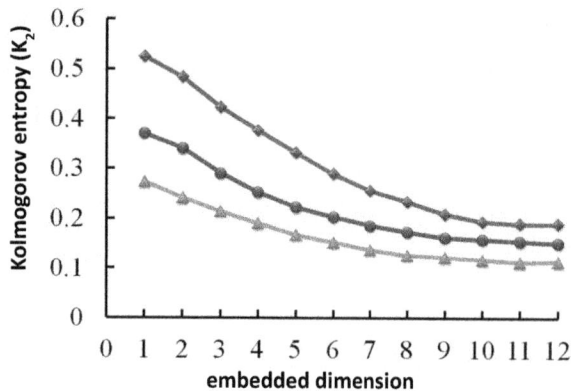

Figure 16: K_2 as a function of m.

Calculation of Kolmogorov Entropy

Using the calculation results of correlation dimension, Kolmogorov entropy can be estimated by Equation 36. Figure 16 shows K_2 variations as a function of embedded dimension *(m)*, in which from top to bottom are micro-crack amount time series, crack tip circumferential stress time series, and crack tip axial stress time series.

As it is shown in Figure 16, a common characteristic of Kolmogorov entropy in three time series is that K_2 gradually decreases and intends to be stable with increasing embedded dimension m. The stable value K_2 needed are 0.179, 0.153, and 0.112, respectively in micro-crack amount time series, crack tip circumferential stress time series, and crack tip axial stress time series.

Calculation and Analysis of 10 Pairs of Experiments

According to the results obtained, we continue to study coal-rock fracturing crack evolution by carrying out 10 experiments and observing the trend Kolmogorov entropy changed with characteristic indicators; the results are shown in Figure 17.

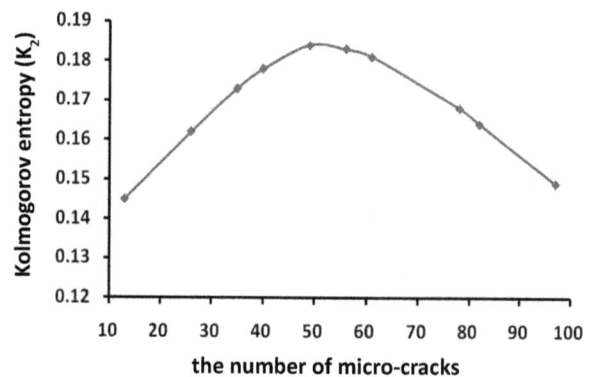

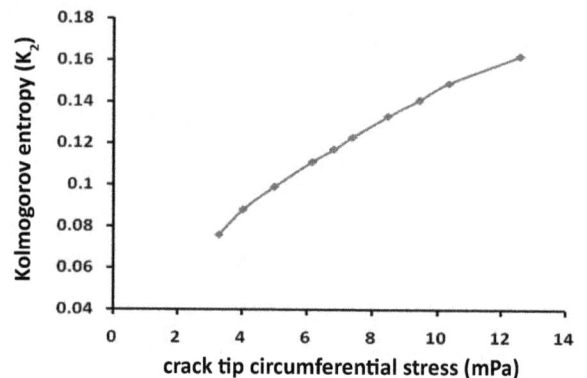

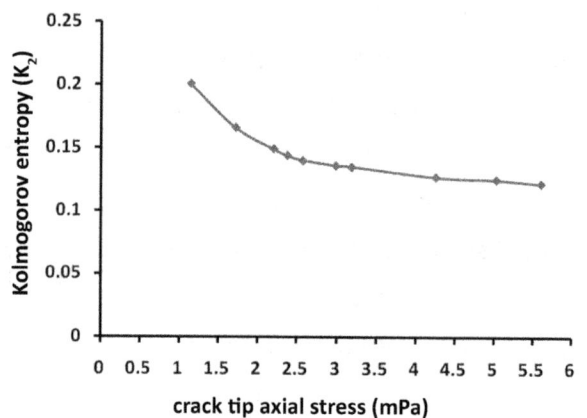

Figure 17: Kolmogorov entropy as a function of characteristic indicators.

It is concluded from the analyses and calculations mentioned above that the correlation dimensions are fractional values, and the maximum Lyapunov index and Kolmogorov entropy are all greater than zero. All these factors meet the conditions of chaos existence, which indicates that crack evolution has a chaotic characteristic. The calculation results provide scientific basis for the evolution and instability of coal-rock fracturing cracks based on the chaos theory.

CONCLUSION

This paper deals with coal-rock fracturing crack evolution based on chaos theory, which is a cutting-edge topic both at home and abroad, and has a great academic value and theoretical significance. The results obtained are as follows:

1. In this paper, based on the model of micro crack evolution developed and by considering the dynamic and static process of fracturing, we derived the expressions of the mechanical distribution characteristics and the numbers of crack evolution of the crack tip.

2. Based on the chaos theory of nonlinear dynamics method, we discussed the evolution behavior of cracks, micro crack amounts, and circumferential stress, and axial stress at crack tip of coal-rock fracturing crack evolution process were used as the characteristic indicators. Correlation dimension, Lyapunov index, and Kolmogorov entropy were introduced as the chaos characteristic quantities. Through the results of 10 experiments, the correlation dimensions were obtained to be fractional values, and the maximum Lyapunov index and Kolmogorov entropy were all greater than zero; these results indicated that coal-rock fracturing crack evolution and the crack network formation process had chaos characteristics.

3. Based on the judgment of chaos, the larger the Kolmogorov entropy is, the higher the crack evolution degree is, and the more complicated the crack system becomes. We calculated the Kolmogorov entropy of 10 experimental data, and the results showed that as the radial stress increased, the Kolmogorov entropy decreased, and a lower degree of chaos was obtained; by increasing the circumferential stress the Kolmogorov entropy increased, and a higher degree of chaos was achieved; as the number of micro cracks evolution increased, the Kolmogorov entropy rose up to a critical value, and it then dropped, which meant that the chaotic degree increased before they were reduced. All in all, in the process of fracturing, large main cracks are tended to form when radial stress increases, while bifurcation micro cracks are created when circumferential stress increases. At the beginning, the crack system becomes more complicated with cracks evolution, and when the evolution reach a degree, cracks are combined with each other, and the crack system becomes steady.

ACKNOWLEDGMENTS

This work is financially supported by The National Natural Science Foundation of China (51404073), China Postdoctoral Science Foundation Funded Project (2014M550180), Heilongjiang Province Postdoctoral Science Foundation Funded Project (LBH-TZ-0503), Youth Science Foundation of Northeast Petroleum University (2013NQ105), The Scientific Research Fund of Heilongjiang Provincial Department of Education (12541090), PetroChina Innovation Foundation(2013D-5006-0209), and Academic Backbone of Heilongjiang Province University Youth Support Program (1253G011).

REFERENCES

1. Junfeng G., Wei T., and Xueqin L., "Optimization of Hydraulic Fracturing Technique in CBM Well in Qinshui Basin," *China Coalbed Methane*, **2011**, *8*, 25-29.
2. Shuheng T., Baocun Z., and Zhifeng Y., "Effect of Crustal Stress on Hydraulic Fracturing in Coalbed Methane Wells," *Journal of China Coal Society*, **2011**, *36*, 65-69.
3. Zhigang L., Shengli F., and Xiaoming W., "Research on Mechanical Property Test and Mechanism of Hydraulic Fracture of Gas Well in Coal Beds," *Petroleum Drilling Techniques*, **2000**, *28*, 10-13.
4. Olovyanny A. G., "Mathematical Modeling of Hydraulic Fracturing in Coal Seams," *Journal of Mining Science*, **2005**, *41*, 61-67.

5. Lekontsev Y. M. and Sazhin P. V., "Application of the Directional Hydraulic Fracturing at Berezovskaya Mine," *Journal of Mining Science*, **2008**, *44*, 253-258.

6. Tonglin L., "A Preliminary Research on Development of Crack in Coal Seams by Hydraulic Fracture," *Journal of China University of Geosciences*, **1994**, *19*, 537-545.

7. Bingxiang H., Qingying Ch., and Changyou L., "Analysis of Mesoscopic Structure Destroy for Coal Rock Mass by Crack Hydraulic Pressure," *Journal of Hunan University of Science and Technology (Natural Science Edition)*, **2009**, *24*, 1-4.

8. Chengcheng L. and Yishan P. "Numerical Simulation on Coal Hydraulic Fracturing Initiation Pressure and Crack Propagation under Different Ground Stresses," Innovation and Practice on Prevention and Control of Coal Mine Bumps, 2013 High-end BBS.

9. Jinsu D., Qiming T., and Youman G., "Investigation of Behavior of Saturated Foundation Soil under Footing Pressure by Hydraulic Gradient Model Test," *Yantu Gongcheng Xuebao*, **1994**, *16*, 8-20.

10. Mian Ch., Fei P., and Yan J., "Experiments and Analysis on Hydraulic Fracturing by A Large-Size Triaxial Simulator," *Chinese Journal of Rock Mechanics and Engineering*, **2000**, *19*, 868-872.

11. Guangzhe D., Shibing W., and Bingxiang H., "Research on Behavior Character of Crack Development Induced by Hydraulic Fracturing in Coal-rock Mass," *Chinese Journal of Rock Mechanics and Engineering*, **2004**, *23*, 3489-3493.

12. Peng T. and Manchao H., "Study of Chaos Mechanics Behavior of Soft Rock in Mines," *Mine Pressure and Roof Management*, **1997**, *1*, 32-35.

13. Guangzhi Y., Xuefu X. and Jiang X., "Research on Bifurcation and Chaos Characteristics of Mesoscopic Fracture Process of Rock," *Journal of Chongqing University (Natural Science Edition)*, **2000**, *23*(2), 56-59.

14. Guangzhi Y., Gun H., Gaofei D., and Guowen S., "Bifurcation and Chaos Analysis of Coal and Rock Damage under Uniaxial Compression Based on CT Values," *Rock and Soil Mechanics*, **2006**, *27*(9), 1465-1470. (in Chinese)

15. Guangzhi Y., Gaofei D., and Wenli P., "Study on Rock Burst using Stick Slip Model," *Rock and Soil Mechanics*, **2005**, *26*(3), 359-364.

16. Hui Z., "Study on Chaotic Features and Nonlinear Forecasting Theory of Mining Quake," *Chinese Journal of Rock Mechanics and Engineering*, **2000**, *19*(6), 813.

17. Weiyuan S. and Yishan P., "Forecast and Chaos Model of Rock Burst," *Journal of China Coal Society*, **2001**, *26*(1), 26-30.

18. Chuanxiao L., "Research on Strength Test of Sandstone with MTS and its Chaotic Features in Different Stages," Rock and Soil Mechanics, **2004**, *25*(12), 1910-1914.

19. Zhixiang L. and Xibing L., "Chaotic Kinetics of Acoustic Emission in Rock Mass," *Journal of Central South University (Science and Technology)*, **2004**, *35*(4), 671-675.

20. Xibing L. and Zhixiang L., "Research on Chaos and Intelligent Identification of Acoustic Emission in Rock Mass," *Chinese Journal of Rock Mechanics and Engineering*, **2005**, *24*(8), 1296-1300.

21. Fangcai Z., Weijun C., and Jian D., "Chaotic Behavior of Rock Failure under Uniaxial Compression," *Journal of Zhuzhou Institute of Technology*, **2006**, *20*(2), 83-86.

22. Yongdong J., Xuefu X., and Guangzhi Y., "Acoustic Emission, Fractal and Chaos Characters in Rock Stress-strain Procedure," *Rock and Soil Mechanics*, **2010**, *31*, 2413-2418.

23. Chi A., Zhengjun L., and Wanchun Zh., "Study on Chaotic Characteristics of Micro-Crack Damage Evolution under Hydraulic Fracturing on Rock Mass," *Science Technology and Engineering*, **2011**, *11*, 2094-2096.

24. Hui T., Xiaoping M., and Meiying Q., "Chaos Characteristic Analysis of Microseism Time Series of Rock Burst," *Safety in Coal Mines*, **2012**, *43*, 140-143.

25. Wolf A., Swift J. B., and Swinney H. L., "Determining Lyapunov Exponents from A Time Series," *Physica D: Nonlinear Phenomena*, **1985**, *16*, 285-317.

26. Zili X., Yiyang W., and Jiliu Zh., "C-C Average Method for Estimating the Delay Time and the Delay Time Window of Nonlinear Time Series Embedding," *Journal of Sichuan University (Engineering Science Edition)*, **2007**, *39*, 151-155.

4

Source Rock Evaluation of the Cenomanian Middle Sarvak (Ahmadi) Formation in the Iranian Sector of the Persian Gulf

Maryam Mirshahani*, Mohammad Kassaie, and Arsalan Zeinalzadeh

Geochemistry Group, Faculty of Upstream Petroleum Industry, Research Institute of Petroleum Industry (RIPI), Tehran, Iran

ABSTRACT

The middle Sarvak formation (Cenomanian) is one of the stratigraphic units of the Bangestan group in the south of Iran. This formation is stratigraphically equivalent to the Ahmadi member of Kuwait and Iraq. There is geochemical evidence that indicates this unit has a high level of organic richness and can be a possible source rock in various locations. This study focuses on the organic geochemistry of the middle Sarvak formation in the Persian Gulf region. Rock Eval pyrolysis, organic petrography, and kerogen elemental analyses were used in order to evaluate the thermal maturity and determine the kerogen type of the middle Sarvak formation. The results of this study show that middle Sarvak formation has entered the oil window in eastern and western parts of the Persian Gulf, but it is immature in the central parts. This regional pattern of organic maturation is probably a consequence of the regional uplift in the Qatar-Fars arch area. The higher maturities in the eastern and western parts are, on the other hand, attributed to its greater depth of burial. The results of the screening analyses by the Rock-Eval pyrolysis in parallel with the results of maceral and elemental analyses show that the kerogen type in the middle Sarvak formation is mainly a mixture of Types II and III.

Keywords: Middle Sarvak Formation, Rock Eval Pyrolysis, Organic Petrography, the Persian Gulf

INTRODUCTION

The Persian Gulf region is known for its prolific petroleum systems owing to the large number of active source rocks. The middle Sarvak formation is considered as a candidate source rock in the studied area based on geochemical observations. This formation is a part of the Bangestan group and is stratigraphically equivalent to the Mauddud, Ahmadi, Rumaila, and Mishrif formations of the southern Persian Gulf region [1]. The middle Sarvak formation is present throughout the Persian Gulf area. A number of geological and geochemical studies on the Sarvak formation in the south of Iran, particularly in the Zagros area, have been carried out [2-5].The goal of this study is to evaluate the source rock potential of the middle Sarvak formation in the Persian Gulf area.

Geological Setting

The Persian Gulf basin encompasses a thick sedimentary succession with alternating clastic,

*Corresponding author

Maryam Mirshahani
Email: mirshahanim@ripi.ir

carbonate, and evaporite sediments, which makes the area particularly prolific for hosting large hydrocarbon deposits. The oldest sediments in the area are believed to be the evaporites, shales, and dolomites of the Late-Precambrian Hormuz series [6]. Generally, there is little data about the sedimentary history of the Lower Paleozoic in the Persian Gulf region, and the sedimentary record comprises mostly of shales and sandstone with rare carbonates in the Devonian and Early Carboniferous [7]. During the Permian, carbonate shelf deposits of the Dalan formation were deposited under warm and shallow-water conditions. More arid conditions during Mid-Late Triassic formed the evaporite deposits of the Dashtak formation, which mark the end of the carbonate cycles. The middle Jurassic sediments mainly consist of normal marine organic-rich carbonates (the Surmeh formation) [8] which are capped by extensive evaporite deposits (Hith formation) of Tithonian age [9, 10]. During Cretaceous, three main stratigraphic sequences [11] were recorded in the Persian Gulf area: the Lower Cretaceous deposits of the Fahliyan, Gadvan, and Dariyan formations; the middle Cretaceous sediments comprising of the Kazhdumi and Sarvak formations; and the Upper Cretaceous deposits of the Ilam, Laffan, and Gurpi formations [12, 13]. A regional unconformity marks the end of the Cretaceous and the boundary between the Late Cretaceous and Early Tertiary sediments [14]. Orogenic folding of the adjacent Zagros during the Late Tertiary resulted in rapid uplift, extensive erosion, and the formation of a thick clastic wedge (Figure 1).

The Paleozoic structural evolution of the area generally took place along regional basement highs [15]. Recent deformation episodes, however, included a Late Cretaceous event (producing NNE-SSW trending faults) and a Late Cenozoic Zagros Orogeny event (reactivating previous folds and causing a new set of

NW-SE trending folds) [16]. Apart from these events, most of the structures in the Persian Gulf basin are affected by episodic salt movements to varying extents [17].

The morphology of the Persian Gulf is highly affected by the Qatar-Fars Arc [18]. The Qatar-Fars Arc is a first-order structure that was created in the central Persian Gulf following the tectonic movements during the Late Precambrian to Early Cambrian in the region (Figure 2). It is a very large (over 100 km wide and 300 km long) regional gentle anticline [19]. According to offshore seismic data in the study area, this structure has a northeast–southwest direction in the Iranian sector of the Persian Gulf and continues southwards to the Qatar peninsula [20]. As Figure 3 demonstrates, the thicknesses of the Pabdeh, Gurpi, and Kazhdumi formations and Ahmadi member significantly decrease toward the central parts of the study area with a noticeable thinning which can be due to the effect of the Qatar-Fars Arc Paleohigh during depositional time [14].

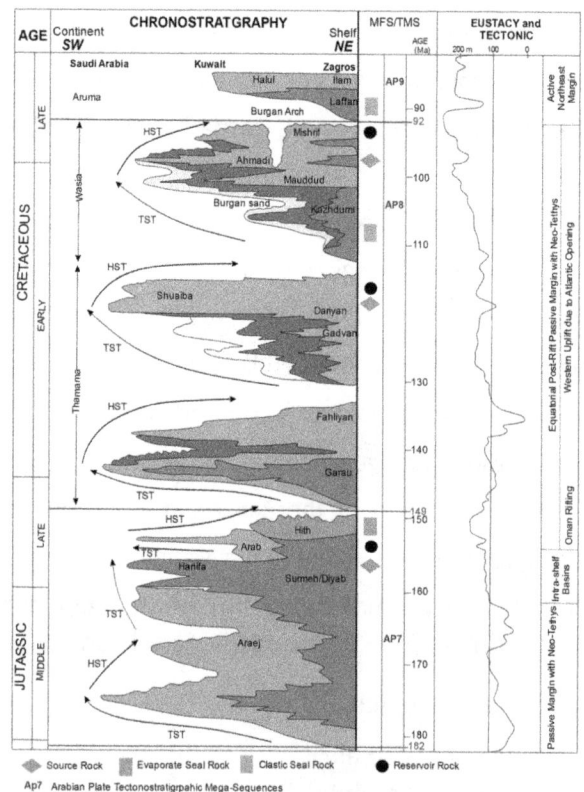

Figure 1: A generalized stratigraphic column of the study area [24].

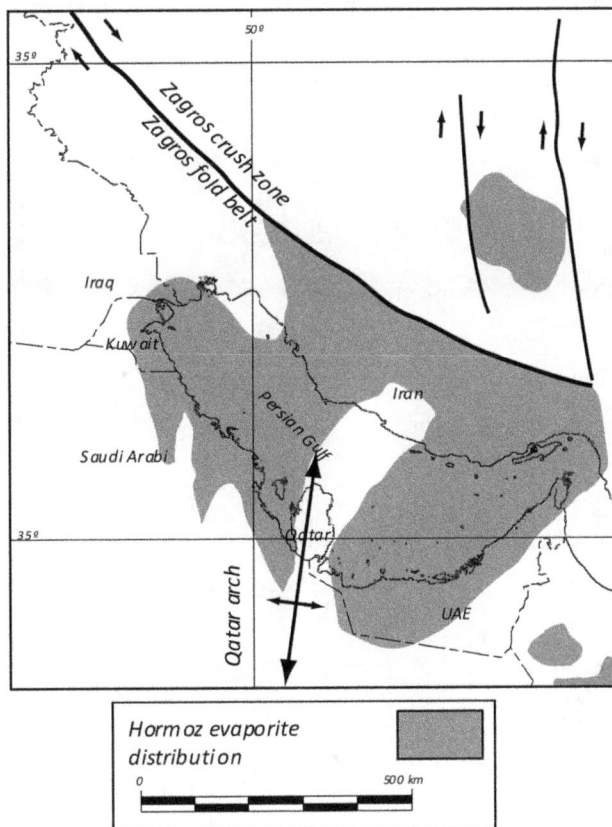

Figure 2: Location of the Qatar-Fars Arc and distribution of Hormoz Salt in the study area [21].

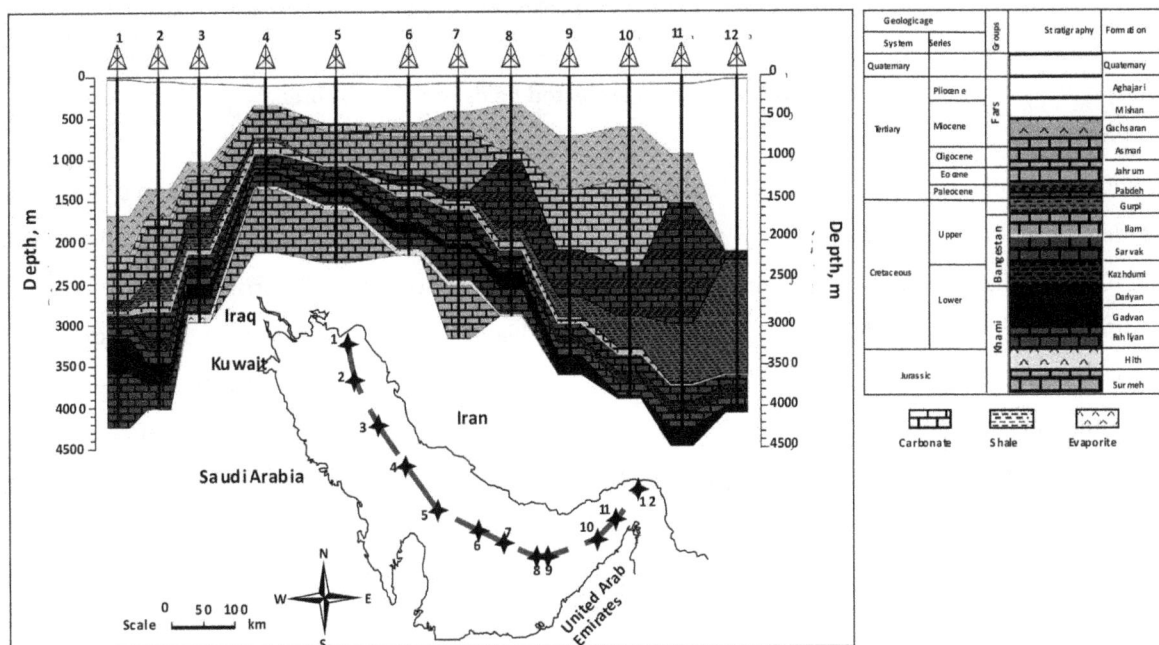

Figure 3: Lithostratigraphic cross section in the Iranian sector of the Persian Gulf through Jurassic to Quaternary (modified after [23]).

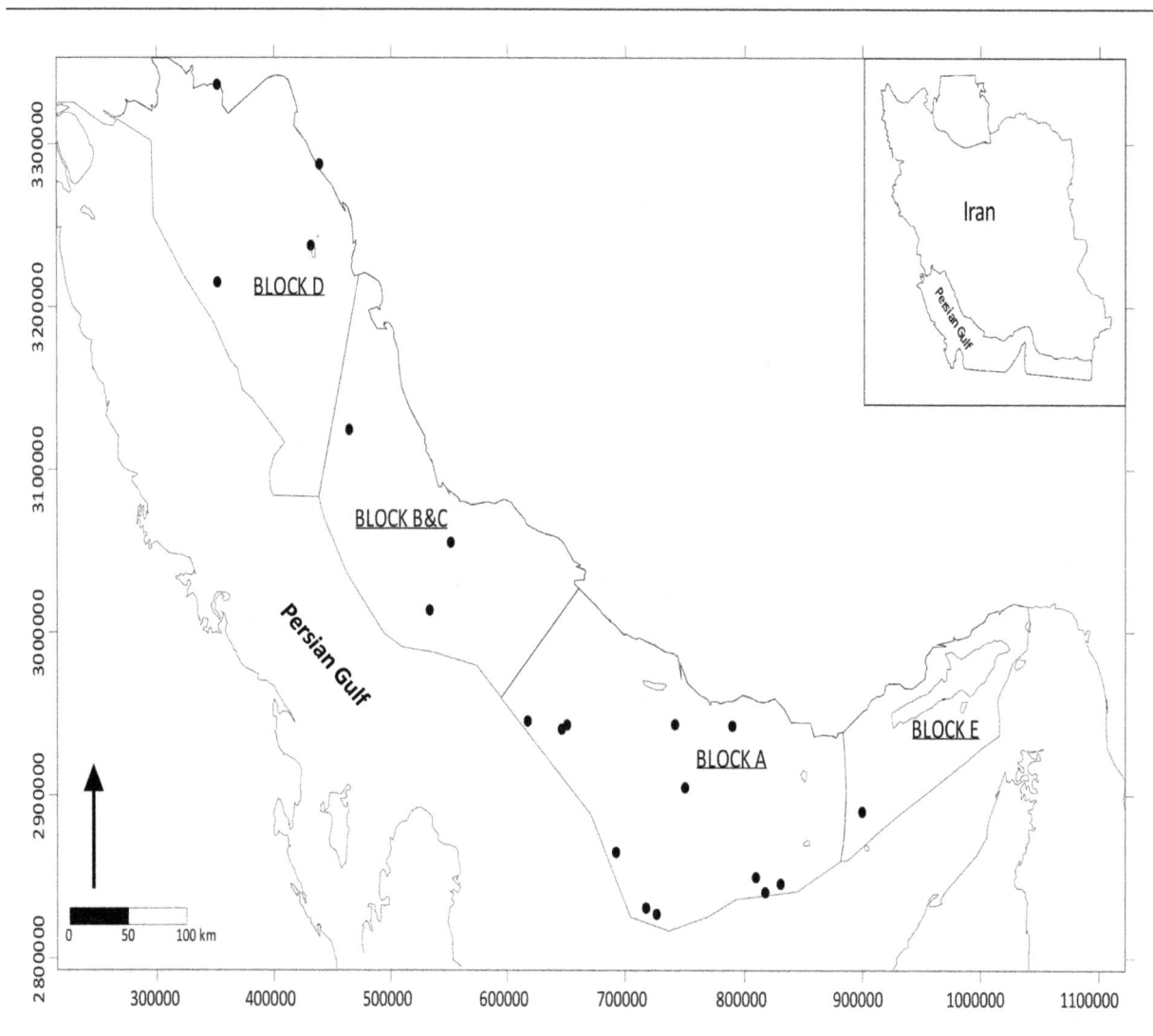

Figure 4: Location map and selected fields of the study area.

The sediments of the middle Sarvak formation were deposited on a passive margin of the Neo-Tethys Ocean. This formation is divided into four depositional members: the Mauddud, bituminous shaley limestone, Khatiyah in the central and western parts of the Persian Gulf, Ahmadi with shaley facies in the northern Persian Gulf, and the Mishrif reefal limestone [1].

The middle Sarvak is present in all wells drilled in the Persian Gulf. The thickness of the formation varies from a minimum 41 meters up to a maximum of 223 meters in the northwestern Persian Gulf [1].

The Ahmadi member is a continuation of the middle Sarvak in the northwestern Persian Gulf (towards Kuwait). Lithologically, it consists of 30 to 60 meters of highly eroded shale with intercalation of thin limestone beds. Most of the present Persian Gulf area was beyond the deltaic influence, and until the deposition of the Ahmadi shales, continued to receive sub-littoral to shallow marine carbonates [25]. The Ahmadi member also consists of fossiliferous limestones. The shale and limestone of the Ahmadi member were deposited during Early to middle Cenomanian in open-marine conditions on the outer shelf at depths between 100 to 200 meters [1]. For the purpose of this study, in order to evaluate source rock potentiality of middle Sarvak formation, cutting samples from a total of 18 wells through the Persian Gulf area have been analyzed with organic geochemical methods. Figure 4 shows the location map of these wells in the Persian Gulf. Four study blocks were delineated to present the northwest (Block A), west-central (Block BC), east central (Block D), and east parts of the Iranian sector of the Persian Gulf.

EXPERIMENTAL PROCEDURES

Materials and Methods

In order to evaluate middle Sarvak formation for hydrocarbon source rock potentiality, Rock-Eval pyrolysis results were used in combination with organic-petrography methods. The latter included vitrinite reflectance measurements, visual kerogen inspection, thermal alteration determination, and kerogen elemental analysis following generally published procedures.

A total of 77 cutting samples were selected for analysis. Cutting cheeps were cleaned and pulverized prior to analysis. Vinci Technologies' Rock-Eval 6 instrument (AGH University of Poland) and Rock-Eval 2 unit (Research Institute of Petroleum Industry of Iran) equipped with a total organic carbon module (TOC) were used for this study.

Based on the Rock-Eval screening analyses, nine samples were selected for kerogen extraction and subsequently used for visual kerogen analyses. In order to obtain kerogen extracts, selected rock samples were crushed to an average particle size of approximately 1 mm. Concentrated HCl and HF were successively added to the pulverized cutting samples to remove carbonate and silicate materials respectively. The samples were washed

in distilled water repeatedly until neutrality (pH=7) was attained, followed by the flotation of the isolated organic matter (kerogen) within zinc-bromide solution (specific gravity=2) to remove residual inorganic matter.

According to the results obtained from the screening analyses, a total of 36 samples were selected for vitrinite reflectance measurements. Vitrinite reflectance measurements were carried out on both polished whole rock block and kerogen concentrate samples by using a Leitz-MPV-SP photometer microscope at RIPI. The measurements were carried out in a random mode according to the ASTM standard method [26].

Nine polished blocks and equal number of thin sections were prepared from the extracts for the identification of kerogen type and the determination of thermal alteration index (TAI).

The kerogen extract samples were used for the elemental analyses to determine the elemental composition of C, H, N, O, and S. The elemental composition of the isolated kerogens (C, H, N, and S) was determined with a Carlo ErBa EA 1108 elemental analyzer at AGH University of Poland.

RESULTS AND DISCUSSION

Source rock characteristics of the middle Sarvak formation were performed based on a geochemical combination. Special emphasizing was on determining the type, amount, maturity, and generative potential of the contained organic matter.

Source Rock Potentiality

The results of Rock-Eval screening analyses are presented in Table 1.

Table 1: Results of Rock-Eval analyses for middle Sarvak formation.

Block	Well	Depth(m)	TOC	T_{MAX}(°C)	S1	S1/TOC	S2	S3	PI	HI	OI
A	1	1805	1.93	416	1.30	0.67	10.12	1.24	0.11	524	64
		1509	0.54	424	0.53	0.98	1.96	1.88	0.21	363	348
	2	1329	0.17	434	0.07	0.41	0.16	0.70	0.29	94	412
	3	1650	0.70	430	0.95	1.36	1.78	2.88	0.35	254	411
	4	2446	2.71	435	3.47	1.28	13.95	0.82	0.20	515	30
		2488	2.75	434	1.67	0.61	16.17	1.05	0.09	588	38
		2492	2.10	432	1.56	0.74	11.76	1.01	0.12	560	48
		2516	1.87	435	1.64	0.88	10.41	0.84	0.14	557	45
		2518	1.83	436	1.76	0.96	10.01	0.88	0.15	547	48
		2524	2.19	433	2.02	0.92	11.53	1.01	0.15	526	46
		2534	2.54	433	3.45	1.36	12.98	1.22	0.21	511	48
		2536	2.80	433	2.97	1.06	14.17	1.37	0.17	506	49
		2544	3.77	431	3.20	0.85	19.62	1.05	0.14	520	28
		2548	4.38	436	2.86	0.65	25.21	0.93	0.10	576	21
		2550	3.66	436	2.46	0.67	19.72	0.97	0.11	539	27
		2564	1.28	437	2.06	1.61	4.88	1.12	0.30	381	88
		2566	3.13	436	2.67	0.85	16.11	0.95	0.14	515	30
		2568	2.98	436	2.57	0.86	15.20	0.97	0.14	510	33
		2570	2.06	435	2.71	1.32	9.55	0.92	0.22	464	45
		2572	4.13	438	3.31	0.80	23.06	0.84	0.13	558	20
		2572	6.45	436	4.40	0.68	38.26	0.90	0.10	593	14
		2574	5.57	430	4.32	0.78	33.19	0.87	0.12	596	16
	5	1555	0.33	419	0.05	0.15	0.11	1.09	0.33	33	330
	6	1408	1.89	412	5.23	2.77	6.08	2.12	0.46	322	112
		1612	0.16	418	0.13	0.81	0.24	0.34	0.35	150	213
		1658	0.33	420	0.31	0.94	0.71	0.62	0.30	215	188
		1710	0.08	347	0.11	1.38	0.10	0.24	0.52	125	300
		1311	3.28	411	0.45	0.14	22.18	2.02	0.02	676	62
		1348	1.67	413	0.51	0.31	9.70	2.10	0.05	581	126
		1378	7.13	410	1.94	0.27	49.86	3.17	0.04	699	44
		1394	4.03	409	1.16	0.29	26.57	3.31	0.04	659	82
		1414	2.40	410	0.62	0.26	15.43	2.58	0.04	643	108
	7	2121	1.37	422	0.41	0.30	7.98	0.53	0.05	582	39
		2137	2.68	424	1.19	0.44	16.94	0.70	0.07	632	26
	8	1100	0.08	418	0.10	1.25	0.26	0.50	0.28	325	625
		1130	0.11	427	0.09	0.82	0.15	0.76	0.38	136	691
		1035	0.48	406	0.31	0.65	0.29	3.13	0.51	60	652
		1040	0.62	378	0.42	0.68	1.54	3.22	0.21	248	519
	9	1862	1.49	419	3.99	2.68	2.26	2.97	0.64	152	199
	10	1756	6.54	410	3.37	0.52	36.64	2.24	0.08	560	34

Block	Well	Depth(m)	TOC	T_{MAX}(°C)	S1	S1/TOC	S2	S3	PI	HI	OI
B and C	11	2131	1.08	415	2.85	2.64	1.23	1.08	0.70	114	100
		2137	0.75	394	1.56	2.08	0.85	1.65	0.65	113	220
		2173	0.37	422	0.65	1.76	0.87	1.16	0.43	235	314
		2182	1.28	414	0.93	0.73	5.06	1.34	0.16	395	105
		2188	1.35	421	1.12	0.83	4.49	1.59	0.20	333	118
		2201	1.16	424	1.05	0.91	4.23	1.79	0.20	365	154
	12	930	0.83	421	2.61	3.14	0.38	1.79	0.87	46	216
		1056	0.25	**	0.12	0.48	0.04	1.35	0.75	16	540
		1058	0.30	406	0.10	0.33	0.13	1.69	0.44	43	563
		1060	0.54	427	0.17	0.31	0.40	1.70	0.30	74	315
		1062	0.48	419	0.22	0.46	0.22	1.85	0.50	46	385
		1064	0.32	409	0.12	0.38	0.02	1.52	0.86	6	475
		1072	0.40	416	0.11	0.28	0.05	1.51	0.69	13	378
		1094	0.51	404	0.11	0.22	0.12	1.60	0.47	24	314
		1106	0.51	431	0.18	0.35	0.92	0.63	0.16	180	124
		1108	0.43	428	0.20	0.47	0.69	0.72	0.23	160	167
		1110	0.39	433	0.21	0.54	0.57	0.66	0.27	146	169
	13	1846	0.30	**	0.11	0.37	0.21	1.58	0.34	70	527
		1890	0.29	411	0.14	0.48	0.33	1.28	0.29	114	441
D	14	3539	0.17	441	0.26	1.53	0.20	0.63	0.57	118	371
		3619	0.31	440	0.29	0.94	0.66	0.19	0.31	213	60
		3695	0.10	451	0.19	1.90	0.09	0.32	0.68	90	320
		3790	0.24	440	0.91	3.79	0.40	0.27	0.69	167	112
	15	2551	0.87	436	2.58	2.97	3.96	1.33	0.39	455	153
		2615	1.11	430	3.89	3.50	3.20	1.18	0.55	288	106
	16	3578	0.38	430	0.41	1.08	1.03	0.70	0.28	271	184
		3674	0.15	425	0.26	1.73	0.07	0.67	0.79	47	447
		3778	0.13	426	0.27	2.08	0.00	0.72	1.00	0	554
		3852	0.22	428	0.55	2.50	0.22	0.75	0.71	100	341
	17	2260	0.44	432	0.14	0.32	0.32	1.11	0.30	73	252
		2380	0.38	435	0.16	0.42	0.52	0.73	0.24	138	193
		2420	3.86	421	0.96	0.25	14.16	2.73	0.06	367	71
		2450	1.34	425	0.35	0.26	5.42	1.16	0.06	404	87
		2480	0.43	432	0.12	0.28	0.47	0.66	0.20	110	154
		2490	0.98	435	0.16	0.16	1.11	1.21	0.13	113	124
E	18	3550	1.13	429	0.58	0.51	5.16	0.66	0.10	457	58

Peters and Cassa [26] presented standard guidelines for evaluating organic richness, quality, and maturity of source rock based on pyrolysis parameters in which a TOC value of 0.5 wt.% is considered as the base limit for an effective source rock. Based on this standard, the TOC range of middle Sarvak formation exhibits variable changes with a lateral alteration in the depositional environment conditions.

The middle Sarvak formation in block A from the eastern Persian Gulf area has a TOC content of 0.08-7.13 wt.%. These values are consistent with source rocks that may have poor to excellent source rock potential with S2 ranging from 0.1 to 49 (mg HC)/(g. rock) at an average S2 value of 12 (mg HC)/(g. rock) [27]. B and C block in the central part of the study area has a TOC content of 0.25-1.35 wt.%. The middle Sarvak formation in this area represents a poor to fair hydrocarbon potential with respect to organic concentration. Moving to the western part (block D), it becomes evident that the middle Sarvak formation in this area is fair to good in terms of oil generation potential (TOC content of 0.1-3.86 wt.% with an average value of 0.69 and an S2 of 0.07-14.16 (mg HC)/(g. rock) at an average S2 value of 2.11 (mg HC)/(g. rock)) (Table 1). Source rock potentiality of middle Sarvak formation was also estimated from the HI versus TOC and S1+S2 versus TOC plots from the Rock-Eval pyrolysis output. These plots indicate variable source rock quality for the middle Sarvak formation in the sampled areas. According to this figure, the data points representing samples from block B and C indicate poor to fair source quality. Poor to excellent oil source quality is represented for Blocks A and D located in the northwest (Figure 5 and Figure 6). As a result, in the regional context, TOC and source rock potentiality variation of the middle Sarvak formation are in satisfactory agreement with the Persian Gulf basin morphology. As represented in the map of Figure 2, Qatar-Fars Arc played an important role in deposition and preservation of organic matter. Qatar-

Fars Arc Paleo-high divided the Persian Gulf basin into two distinct basins, namely one in the west and another in the east of Qatar-Fars Arc. The amounts of TOC increase from central Persian Gulf towards northwestern direction in the western basin and northeast direction in the eastern basin.

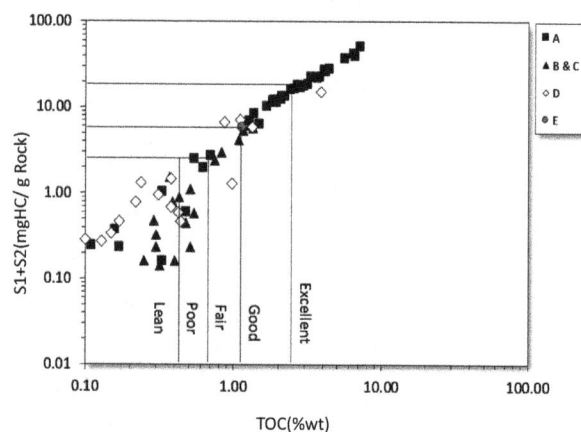

Figure 5: S1+S2 versus TOC in the middle Sarvak formation [27].

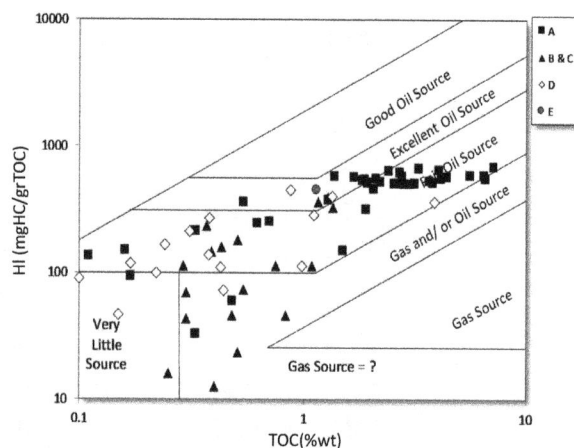

Figure 6: HI versus TOC in the middle Sarvak formation [27].

Kerogen Type

The type of organic matter present in the source rocks can be evaluated based on the plot of S2 (remaining hydrocarbon potentiality) verses TOC and modified Van Krevelen diagrams of HI versus temperature of maximum generation (Tmax) and OI versus HI. In the modified Van Krevelen diagram of HI-OI and HI-Tmax and S2 versus TOC (Figure 7, Figure 8, and Figure 9), most of the studied samples fall in the zone of Type II and mixed Types II-III kerogen. Mixed Type II/III kerogen may originate from the mixtures of terrigenous and marine organic matter with varying oil and gas generation potentials. This type of organic matter may also originally be marine Type II organic matter, which has partially been oxidized during deposition. Additional support is provided by the chemical and optical studies of the organic matter. These methods provide better means for defining the type of organic matter in the studied samples. The optical analysis of organic matter was conducted on selected samples from both D and A blocks. These results revealed that liptinite is the major maceral constituent inside the middle Sarvak formation (Figure 10). This result is in agreement with the higher amounts of amorphous particles in the studied samples under transmitted light. The elemental analysis data, in combination with the Rock-Eval data, support the optical methods result. All of the studied sample have higher amounts of elemental hydrogen (and similarly higher HI values), implying that a major proportion of the organic matter available in the studied samples is of Type II (Table 2). The results from the elemental analysis of selected samples are displayed in a typical Van Krevelen diagram by plotting H/C versus O/C ratios. Figure 11 shows the Van Krevelen cross-plot for the organic matter of the analyzed samples. The occurrence of uplift in the central part of the Persian Gulf resulted from the presence of the Qatar Arch provided a shallower sea in this part of the study area. Consequently, organic matters were exposed to more oxidizing activities resulting in relatively poor preservation of organic matter in the central Persian Gulf compared to the other parts. As a result, the organic matter quality of middle Sarvak formation shows an increase in the potentiality of sedimentary environment for organic matter accumulation and preservation by moving from central part (block B and C) to the west (block D) and east (block A) part of the Persian Gulf.

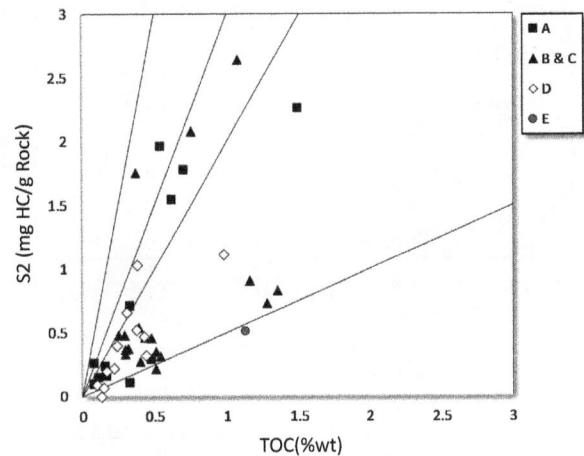

Figure 7: S2 versus TOC in middle Sarvak [29].

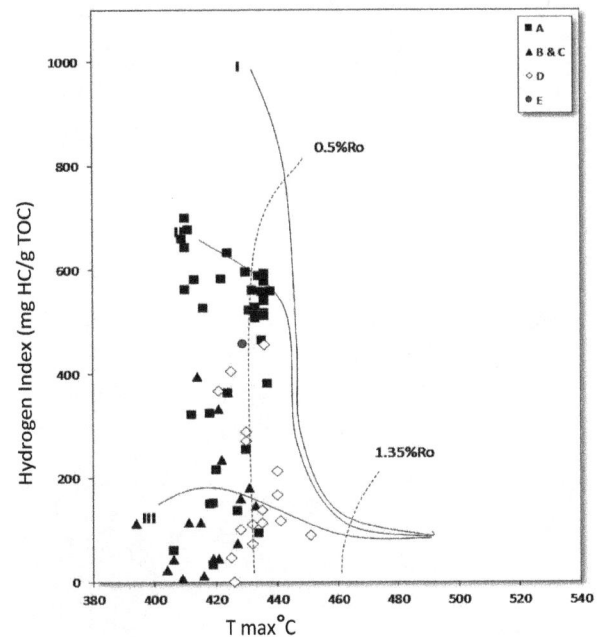

Figure 8: HI versus Tmax in middle Sarvak [29].

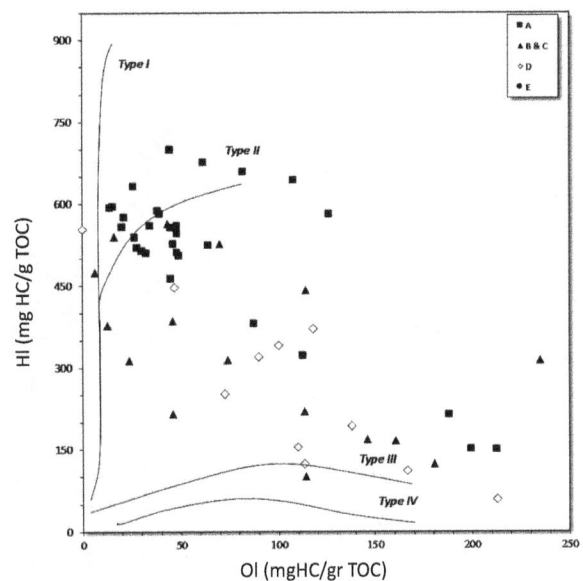

Figure 9: HI versus OI in middle Sarvak [29].

Table 2: Determination of kerogen type based on chemical and optical properties of middle Sarvak formation.

General Information			Chemical methods				Optical Methods							Kerogen Type
Elemental Analyze			Rock-Eval		Reflected Light		Transmitted Light							
Block	Well	Sample No. Interval	H/C	O/C	HI	OI	Vit.	Lip.	Iner.	Amor.	Herb.	Woody	Coaly	
D	7	S-1034-1035	1.13	0.17	386	79	30	50	20	65	0	15	20	II
A	4	S-637-638	1.3	0.03	595	15	30	60	10	75	5	10	10	II
A	4	S-645	1.4	0.03	576	21	30	50	20	80	0	15	5	II
A	4	S-653-654	1.23	0.05	551	34	35	60	5	70	0	20	10	II
A	6	S-75-79	1.32	0.11	655	84	30	60	20	85	0	5	10	II
A	6	S-81-86	1.2	0.14	672	71	20	65	15	80	0	10	10	II
Average			1.3	0.1	572.4	50.7	29	58	15	76	1	13	11	

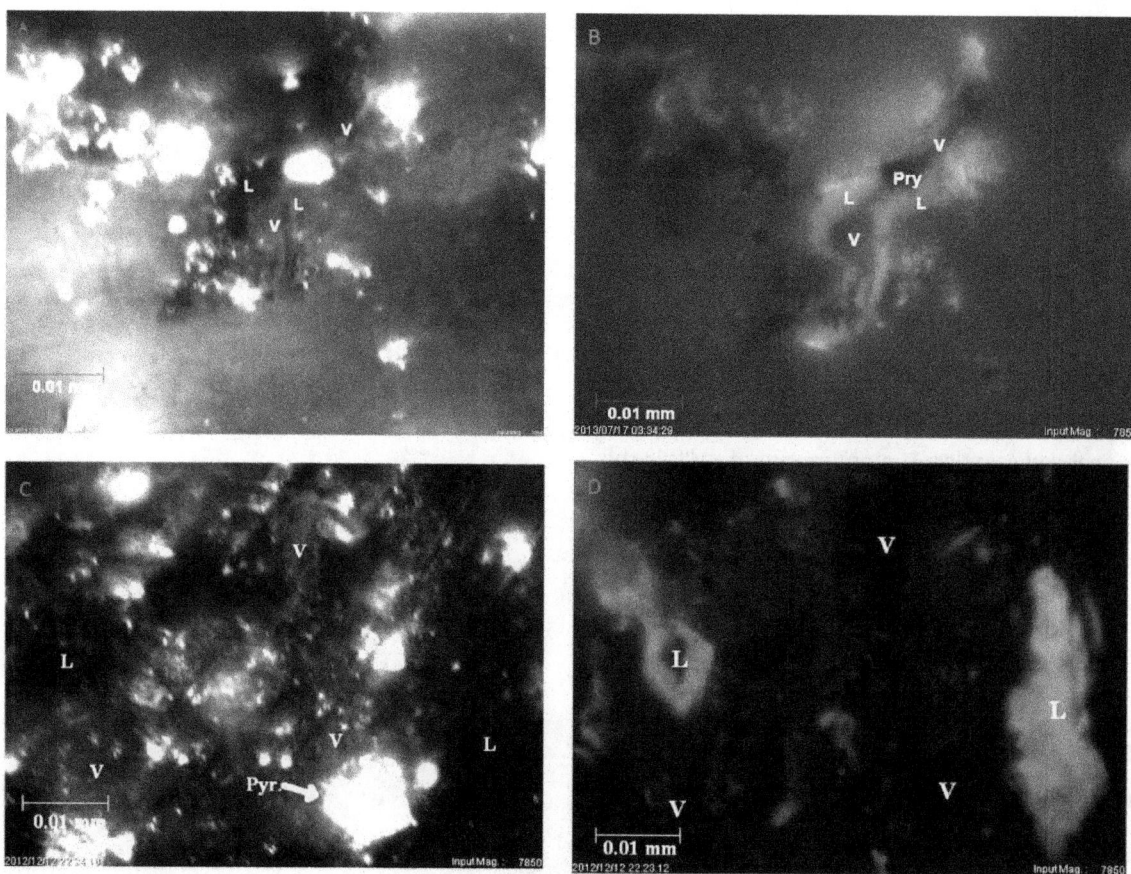

Figure 10: Photomicrographs of macerals in the middle Sarvak formation; A and C are reflected light, and B and D are UV; (V: Vitrinite, L: Liptinite).

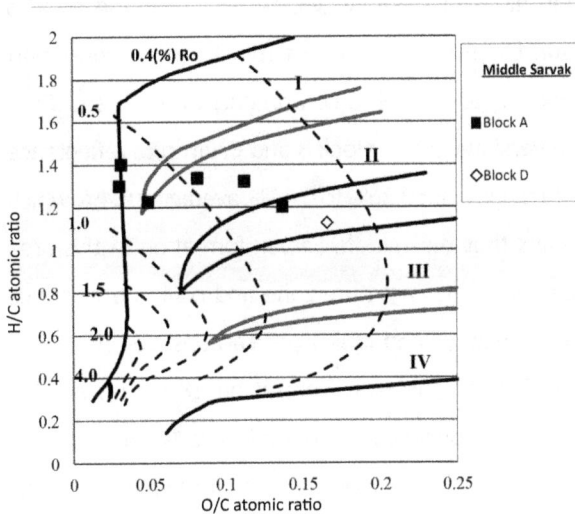

Figure 11: Bivariate plot H/C versus O/C for middle Sarvak formation [30].

Organic Matter Thermal Maturity

The level of maturity is evaluated using two main factors, including petrographical observation and thermal indicators. Vitrinite reflectance and palynofacies analysis are the most reliable petrographical indicators. Meanwhile, Tmax obtained from Rock-Eval pyrolysis, is a good thermal indicator of organic maturity [29, 28]. The extent of organic maturation of the middle Sarvak formation was evaluated by using several important parameters. Tmax and production index (PI) were used for this purpose. In addition, petrographic data such as vitrinite reflectance measurements and thermal coloration studies (both fluorescence and transmitted light microscopy) were applied. The bottom of oil generation is considered to be equivalent as a vitrinite reflectance of 0.6% (Tmax =435 °C), and its peak is regarded to be around 0.65-0.9% VRo (Tmax ranges from 445 °C to 450 °C) [26].

The plot of Tmax versus HI for 4 blocks over the study area illustrates different levels of maturity for middle Sarvak formation (Figure 8). As shown in this figure and Table 1, samples obtained from block A in the eastern part of Persian Gulf have a Tmax value ranges

between 406 and 438 °C with an average value of 425 °C. This evidence indicates that the middle Sarvak formation in block A is located at early oil generation window. Meanwhile, the level of maturity considerably decreased according to Figure 8, moving towards central part. In block B and C from the central part of the Persian Gulf, Tmax ranges from 394 °C to 433 °C with an average value of 417 °C. Regarding Peters's classification [26], the middle Sarvak formation in this area falls within immature zone and could not produce commercial gas and liquid hydrocarbons. In block D, the western part of the study area, Tmax value ranges between 421 °C and 451 °C with an average value of 435 °C, which means the middle Sarvak formation is located in the early stage of oil window maturation level.

The level of organic matter maturity was also estimated from the Tmax versus PI plot from the Rock-Eval pyrolysis output [28]. According to this diagram, Tmax of oil generation zone ranges between 435 °C and 460 °C, and its PI values lie between 0.1 and 0.4. The average PI value for the middle Sarvak formation varies in different blocks in the study area (Figure 12 and Figure 13). These figures indicate that this formation is thermally mature in block A and D and is placed in a nearly hydrocarbon generation zone. However, the middle Sarvak formation in block B and C from the central part of the Persian Gulf is not located in the hydrocarbon generation zone. Meanwhile, these figures (Figure 12 and Figure 13) can be used to identify the type of hydrocarbon produced. According to these relations, the samples from block B and C are thermally immature, and the hydrocarbons are nonindigenous; nevertheless, in A and D blocks, samples are nearly thermally mature, and the hydrocarbons are considered to be most indigenous (PI ranges from 0.1 to 0.4).

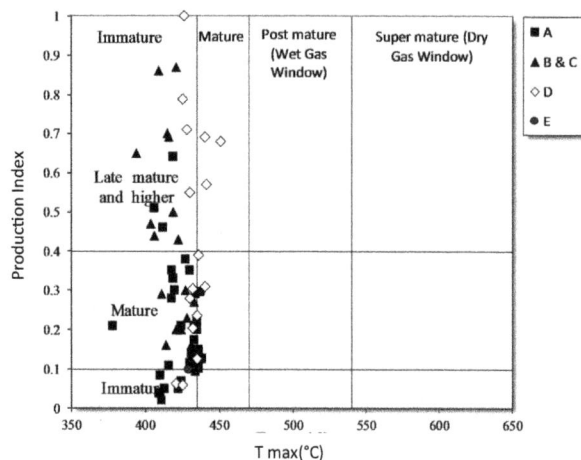

Figure 12: PI versus Tmax diagram in middle Sarvak formation [31].

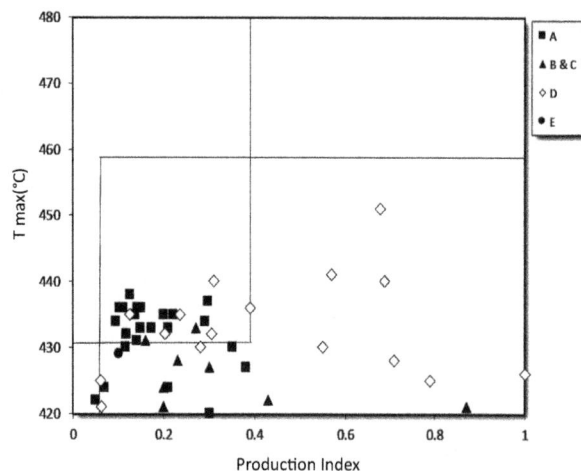

Figure 13: Tmax versus PI diagram in middle Sarvak formation [32].

The maturity of the middle Sarvak formation is also evaluated by using vitrinite reflectance measurements. Vitrinite reflectance data for the studied formation over the Persian Gulf area is shown in Table 3. The range of vitrinite reflectance data in the middle Sarvak formation is between 0.45% and 0.6% in Block A, with an average value of 0.52%, which means this formation located in the early stage of oil generation window. In Block D, vitrinite reflectance data ranges from 0.51% to 0.57 % with an average of 0.54%. This parameter indicates that the middle Sarvak formation in the western part of the Persian Gulf falls within the oil window maturation level. Meanwhile, the level of

maturity considerably decreased, according to VRo data, by moving from east (block A) and west part (block D) towards the central part (block B and C) of the Persian Gulf. In block B and C, vitrinite reflectance data range from 0.36% to 0.44%, averaging 0.4%, which means that the middle Sarvak formation in this area falls within an immature zone and could not produce commercial gas and liquid hydrocarbons. Iso-reflectance map for this formation based on 18 measurements over the study area is shown in Figure 14. According to this figure, it can be concluded that the middle Sarvak formation has remained immature in the central parts, but it has reached the oil generation stage in the western and eastern parts of the Persian Gulf. The comparison between VRo map and burial depth map of the formation (Figure 15) illustrates a good correlation between these two parameters, which indicates a role of burial depth in the thermal maturity of the formation. The burial depth of the middle Sarvak formation is an important variable which is considered to affect geologically related properties such as thermal maturity (Figure 15). The figure proposes an increase in the central parts of the study area (Qatar-Fars Arc) toward the western and eastern parts (from about less than 1000 m to more than 3000 m). The thickest zones (blocks A and D) represent more accommodation space (i.e. deeper basins). Accordingly, trough zones are more susceptible to the maturation of the middle Sarvak formation, which is in agreement with the thermal indicator spatial variation in the Persian Gulf area. Thermal alteration index (TAI) measured based on the kerogen color is also consistent with these findings. In this study, the color of the organic matter inside the middle Sarvak formation varies from light brown to medium brown (Figure 16) with corresponding TAI values ranging from 2 to 2+ (Table 4). Generally, these results show an immature to early mature zone for the studied samples.

Table 3: Vitrinite reflectance data (Ro%) of the Sarvak formation in the Persian Gulf.

Block	Well	Formation	Sample No.	Depth	Number of Reading	Vitrinite Reflectance (%)		
						Min	Max	Mean
B and C	11	Middle Sarvak	FC-1-17	2130.55	5	0.3	0.66	0.44
B and C	11	Middle Sarvak	FC-1-19	2173.22	5	0.33	0.57	0.40
B and C	11	Middle Sarvak	FC-1-22	2200.66	3	0.27	0.53	0.38
A	2	Middle Sarvak	S-587	1329	11	0.36	0.54	0.46
A	4	Middle Sarvak	S-654	2446	16	0.46	0.69	0.54
A	4	Middle Sarvak	S-653	2488	8	0.42	0.62	0.53
A	4	Middle Sarvak	S-652	2492	8	0.43	0.63	0.52
A	4	Middle Sarvak	S-649	2524	11	0.44	0.62	0.54
A	4	Middle Sarvak	S-647	2536	10	0.45	0.62	0.52
A	4	Middle Sarvak	S-645	2548	10	0.48	0.67	0.55
A	4	Middle Sarvak	S-644	2550	9	0.48	0.65	0.55
A	4	Middle Sarvak	S-641	2568	11	0.47	0.71	0.57
A	4	Middle Sarvak	S-639	2572	10	0.47	0.71	0.6
A	4	Middle Sarvak	S-637	2574	11	0.46	0.72	0.6
B and C	13	Middle Sarvak	S-911	1846	9	0.31	0.43	0.36
D	17	Middle Sarvak	S-1037	2260	14	0.43	0.6	0.51
D	17	Middle Sarvak	S-1035	2420	8	0.47	0.68	0.55
D	17	Middle Sarvak	S-1034	2450	11	0.48	0.7	0.56
D	17	Middle Sarvak	S-1032	2490	13	0.48	0.7	0.57

Table 4: Range and mean of Ro% and TAI data for the middle Sarvak formation in different blocks.

Block	Number of Samples	Range of Ro	Average Ro (%)	TAI	Maturity
A	11	0.46-0.6	0.54	+2 to 2	Early Mature
B and C	4	0.36-0.44	0.39	-	Immature
D	4	0.51-0.57	0.54	+2	Early Mature
E	No sample	-	-	-	-

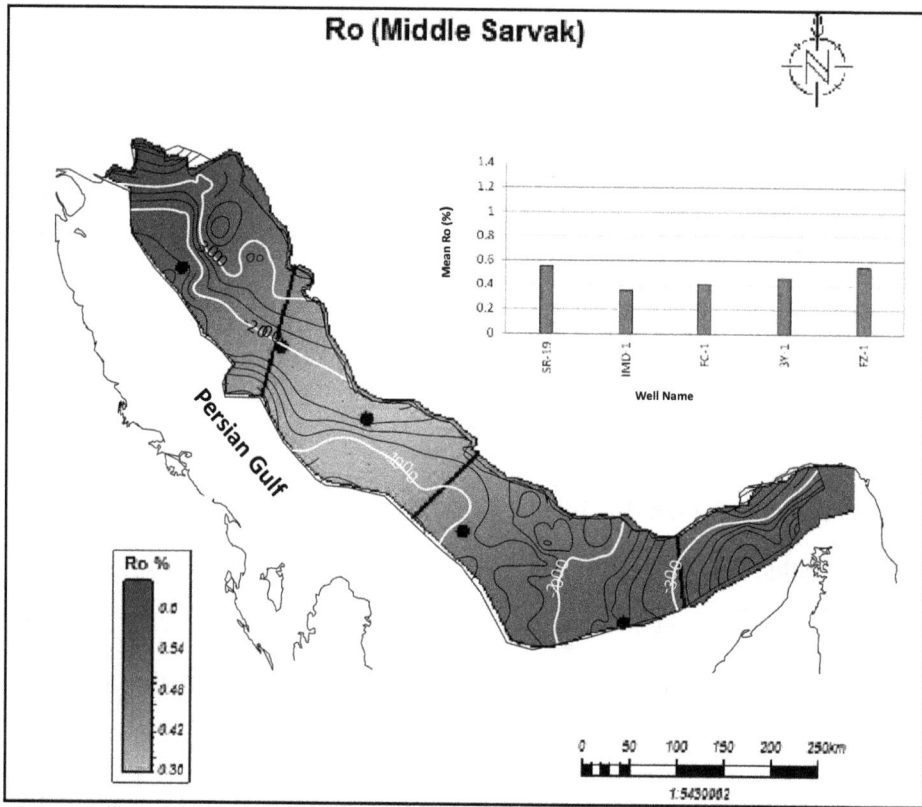

Figure 14: Regional variation of Ro data for the middle Sarvak formation in the Persian Gulf.

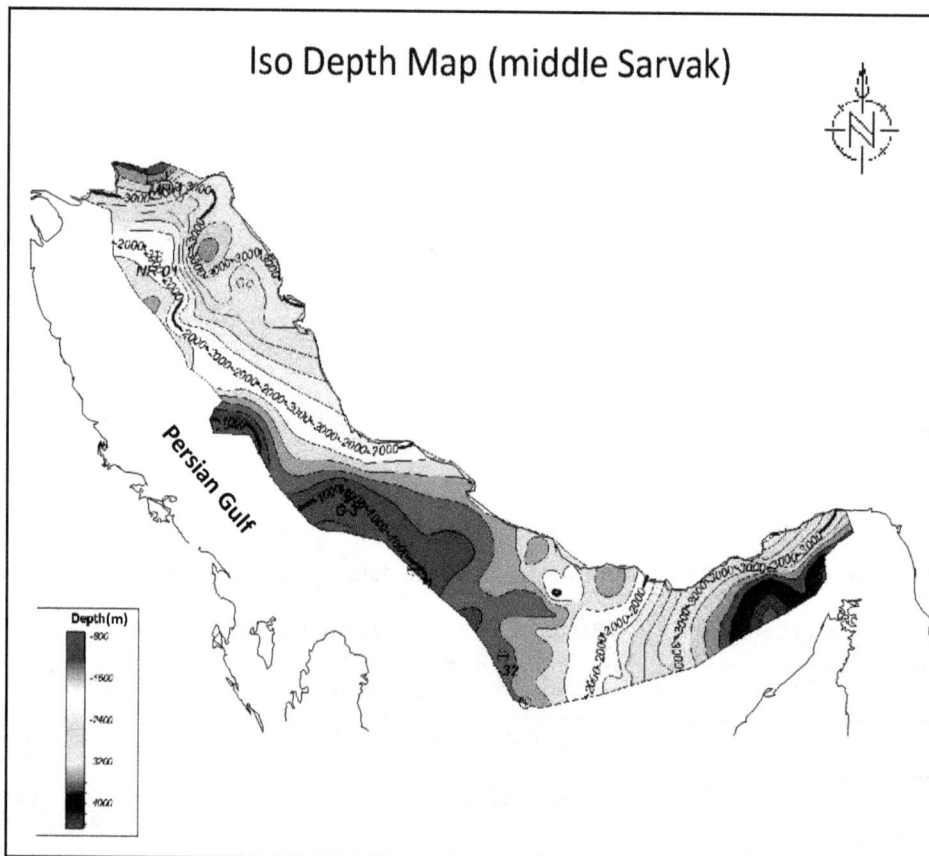

Figure 15: Depth map of the middle Sarvak formation; Qatar-Fars Arc Paleo-high represents the minimum depth for the middle Sarvak formation.

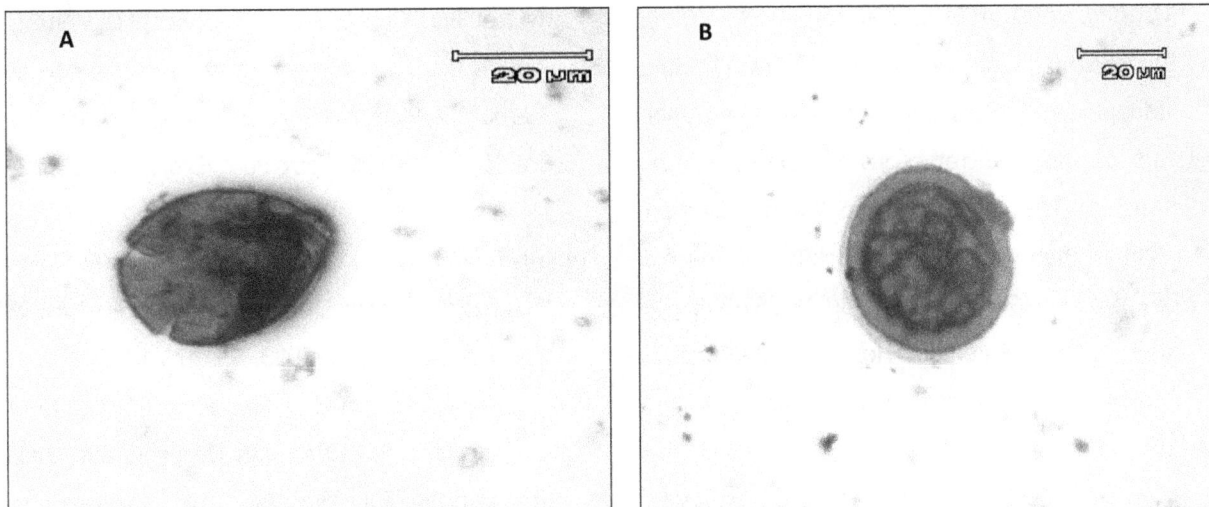

Figure 16: Organic matter of the middle Sarvak formation under transmitted light.

CONCLUSIONS

In this study, the source rock potential of 76 cutting samples from the middle Sarvak formation in the Persian Gulf were investigated. This investigation was based on the gathered data such as source rock richness, quality, distribution, and maturity which were obtained from Rock-Eval, vitrinite reflectance, and palynofacies analyses.

In summary, based on the Rock-Eval pyrolysis results, the variation of organic matter quality- and quantity-dependent parameters indicates Type II/III kerogen in the western and eastern blocks (A and D) and Type III kerogen in block B and C . The result of this study shows that middle Sarvak formation is immature in the central area of the Persian Gulf. The existence of Qatar-Fars Arc causes uplift in this region and can be a reason for immaturity in this part of the study area. In contrast to the central part, the middle Sarvak formation has entered to oil window in the eastern and western areas. The overall asymmetric nature of the basin in combination with shallower basement in the central parts with respect to the eastern and western regions resulted in the different trend of maturity in the study area. The higher maturity in the eastern and western regions can be attributed to the increasing burial depth and thickness of this formation toward the northwest and east.

ACKNOWLEDGMENTS

This work was conducted as a part of Pearl Program Research which is shared project between Iranian Offshore Oil Company (IOOC) and RIPI. The authors gratefully acknowledge Iranian Offshore Oil Company IOOC staff, especially Dr. Ali Chehrazi for the financial support of the project. Also, we appreciate Prof. Kotarba for assistance with the Rock Eval pyrolysis at RIPI and AGH university of Poland respectively.

REFERENCES

1. Ghazban F., "Petroleum Geology of the Persian Gulf," Tehran University, **2007**.

2. Bordenave M. L. and Burwood R., "Source Rock Distribution and Maturation in the Zagros Orogenic Belt: Provenance of the Asmari and Bangestan Reservoir Oil Accumulations," *Organic Geochemistry*, **1990**, *16*(1), 369-387.

3. Bordenave M. L. and Huc A. Y., "The Cretaceous Source Rocks in the Zagros Foothills of Iran: An Example of a Large Size Intracratonic Basin," *AAPG Bulletin*. 77. CONF-930306, **1993**.

4. Bordenave M. L. "The Middle Cretaceous to Early Miocene Petroleum System in the Zagros Domain of Iran, and its Prospect Evaluation," *AAPG Annual*

    Meeting, Houston, Texas, **2002**.

5. Beiranvand B., Ahmad. A., and Sharafodin M., "Mapping and Classifying Flow Units in the Upper Part of the Mid-cretaceous Sarvak Formation (Western Dezful Embayment, SW Iran) Based on a Determination of Reservoir Rock Types," *Journal of Petroleum Geology*, **2007**, *30*(4), 357-373.

6. Kent P., "The Emergent Hormuz Salt Plugs of Southern Iran," *Journal of Petroleum Geology*, **1979**, *2*, 117-144.

7. Murris R. J., "Middle East-Stratigraphic Evolution and Oil Habitat," *ABSTRACT. AAPG Bulletin*, **1981**, *65*, 1358-1358.

8. Alsharhan A., "Oxfordian-Kimmeridgian Diyab Formation as a Major Source Rock Unit in Southern Arabian Gulf," *The Society for Organic Petrology (TSOP)*, 18th Annual Meeting, **2001**

9. Alsharhan A. and Magara, K., "The Jurassic of the Arabian Gulf Basin: Facies, Depositional Setting and Hydrocarbon Habitat," *Global Environments and Resources*, **1994**, *17*, 397-412.

10. Lasemi Y. and Jalilian A., "The Middle Jurassic Basinal Deposits of the Surmeh Formation in the Central Zagros Mountains, Southwest Iran: Facies, Sequence Stratigraphy, and Controls," *Carbonates and Evaporites*, **2010**, *25*(4), 283-295.

11. Alsharhan A. and Kendall C. S. C., "Cretaceous Chronostratigraphy, Unconformities and Eustatic Sea Level Changes in the Sediments of Abu Dhabi, United Arab Emirates," *Cretaceous Research*, **1991**, *12*, 379-401.

12. Van Buchem F. S., Razin P., Homewood P. W., Oterdoom W. H., and Philip J., "Stratigraphic Organization of Carbonate Ramps and Organic-rich Intrashelf Basins: Natih Formation (Middle Cretaceous) of Northern Oman," *AAPG Bulletin*, **2002**, *86*, 21-53.

13. Vincent B., van Buchem F. S., Bulot L. G., Jalali M., et al., "Depositional Sequences, Diagenesis and Structural Control of the Albian to Turonian Carbonate Platform Systems in Coastal Fars (SW Iran)," *Marine and Petroleum Geology*, **2015**, *63*, 46-67.

14. Alsharhan A. and Nairn A., "Carbonate Platform Models of Arabian Cretaceous Reservoirs," *AAPG Special Publications*, **1993**, 56, 173-184.

15. Faqira M., Rademakers M., and Afifi A., "New Insights into the Hercynian Orogeny, and their Implications for the Paleozoic Hydrocarbon System in the Arabian Plate," *GeoArabia*, **2009**, *14*, 199-228.

16. Soleimany B., Nalpas T. i., and Montserrat F. S., "Multidetachment Analogue Models of Fold Reactivation in Transpression: the NW Persian Gulf," *Geologica Acta: an International Earth Science Journal*, **2013**, *11*, 265-276.

17. Jordan J. r., Connolly J. r., and Vest H. A., "Middle Cretaceous Carbonates of the Mishrif Formation, Fateh Field, Offshore Dubai, UAE," Carbonate Petroleum Reservoirs, *Springer*, **1985**, 425-442.

18. Aali J., Rahimpour-Bonab H. and Kamali M. R., "Geochemistry and Origin of the Worlds Largest Gas Field from Persian Gulf, Iran," *Journal of Petroleum Science and Engineering*, **2006**, *50*, 161-175.

19. Ziegler M. A., "Late Permian to Holocene Palynofacies Evolution of the Arabian Plate and its Hydrocarbon Occurrences," *Journal of Geoarabia*, **2001**, *6*, 445-504.

20. Perotti C. R., Bertozzi G., Feltre L., Rahimi M., et al., "The Qatar-South Fars Arch Development (Arabian Platform, Persian Gulf): Insights from Seismic Interpretation and Analogue Modelling," INTECH Open Access Publisher, **2011**.

21. Ghazban F. and Al-Aasm I. S., "Hydrocarbon-induced Diagenetic Dolomite and Pyrite

Formation Associated with the Hormoz Island Salt Dome, Offshore Iran," *Journal of Petroleum Geology*, **2010**, *33*, 183-196.

22. Mashhadi Z. S., Rabbani A. R., Kamali M. R., Mirshahani M., et al., "Burial and Thermal Maturity Modeling of the Middle Cretaceous–Early Miocene Petroleum System, Iranian Sector of the Persian Gulf," *Petroleum Science*, **2015**, *12*, 367-390.

23. Sharland P. R., Archer R., Casey D., Davies,R., et al., "Arabian Plate Sequence Stratigraphy," *GeoArabia, Journal of the Middle East Petroleum Geosciences*, **2013**, *18*, 56-74.

24. Mina P., Razaghnia M. T. , and ParanY., "Geological and geophysical studies and exploratory drilling of the Iranian continental shelf–Persian Gulf," *7th World Petroleum Congress*, World Petroleum Congress, **1967**.

25. Taylor G. H., "Organic Petrology: Berlin," *Gebrüder Borntraeger*, **1998**.

26. Peters K. E. and Cassa M. R., "Applied Source Rock Geochemistry," Chapter 5: Part II, *Essential Elements*, **1994**, 93-120.

27. Kenneth E. P., Walters C. C., and Moldowan J. M., "The Biomarker Guide: Biomarkers and Isotopes in the Environment and Human History," Cambridge University Press, **2005**.

28. Hunt J. M., "Petroleum Geology and Geochemistry," Freeman and Company, New York, **1996**.

29. Krevelen D. W., "Coal-typology, Chemistry, Physics, Constitution," *Elsevier Science and Technology, 3*, **1961**.

30. Magoon L. B. and Wallace G. D., "The Petroleum System-from Source to Trap," *AAPG Memoir*, **1994**, *60*, 3-24.

31. Langford F. F. and Blanc-Valleron M. M., "Interpreting Rock-Eval Pyrolysis Data Using Graphs of Pyrolizable Hydrocarbons vs. Total Organic Carbon," *AAPG Bulletin*, **1990**, *74*(6), 799-804.

5

An Experimental Investigation of Feasibility of Gas Huff and Puff for Recovering Crude Oil with Different Viscosities

Ping Jiang[1], Lei Zhang[1], Jijiang Ge[1]*, Guicai Zhang[1], Yufeng Yuan[2], Yang Wang[1], and Haihua Pei[1]

[1] College of Petroleum Engineering, China University of Petroleum (East China), Qingdao 266580, China
[2] Engineering Technology Research Institute of Jiangsu Oilfield, China Petroleum & Chemical Corporation, Yangzhou 225009, China

ABSTRACT

Investigating multi cyclic gas (N_2 and CO_2) huff and puff have been conducted by long sandpack simulations with four different crude oils from Jiangsu oilfield. According to the production differential pressure and the production regularity during huff and puff, the production process was divided into three sections; the first one is free gas production section; the second section is low viscous oil production section, and the third one is gas driving exploitation section. According to the results, the production decreases with the increase of cycle times at a certain backpressure, and the effect of multi-cycle huff and puff is improved as the backpressure decreases. In the same cycle, the oil production increases as the backpressure declines. In the first huff and puff cycle, positive synergistic effects are generated while CO_2 and N_2 are being mixed, and there is an optimum ratio between them. The amount of CO_2 in the optimum ratio increases as the backpressure rises. The effect of single gas is better than that of mixed gas in multi-cycle huff and puff, and the effect of CO_2 is better than that of N_2. The effect of huff and puff is not influenced by the injection mode. At a high backpressure, the huff-and-puff effect becomes better as the oil viscosity increases, but at low backpressure, it becomes worse with an increase in oil viscosity. The huff-and-puff effect also gets worse as the temperature rises. The oil recovery degree is improved as the CO_2 injection amount increases, but there is an optimal oil exchange ratio. The depressurization process should be neither too quick nor too slow in the CO_2 huff-and-puff process. The cyclic oil production is improved with the increase of injection rate while the average injection rate is lower than 90 L/min. The equilibrium time of CO_2 is shorter than that of N_2. The economic return of N_2 huff and puff is better than that of CO_2 at high producing pressure drop, and light oil has higher economic returns than heavy oil.

Keywords: Depressurization, Free Gas Production, Huff-and-Puff Effect, Injection Mode, Oil Reservoirs.

*Corresponding author
Jijiang Ge
Email: gejijiang@163.com

INTRODUCTION

The investigation of CO_2 huff-n-puff applicability to the enhanced recovery of light oil began in 1984. It accounts for nearly 60 percent of EOR production in the United States. Cyclic injection could be the good EOR option for small or discontinuous reservoirs because the single-well process does not demand well-to-well displacement. The application of huff and puff process has been tested as a means of implementing variety of enhanced oil recovery processes such as CO_2 and hydrocarbon solvent injection in conventional oil reservoirs. Huff and puff process improves oil recovery through oil swelling, hydrocarbon extraction, viscosity reduction, and relative permeability effects [1, 2]. Information on the performance of CO_2 huff-n-puff in these conditions is provided in a laboratory core flood study, case histories, field-test evaluations, and a numerical simulation. Tang et al. [3] used gas huff-n-puff in a case study in Sudan, in which they found the production rate could be optimized using a nodal analysis. Rubin [4] simulated non-Darcy flow in stimulated fractured shale reservoirs. Wan et al. [5] first evaluated the huff and puff process in a shale oil reservoir using the Rubin's simulation method to set the fracture in the numerical model. He found out that huff-n-puff could increase shale oil recovery by 29% more than the primary gas flooding method. Sheng and Chen [6] also performed a comparison study to compare huff-n-puff with gas flooding. They found out that lower bottom-hole pressure leads to a relative higher oil recovery. Yu et al. [7, 8] also built a fracture model to simulate CO_2 huff-n-puff in Bakken tight oil reservoirs. They compared the CO_2 flooding in two horizontal wells

with CO_2 huff-n-puff in a horizontal well. Then, they analyzed the influences of parameters such as an injection rate, time, the number of cycle, and CO_2 diffusivity on well performance. They claimed that the most important parameter is CO_2 injection rate which is followed by CO_2 injection time, number of cycle, and CO_2 diffusivity. Chen et al. [9] used the compositional models of Bakken formation to simulate CO_2 huff and puff which was stimulated in both homogenous and heterogeneous reservoirs. In their model, the huff-and-puff process is from 300 to 1000 days, which might be too long for industrial production. Vinassa et al. [10] applied CO_2 huff-n-puff injection in Chattanooga shale formation. They found out that the injection rate and injection time have the largest impact on incremental oil recovery. They thought that the injection time is directly related to the injected CO_2 volume and the changes in reservoir pressure and fluid properties during huff-n-puff. However, in the industrial production, the well is constraint to not only the injection rate and injection time, but also the maximum injection pressure. The performance of gas huff-n-puff will be influenced by many operation parameters such as the injection time and rate, production time and rate, soaking time, and so on.

During this low oil price period, researchers have gradually applied cyclic gas injection recovery to unconventional resources in recent years. Now, it is a big issue to enhance the oil recovery using a more efficient approach in a low permeability reservoir especially the unconventional oil reservoirs, and it has been proven to be an effective recovery method to enhance oil recovery in laboratory experiments for shale oil production [11,12,13]. Meng et al.

[14] conducted an experimental study about the effect of huff-n-puff gas injection on enhancing condensate recovery in shale gas reservoirs. They analyzed the parameters such as huff-n-puff cycles and soaking time and found out that soaking time slightly affects condensate recovery. Yu and Sheng [15] analyzed the effect of pressure depletion time on shale oil recovery and found out that the oil recovery increased as the pressure depletion rate rose. Gamadi et al. [16, 17] compared N_2 with CO_2 huff–n-puff performances on shale core plugs in different regions such as Barnett, Marcos, and Eagle Ford. He claimed that the huff-n-puff method could enhance oil recovery by 10 to 50% in a laboratory study. They also tested the influence of soaking period on ultimate recovery factor and found out that the longer the shut-in-period was, the more the ultimate recovery became.

While the sequential steam huff-and-puff technique can significantly increase the oil recovery from heavy oil reservoirs, cyclic CO_2 injection has also been proposed as an alternative to cyclic steam stimulation for the heavy oil reservoir. Moreover, in China, gas huff-n-puff was widely used in heavy oil reservoirs such as in Liaohe and Henan oilfield with oil viscosity higher than 10000 mPa.s in reservoir conditions. Recently, petroleum operators have shown increasingly an interest in taking CO_2 huff-n-puff process as a preferred option to extract light oil in low-pressure and low-permeability reservoirs [18]. The problem of CO_2 huff-and-puff technique application was lack of CO_2 source in China. In addition, the precipitation of scale during immiscible CO_2 injection is another challenge encountered during CO_2 huff-n-puff process. Evidence of such a problem was seen in

Crooks Gap field, Wyoming, during immiscible CO_2 injection [19]. Dissolution of CO_2 in the formation water results in the formation of carbonic acid, which in turn can dissolve formation minerals during injection and soak period; also, this process improves formation permeability [20]. However, if calcium carbonate precipitates near the wellbore, this will lead to reduced productivity.

N_2 has been successfully used as the injection fluid for the recovery of crude oil, with the advantages of being more cost-effective than CO_2 or hydrocarbon-gas and being non-corrosive; in addition, N_2 becomes an economical alternative for oil recovery using a gas miscible displacement (Hudgins et al., 1990). Lots of literature discussed the recovery of crude oil by N_2 injection into fractured reservoirs based on the findings from the experimental and simulation studies [21-23]. The very good agreement between experimental and numerical simulation results showed that N_2 was a feasible injection gas for recovering crude oil by gravity floods. Also, the performance of N_2 and enriched N_2 injections, which were applied as the secondary and tertiary oil recovery processes, was tested and analyzed.

The objective of this study is to investigate the influence of multi cyclic gas (N_2 and CO_2) huff and puff on enhancing oil recovery. First, oils with different viscosities (from light oil to heavy oil) were utilized. Second, the same oil was used to study the effect of mixing N_2 with CO_2 at different ratios. Third, long sandpacks were used instead of short cores so as to truly reflect the influence of gas diffusion and oil saturation on huff and puff. We hope that this work may expand the application of gas huff and puff.

EXPERIMENTAL PROCEDURES

Materials

The properties of crude oils and formation waters from Jiangsu oilfield are shown in Table 1.

The viscosity-temperature curves of the oils are shown in Figure 1. The mainly used gases in this paper are CO_2 (purity of 98.1%) and N_2 (purity of 99.9%). CO_2 is the greenhouse gas which comes from the power plants near the reservoir, and N_2 is a commercial product from gas producers.

they were filled in the sandpack. The permeability of the sandpack ranges from 90 to 200 mD.

Table 1: Oil and water properties of the tested oilfield

Crude oil	Density (formation condition), g/cm³	Viscosity (50°C), mPa·s	Viscosity (formation condition), mPa·s	Formation temperature, °C	Average permeability, ×10⁻³µm²	Chloride content, mg/L	Salinity, mg/L	Water type
Tai5	0.8167	46.0	4.89	100	75.4	23487	50245	Na_2SO_4
Zhen177	0.6850	14.8	1.60	93	77.4	4017	8631	$NaHCO_3$
Xu27	0.8559	192.0	10.0	87	11.1	3752	11027	$NaHCO_3$
Wei5	0.9019 -0.9062	562.0	218.9 -292.3	60-70	192.3	8693	16264	$NaHCO_3$

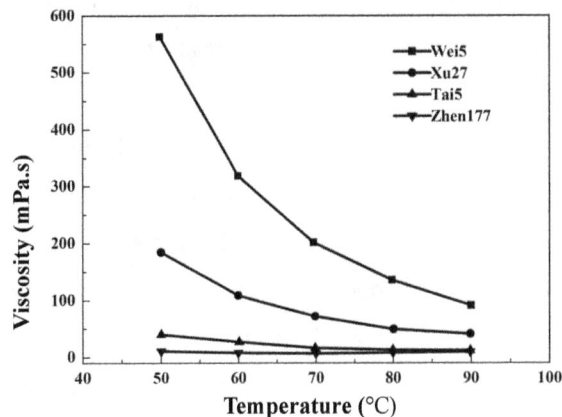

Figure 1: Viscosity-temperature relationship of the crude oils from Jiangsu Oilfield.

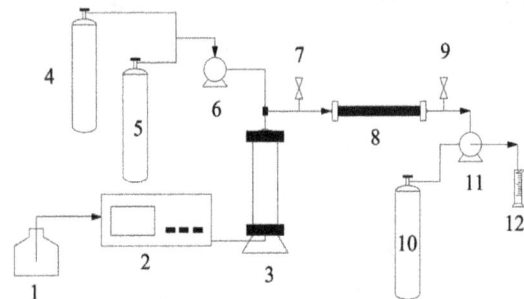

Figure 2: A schematic of the experimental setup; 1) Water source; 2) Pump; 3) Intermediate container; 4) CO_2 cylinder; 5) N_2 cylinder; 6) Backpressure valve; 7) Valve; 8) Sandpack; 9) Valve; 10) N_2 cylinder; 11) Backpressure valve; 12) Cylinder.

Apparatus

The schematic of an experimental set-up is shown in Figure 2. The sandpacks used in the experiment were 1500 mm in length and 45 mm in inner diameter. Quartz sands with a mesh of 20 to 70 were dried after they had been cleaned with distilled water, and then

Water/Oil Saturation

Firstly, the sandpack was saturated with simulated formation water which has the same salinity with the formation water, and the pore volume was calculated. Then, it was put in the thermostatic water bath, and the permeability was measured at a flow rate of 1.0 mL/min.

After the water saturation, the sandpack was saturated with crude oil at a flow rate of 1.0 mL/min at the formation temperature. The process was stopped until there was no water production at the outlet end, and then the oil saturation as well as the irreducible water saturation was calculated.

The Depressurization Process

The outlet valve of the sandpack was closed when the oil saturation was finished, and the inlet was closed when the pressure reached 20 MPa after the oil saturation. The pressure of the back-pressure valve at the outlet was increased to 20 MPa with gas right before the outlet valve was opened, and it was then reduced gradually to atmospheric pressure (in each step, the pressure was reduced by 2 MPa until there was no production). The liquid production, the water production, and the oil production were recorded.

The Gas Huff-and-Puff process
Gas Injection

The gas was injected into the intermediate container (at 60 °C) with a piston by a gas booster pump until the pressure reached 20 MPa. The valve of the intermediate container was opened, and the gas flowed into the sandpack that had been depressurized. The piston was displaced by water with a high pressure piston pump in order to keep the gas pressure at 20 MPa, and this equilibrium state was maintained for 6 hours (the pressure was increased repeatedly during this process). The injected gas volume was calculated through measuring the volume of the water that had been injected into the intermediate container.

Huff and Puff

The depressurization development was carried out as noted below.

Multi-cycle Huff and Puff

Step (1) and step (2) were repeated to conduct the multi-cycle huff and puff.

RESULTS AND DISCUSSIONS
The Production History Curve of Depressurization Development

The production history curve of the gas huff and puff depressurization production was studied with Zhen177 crude oil at 93 °C as shown in Figure 3. It shows that the curve can be divided into three sections as follows:

Section 1 (Pressure Reduces from 20 to 14 MPa): Free Gas Production Section

In this section, the main production is free gas, with little oil. It is because the flow channel for crude oil is occupied by the early gas production, and thus the flow of crude oil is hindered. The viscosity of free gas is much lower than that of crude oil, and gas has higher saturation and relative permeability near the outlet of the core, making the flow resistance of the gas lower than that of the oil; this leads to the priority production of gas.

By comparing the mixed gas of N_2-CO_2 with a different composition, it shows that the free gas production stage is extended with the increase of N_2 content. This is because N_2 has a lower solubility in crude oil as well as a smaller molecular weight compared to CO_2, and thus the free gas has a larger volume at the same pressure.

Section 2 (Pressure Drops from 14 to 5.0 MPa): Low-viscosity Crude Oil Production Section

The main production in this section is crude oil, accompanied with the discontinuous production of free gas. The oil production in this section accounts for 35-50% of the total oil production. The viscosity and the flow resistance of crude oil are reduced because of the dissolution of gas. Meanwhile, the volume expansion and the increase of driving energy change the immobile oil into movable oil.

The oil production in this section rises with higher CO_2 proportion of the mixed gas since CO_2 has a stronger ability to reduce the oil viscosity compared to N_2.

Section 3 (Pressure Drops from 5.0 to 0.1 MPa): Gas Driving Exploitation Section

In this section, the oil production accounts for 50-65% of the total amount. Gases are produced mainly in the form of bubbles or slugs and have two effects on displacing the oil. The first type is free gas driving. When the outlet pressure drops below 5 MPa, the free gas is segmented into clusters by crude oil. Furthermore, the gas expands with a decrease in pressure, resulting in the slug-type flow of oil and gas, which drives the oil to the outlet constantly. Since the volume factor and compressibility of N_2 is higher than that of CO_2, the driving effect of N_2 as free gas is better. Another type is dissolved gas driving. The dissolved gas spills over and then forms bubbles inside the oil at a low pressure. Therefore, the bubbles expand as the pressure decreases, and the crude oil, as well as gas, is produced in the form of foam. CO_2 is more able in dissolved gas driving than N_2 due to its higher solubility.

The Effect of Multi-cycle Gas Huff and Puff at Different Backpressure

The effect of multi-cycle gas huff and puff at different back pressures was studied with Zhen177 crude oil at 93 °C. The results are shown in Figures 4-7.

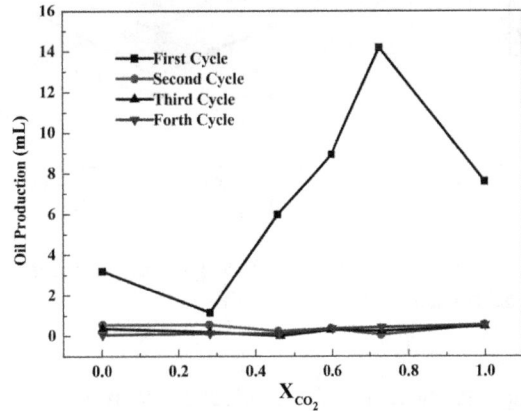

Figure 4: The effect of multi-cycle gas huff and puff at the backpressure of 12 MPa.

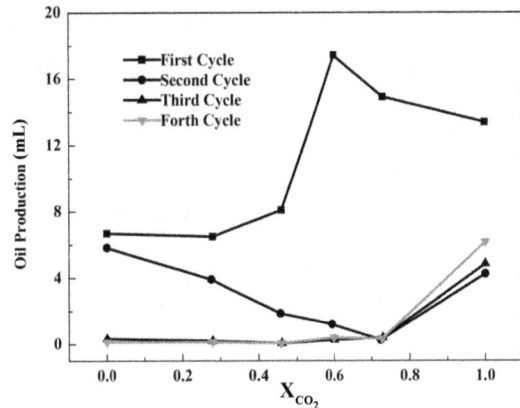

Figure 5: The effect of multi-cycle gas huff and puff at the backpressure of 8 MPa.

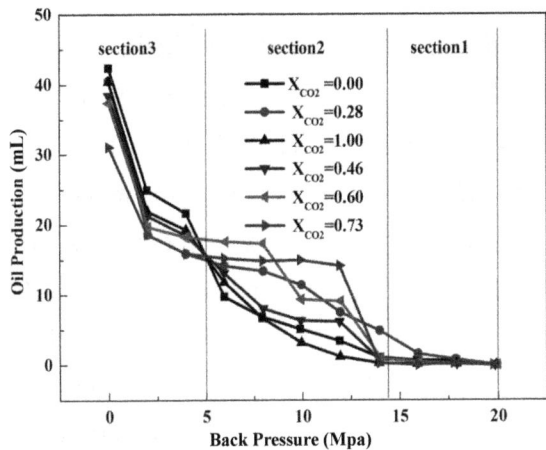

Figure 3: Recovery curve of depressurizing production of gas huff and puff.

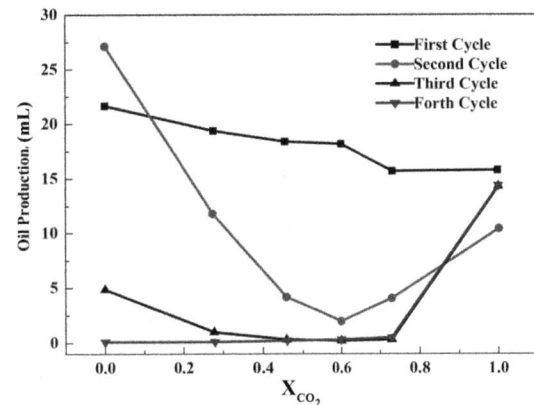

Figure 6: The effect of multi-cycle gas huff and puff at the backpressure of 4 MPa.

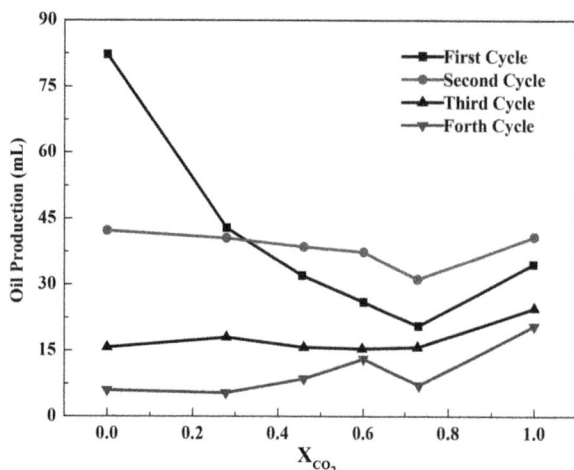

Figure 7: The effect of multi-cycle gas huff and puff at the backpressure of 0.1 MPa.

The figures show that the production decreases with an increase in cycle times at a certain backpressure. This is because the oil saturation decreases with an increase in cycle times. Meanwhile, gas channels are formed in the core when oil saturation is reduced to a certain degree, so the displacement effect of free gas is significantly weakened.

Moreover, the effect of multi-cycle is improved with a decrease in backpressure. It shows that little oil is produced during the 2-4 cycle; furthermore, the backpressure is higher than 12 MPa, which illustrates that the oil production displaced by the free gas mainly depends on the oil saturation near wellbore area. However, when the backpressure is lower than 8 MPa, the oil production obviously corresponds to a decrease in backpressure due to the effect of the dissolved gas driving and viscosity reduction mechanism. Additionally, in the first huff and puff cycle, positive synergistic effects are generated with the mixture of CO_2 and N_2, and there exists an optimum ratio between these two gases. The amount of CO_2 at this optimum ratio increases with a decrease in backpressure. This is because CO_2 has a higher solubility in crude oil than N_2,

making the oil viscosity lower, which is beneficial to increasing the production in the second section (Low-viscosity crude oil production section). N_2 has higher compression ratio and volume expansion ability during the pressure drop process, which contributes to an increase in the production in the third section (gas driving exploitation section). It is the main reason that the oil production of the single N_2 huff and puff is higher than that of the single CO_2 or CO_2-N_2 mixture at low backpressure in the first and second cycles. Also, the effect of single gas is better than that of mixed gas in multi-cycle huff and puff, and the effect of CO_2 is better than that of N_2. Since the oil saturation near wellbore is reduced during the multi-cycle huff and puff, the effect of the first two sections becomes less, and the increment of oil production mainly relies on the gas driving period. However, free gas exists continuously in the porous medium during the depressurizing process due to the oil saturation reduction, making the gas clusters dispersed in crude oil become less, and the effect of oil displacement by gas expansion is weakened. Therefore, the main mechanism of increasing oil production is dissolved gas driving.

The Influence of Injection Mode

The influence of injection mode on huff-and-puff effect was studied with Tai5 crude oil and formation water at 95 °C. In this experiment, the molar ratio of CO_2 and N_2 was 1:1. The shut in time was 6 hrs and the equilibrium pressure of mixed gas was 20 MPa. Five kinds of injection modes were studied: Pattern a: N_2 was injected first, followed by CO_2. Pattern b: Half of CO_2 was injected first, followed by half of N_2; then, the other half of CO_2 was injected, followed by the other half of N_2.

Pattern c: Half of N_2 was injected first, followed by half of CO_2; then, the other half of N_2 was injected, followed by the other half of CO_2.

Pattern d: Half of N_2 was injected first, followed by CO_2; then, the other half of N_2 was injected.

Pattern e: Half of CO_2 was injected first, followed by N_2; then, the other half of CO_2 was added.

The results are listed in Table 2. It shows that oil recovery was not influenced by the injection mode. This is because the gas-gas as well as the gas-liquid has achieved an equilibrium state within the shut-in time. When composition, temperature, and pressure are at certain values, the distribution of the gas in gas phase and liquid phase is constant regardless of the injection mode.

Table 2: Influence of injection mode on oil recovery

Injection Mode	a	b	c	d	e
Pore Volume (mL)	576	591	587	563	584
Porosity (%)	32.4	31.8	32.6	33.4	31.9
Permeability ($10^{-3}\mu m^2$)	159.4	173.6	118.6	133.7	164.2
Saturated Oil (mL)	406	411	404	396	417
Oil Saturation (%)	70.5	69.5	68.8	70.3	71.4
Oil Production in Elastic Exploitation Process (mL)	23.7	22.3	21.8	23.4	22.9
Oil Recovery in Elastic Exploitation Process (%)	5.84	5.43	5.40	5.91	5.49
Oil Production in Gas Huff and Puff (mL)	47.3	46.9	44.2	46.5	47.1
Oil Recovery in Gas Huff and Puff (%)	11.65	11.41	10.94	11.74	11.29

The Influence of Oil Viscosity

The influence of oil viscosity was studied at 93 °C. The results are shown in Table 3 and Figure 8.

Table 3: Influence of oil viscosity on huff and puff effect

Crude oil	Oil viscosity (mPa·s)	Oil production at the backpressure of 0.1 MPa (mL)		Oil production at the backpressure of 10 MPa (mL)	
		CO_2	N_2	CO_2	N_2
Zhen 177	5.51	58.55	65.65	9.75	0.25
Tai5	9.61	54.45	61.05	9.07	0.23
Xu27	32.8	48.90	49.00	11.50	2.45
Wei5	81.8	40.50	42.25	14.50	5.15

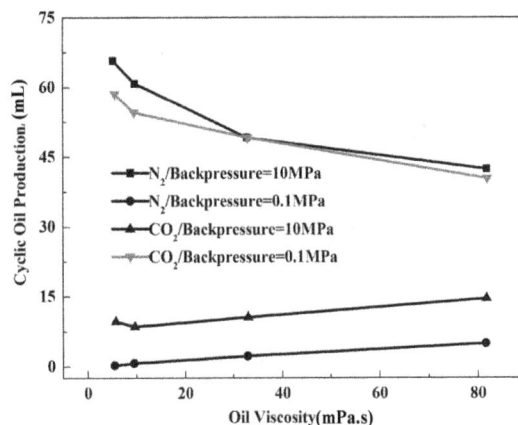

Figure 8: The influence of oil viscosity on cyclic oil production.

At a high backpressure, the oil recovery increases with an increase in oil viscosity, and the effect of CO_2 is superior to that of N_2. However, at a low backpressure, the huff and puff effect becomes worse as the oil viscosity increases, and the effect of CO_2 is about the same as that of N_2. This is because viscosity reduction is the main mechanism of gas huff and puff to improve oil recovery at a high backpressure, and the viscosity reduction efficiency is improved when the oil viscosity increases. Nevertheless, at a low backpressure, gas driving is the main mechanism to improve oil recovery; furthermore, the gas driving efficiency is improved when the oil viscosity decreases, and thus the huff-and-puff effect becomes better.

The Influence of Temperature

The simulated oil was prepared with Xu27 crude oil and diesel oil to study the influence of temperature on the huff-and-puff effect. The component of the simulated oil is shown in Figure 9.

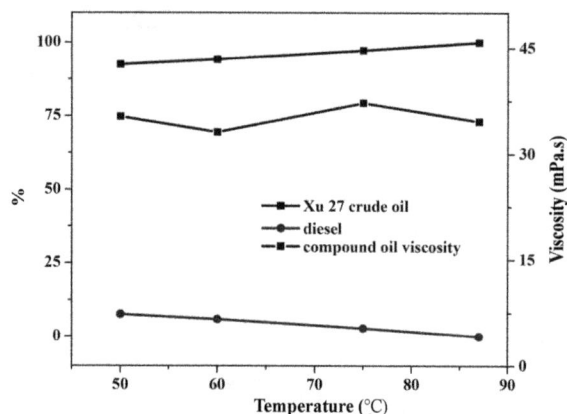

Figure 9: Simulated crude oil viscosity.

The viscosity of the used oil was maintained between 33.1 and 37.2 mPa.s, and the influence of temperature on CO_2 huff-and-puff effect was studied (see Figure 10).

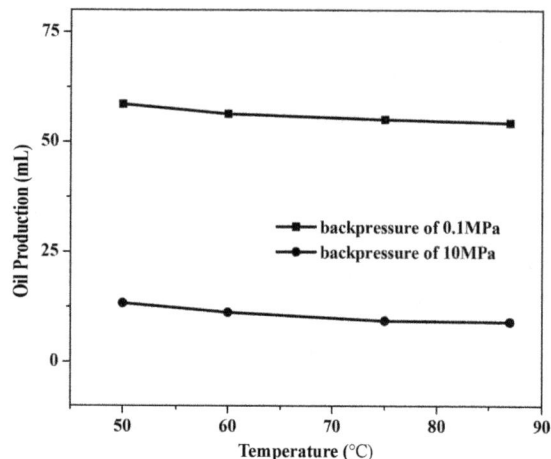

Figure 10: The influence of temperature on oil production.

The cycle production of CO_2 huff and puff decreases as the temperature rises. This is because the solubility of CO_2 in crude oil declines when temperature increases, and thus the viscosity reduction effect and dissolved gas driving effect are weakened, leading to a decline in oil production.

The Influence of Injection Pressure/ Injection Volume

The influence of CO_2 injection volume was studied with Tai5 crude oil at the backpressure of 0.1 MPa at 95 °C. The results are listed in Table 4 and Figure 11.

Table 4: Influence of CO_2 injection amount on huff and puff effect

CO_2 injection pressure (MPa)	20	15	10	5	3
Injection amount (mol)	0.5837	0.3336	0.2247	0.1668	0.1126
Oil production at the backpressure of 0.1MPa (mL)	57.3	49.1	47.5	37.55	21.34

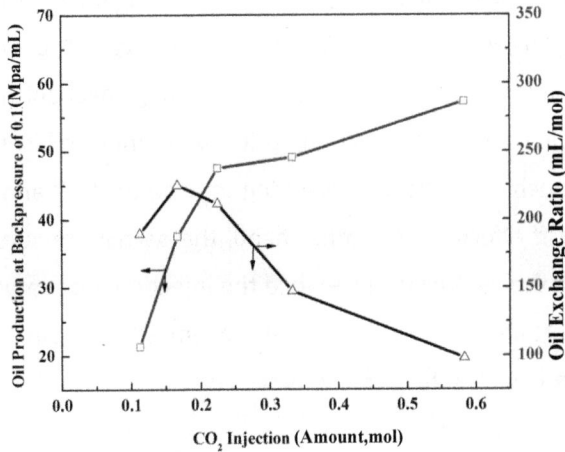

Figure 11: The influence of CO_2 injection amount on huff-and puff-effect.

It shows that the oil recovery is improved as the amount of injected CO_2 increases, while there is an optimal value of the oil exchange ratio. With an increase in the amount of CO_2 injection, the partial pressure of CO_2 in the core rises, leading to an increase in solubility in oil and a reduction in oil viscosity, which enhances the oil production. The incremental oil production of gas huff and puff mainly lies in the gas driving section (the third section), and thus the oil exchange ratio may reach the highest value as long as the conditions of the third section are obtained. In Figure 11, the oil exchange ratio is the highest when CO_2 injection pressure is 5 MPa, and this is corresponding to the result displayed in Figure 3.

The Influence of Depressurization Mode

The influence of depressurization mode was studied with Xu27 crude oil at 87 °C as shown in Table 5. The four kinds of modes are studied as follows:

A: depressurization step: 20-18-16-14-12-10-8-6-4-2-0.1;

B: depressurization step: 20-16-12-8-4-0.1;

C: depressurization step: 20-10-0.1;

D: onetime depressurization: 20-0.1.

Table 5: Influence of depressurization mode

Depressurization method	A	B	C	D
Oil production (mL)	54.45	63.42	56.38	49.22
Recovery (%)	11.71	13.32	11.87	10.61

It shows that the depressurization should be neither too fast nor too slow in the CO_2 huff-and-puff process, with the existence of a reasonable value. If the production pressure drop is too large, CO_2 will have a high fingering speed and make a breakthrough easily, which diminish the driving effect of free gas. If the production pressure drop is too small, gas-liquid stratified flow will occur under the action of gravity, which can also diminish the driving effect of free gas. Therefore, a reasonable production pressure is to ensure that the gas-liquid is produced in the form of slug flow.

The Influence of CO_2 Injection Rate

The influence of CO_2 injection rate was analyzed with the result of the first gas huff and puff applied to Xu27 crude oil at the backpressure of 0.1 MPa and at a temperature of 87 °C. The average injection rate is defined as the time spends for the pressure to reach 20 MPa by injecting CO_2. The results are shown in Tables 6 and 12.

Table 6: Influence of CO_2 injection rate

Injection time (min)	45	90	120	150	200
Average injection rate (L/min)	184	93.6	70.3	55.3	41.5
Oil production (mL)	59.33	54.45	48.27	46.16	45.17
Recovery degree (%)	12.76	11.71	10.60	9.88	9.93

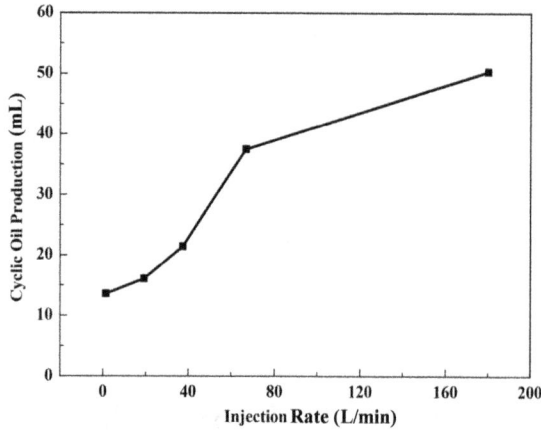

Figure 12: The influence of CO$_2$ injection rate on huff-and-puff effect.

The cyclic oil production is improved with an increase in the average injection rate. However, once the average injection rate is larger than 90 L/min, the cyclic oil production increases slowly as the injection rate rises. This is because the gas is easier to finger into the deep reservoir along the high permeability layers when the injection rate is larger. On the one hand, the viscosity reduction mechanism as well as the gas driving mechanism will have a larger action radius while more oil is in touch with the gas, resulting in a better huff-and-puff effect. On the other hand, the partial pressure of the gas will decline while the injection amount is certain and the action radius becomes larger, which can weaken the effect of viscosity reduction and gas driving.

The Influence of Shut-in Time

The influence of shut-in time was described with the result of the first gas huff and puff applied to Xu27 crude oil at the backpressure of 0.1 MPa and at a temperature of 87 °C. The results are shown in Table 7 and Figure 13.

Table 7: Influence of shut in time on huff and puff effect

Shut in time (hr.)	0	2	4	6	8	10
Recovery of first CO$_2$ huff and puff cycle (%)	3.10	4.32	9.01	10.76	10.58	10.69
Recovery of first N$_2$ huff and puff cycle (%)	1.31	2.16	7.43	11.84	11.99	11.73

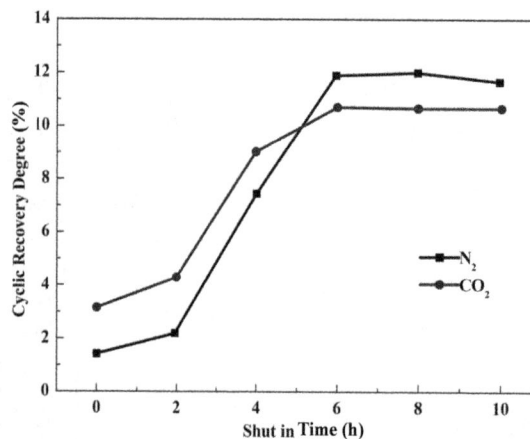

Figure 13: The influence of shut in time on huff-and-puff effect.

During the shut-in time, the gas diffuses into the deep reservoir and reduces the viscosity of the oil. The oil recovery is constant while the dissolution and diffusion processes are researching equalization. It shows that the gas-liquid equilibrium time of CO_2 is shorter than that of N_2.

The Huff-and-puff Results of Different Oils

For different kinds of oils, the relationship between the cost per ton oil and backpressure during the first huff-and-puff cycle is shown in Figure 14 (CO_2 huff and puff) and Figure 15 (N_2 huff and puff).

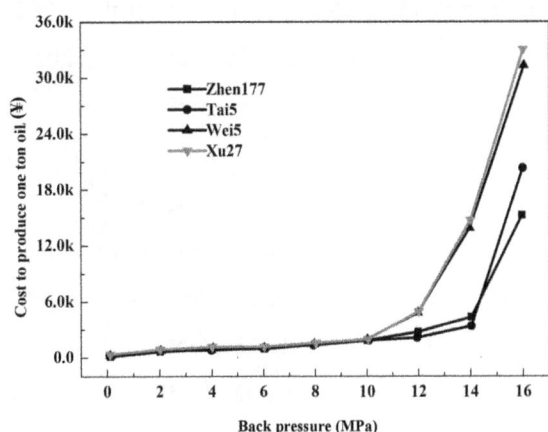

Figure 14: A relationship between the cost per ton oil and backpressure in CO_2 huff and puff.

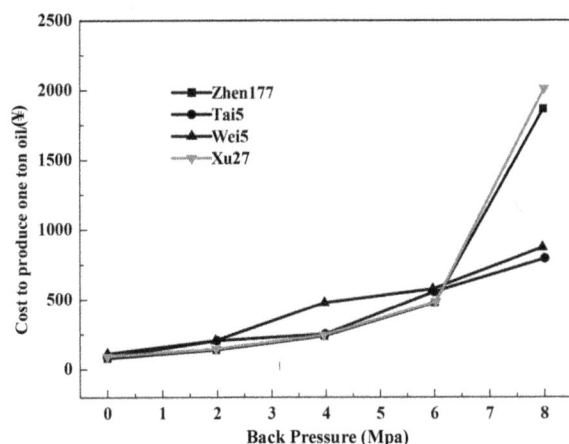

Figure 15: A relationship between the cost per ton oil and backpressure in N_2 huff and puff.

The calculation was conducted by consulting that the price of CO_2 and N_2 are respectively 1000¥ per ton and 2¥ per cubic meter in a standard condition. It can be seen that there is a turning point in the cost-backpressure curve. The turning point is corresponding to the starting point of the second section (as shown in Figure 3), which is the end point of the free gas production section.

The turning point backpressure of the heavy oils (Xu27 and Wei5) is lower than that of the light oils (Zhen177 and Tai5). This is because the mobility ratio between gas and oil rises with an increase in the oil viscosity, leading to the lag of oil production. Meanwhile, the oil with low viscosity is easily displaced by gas at high backpressure.

The turning point backpressure of CO_2 is larger than that of N_2 because CO_2 has the better viscosity reduction ability and enhance the recovery at higher backpressures.

CONCLUSIONS

The recovery curve of the gas huff-and-puff depressurization production can be divided into three sections; Section 1 free gas production (pressure reduces from 20 to 14 MPa), in which the main production is free gas, with little oil; Section 2 low-viscosity crude oil production (pressure drops from 14 to 5.0 MPa), in which the main production is crude oil, accompanied with the discontinuous production of free gas. The oil production in this section accounts for 35-50% of the total amount. Section 3 gas driving exploitation section (pressure drops from 5.0 to 0.1 MPa), in which the oil production accounts for 50-65% of the total amount. Gases are produced mainly in the form of bubbles or slugs. Moreover, the production decreases with an increase in cycle times at a certain backpressure, and the effect of multi-cycle huff and puff is improved with a decrease in backpressure. In the same cycle, the oil production increases as the backpressure declines. In the first huff-and-puff cycle, positive synergistic effects are generated after CO_2 and N_2 are mixed together, and there is an optimum ratio between these two gases. The amount of CO_2 in this optimum ratio increases as the backpressure rises. The effect of single gas is better than that of mixed gas in multi-cycle huff and puff, and the effect of CO_2 is better than that of N_2. Furthermore, the effect of gas huff and puff is not influenced by the injection mode. At high backpressures, the huff and puff effect becomes better as the oil viscosity increases, but at low backpressures, the huff and puff effect becomes worse with an increase in oil viscosity. The effect of gas huff and puff worsens with an increase in the temperature. The oil recovery degree is improved as the CO_2 injection amount increases, but there is an optimal value of the oil exchange ratio. The depressurization development should be neither too quick nor too slow in the CO_2 huff and puff process. The cyclic oil production is improved with an increase in injection rate, while the average injection rate is lower than 90 L/min. The equilibrium time of CO_2 is shorter than that of N_2. Finally, the economic return of N_2 huff and puff is better than that of CO_2 at a high producing pressure drop. Moreover, the light oil has a larger economic return compared to heavy oil which needs a high producing pressure drop.

REFERENCES

1. Momger T. G. and Coma J. M., "A Laboratory and Field Evaluation of the CO_2 Huff-n-Puff Process for Light-Oil Recovery," *SPE Reservoir Engineering*, **1988**, *3*, 1168-1176.

2. Haskin H. K. and Alston R. B., "An Evaluation of CO_2 Huff-n-Puff Tests in Texas," *SPE Reservoir Engineering*, **1989**, *2*(41), 177-184

3. Tang X., Li G., Wang R., and Wu X., "Innovative In-situ Natural Gas Huff and Puff in Same Well Bore for Cost-Effective Development: A Case Study in Sudan," SPE 144836 Presented at *SPE Enhanced Oil Recovery Conference*, Kuala Lumpur, Malaysia, **2011**.

4. Rubin B., "Accurate Simulation of Non Darcy Flow in Stimulated Fractured Shale Reservoirs," paper SPE 132093 presented at *the SPE Western Regional Meeting*, Anaheim, California, **2010**.

5. Wan T., Sheng J. J., and Soliman M. Y., 2013b. "Evaluation of the EOR Potential in Fractured Shale Oil Reservoirs by Cyclic Gas Injection," In Paper SPE 168880 or URTeC1611383 Presented at *the Unconventional Resources Technology Conference Held in Denver*, Colorado, USA, **2013**.

6. Sheng J. J. and Chen K., "Evaluation of the EOR Potential of Gas and Water Injection in Shale Oil Reservoirs," *Journal of Unconventional Oil and Gas Resources*, **2014**, *5*, 1-9.

7. Yu W., Lashgari H., and Sepehrnoori K., "Simulation Study of CO_2 Huff-n-Puff Process in Bakken Tight Oil Reservoirs," SPE 169575 in *SPE Western North American and Rocky Mountain Joint Meeting*, Denver, Colorado, USA, **2014**.

8. Yu W., Al-Shalabi E. W., and Sepehrnoori K., "A Sensitivity Study of Potential CO_2 Injection for Enhanced Gas Recovery in Barnett Shale Reservoirs," *SPE 169012* in *SPE Unconventional Resources Conference*, Woodlands, Texas, USA, **2014**.

9. Chen C., Balhoff M. T., and Mohanty K. K., "Effect of Reservoir Heterogeneity on Primary

Recovery and CO_2 Huff 'n' Puff Recovery in Shale-oil Reservoirs," SPEREE, **2014**, *17*(3), 404-413.

10. Vinassa M., Cudjoe S. E., Gomes J. H. B., and Ghahfarokhi R. B., "A Comprehensive Approach to Sweet-Spot Mapping for Hydraulic Fracturing and CO_2 Huff-n-Puff Injection in Chattanooga Shale Formation," SPE 175952 Presented *at SPE/CSUR Unconventional Resources Conference Calgary*, Alberta, Canada, **2015**.

11. Sheng J. J. and Chen K., "Evaluation of the EOR Potential of Gas and Water Injection in Shale Oil Reservoirs," *Journal of Unconventional Oil and Gas Resources*, **2014**, *5*, 1-9.

12. Sheng J. J., "Enhanced Oil Recovery in Shale Reservoirs by Gas Injection," *Journal of Natural Gas Science and Engineering*, **2015,** *22*, 252-259.

13. Yu Y. and Sheng J., "An Experimental Investigation of the Effect of Pressure Depletion Rate on Oil Recovery from Shale Cores by Cyclic N_2 Injection." SPE-178494 prepared for presentation at *the Unconventional Resources Technology Conference held in San Antonio*, Texas, USA, **2015**.

14. Meng X., Yu Y., and Sheng J. J., "An Experimental Study on Huff-n-Puff Gas Injection to Enhance Condensate Recovery in Shale Gas Reservoirs," SPE 178540 Presented at *Unconventional Resources Technology Conference*, San Antonio, Texas, **2015**.

15. Yu Y. and Sheng J. J., "An Experimental Investigation of the Effect of Pressure Depletion Rate on Oil Recovery from Shale Cores by Cyclic N_2 Injection," SPE 178494 in *SPE Unconventional Resources Technology Conference*, San Antonio, Texas, **2015**.

16. Gamadi T. D., Sheng J. J., and Soliman M. Y., "An Experimental Study of Cyclic Gas Injection to Improve Shale Oil Recovery," SPE 166334 in *SPE Annual Technical Conference and Exhibition*, New Orleans, Louisiana, USA, **2013**.

17. Gamadi T. D., Sheng J. J., Soliman M. Y., Menouar H., et al., "An Experimental Study of Cyclic CO_2 Injection to Improve Shale Oil Recovery," SPE 169142 in *SPE Improved Oil Recovery Symposium*, Tulsa, Oklahoma, USA, **2014**.

18. Zhengxiang W., Jinhua M., Ruimin G., et al., "Optimizing Cyclic CO_2 Injection for Low-Permeability Oil Reservoirs through Experimental Study," *SPE 167193 prepared for presentation at the SPE Unconventional Resources Conference-Canada held in Calgary*, Alberta, Canada, **2013**.

19. Smith L. K., MacGowan D. B., and Surdam R. C., "Scale Prediction during CO_2 Huff-n-Puff Enhanced Recovery, Crooks Gap Field, Wyoming," Paper SPE 21838 presented at *the Rocky Mountain Regional Meeting and Low-permeability Reservoirs Symposium*, Denver, Colorado, **1991**.

20. Crawford H. R., Neill G. H., Bucy B. J., and Crawford P. B., "Carbon Dioxide-a Multipurpose Additive for Effective Well Stimulation," *JPT*, **1963**, *3*(15), 237-242

21. Sayegh S. G., Najman J., and Wang S. T., "Multiple Contact Phase Behavior in the Displacement of Crude Oil with Nitrogen and Enriched Nitrogen," *Jour. Can. Pet. Tech.*, **1987**, *26*(6), 31-39.

22. Sayegh S. G., Wang S. T., and Fosti J. E., "Recovery of Crude Oil by Nitrogen Injection-Laboratory Displacement Data," *Jour. Can. Pet. Tech.*, **1988**, *27*(6), 74-79.

23. Vicencio O. A., Sepehrnoori K., and Miller M. A., "Simulation of Nitrogen Injection into Naturally Fractured Reservoirs," Paper SPE 92110 presented at *the 2004 SPE International Petroleum Conference* held in Mexico Puebla, Mexico, **2004**.

6

An Experimental Investigation into the Effects of High Viscosity and Foamy Oil Rheology on a Centrifugal Pump Performance

Hiep Phan[1]*, Tan Nguyen[1], Eissa Al-Safran[2], Olav-Magnar Nes[3], and Arild Saasen[3]

[1]Department of Petroleum Engineering, New Mexico Institute of Mining and Technology, New Mexico, United States
[2]Kuwait University, Kuwait
[3]Det Norske Oljeselskap ASA, Norway

ABSTRACT

This paper presents the effect of high viscosity oil and two-phase foamy-oil flow on the performance of a single stage centrifugal pump. As a part of this, the applicability of a pipe viscometer to determine the foamy oil viscosity is investigated. Unlike previous studies, a sealed fluid tank is used in a closed flow loop to allow conditioning the two-phase mixture to create foamy-oil. Furthermore, the sealed tank enables the accurate determination of the gas void fraction (GVF) of the foamy-oil under steady state conditions. The results show that up to a GVF of 38%, the foamy-oil improves head performance in comparison with the single-phase pump performance. In addition, the viscosity of foamy oil is found to be similar to the single-phase oil due to the Newtonian behavior of foamy oil. New correction factors have been developed to predict the performance of a single-stage centrifugal pump handling foamy oil.

Keywords: Centrifugal Pump Performance, Two-phase Foamy Oil, Electrical Submersible Pump Performance, Foamy Oil Viscosity

INTRODUCTION

Artificial lift methods are often used in wells with insufficient reservoir pressure to achieve desirable oil production. In such cases, ESP is commonly applied, especially in offshore wells, due to its flexibility, operational temperature range, a relatively small surface control facility, and a high production rate. The electrical submersible pump is a multi-stage centrifugal pump, inheriting the shortcomings of a single-stage centrifugal pump. Many factors; such as, viscosity, gas or solid presence, and rotational speed could greatly deteriorate the pump performance. Furthermore, it is common to have free gas in oil wells. However, the two-phase oil-gas effect on centrifugal pump performance is not fully understood. Most of the research on two-phase centrifugal pump and ESP performance has been performed with water and air [1-5]. A few studies have investigated the viscous effects on two-phase liquid-gas pump performance, but none of them has studied the effects of high viscosity, high gas volume fraction (GVF), and foamy-oil effects on a centrifugal pump performance [5-6].

Foamy oils are characterized as heavy (or high

*Corresponding author

Hiep Phan
Email: hphan@nmt.edu

viscosity) oils with the stable presence of tiny gas bubbles. The foaminess often persists in open vessels for several hours [7], which would suggest that the interaction between these small bubbles be much less energetic compared to a low viscosity mixture such as a water-air mixture. In water-air mixtures, the gas bubbles would quickly coalesce to form bigger bubbles, which would eventually separate from the liquid phase. For solution-gas-drive heavy oil wells, it is often observed that foamy-oil reservoirs exhibit an anomalously high production rate and a primary recovery factor, exceeding the prediction using Darcy's Law [8]. To explain such an unusual outcome, various theories have been proposed; besides, some of theories have been suggesting that the effect be caused by a viscosity difference between the foamy-oil and the single-phase oil. Most recently, Alshmakhy and Maini have used three different types of viscometers (i.e. electromagnetic viscometers, capillary tubes, and sand-packed slimtubes), and concluded that foamy-oil viscosity has been similar to that of a live oil up to a GVF of 40% [9].

The electrical submersible pump can readily generate the foamy-oil as documented by Solano [10]. He noted that the micro-foam existence has also persisted in atmospheric conditions for several hours; moreover, the foamy sample has exhibited a color change compared to the single-phase oil. The occurrence of foamy-oil can be explained based on previous studies on two-phase water-air pump performance. In visualization studies of two-phase flow, upstream stages have been shown to act as "flow conditioners" or "bubble breakers" for downstream stages [11-12]. To improve the understanding of two-phase viscous oil and gas effects on a single-stage centrifugal pump

performance, and to extend the ESP downstream stages performance, the objective of this paper is to gather pump performance experimental data under single-phase viscous flow and two-phase highly viscous foamy oil at different GVF's. At the same time, the study intends to investigate the rheology of foamy oil using the pipe viscometer.

EXPERIMENTAL PROCEDURES

Experimental Facility

A schematic of the facility is presented in Figure 1. The liquid is stored in the reservoir tank with a capacity of 170 liters, which is elevated above the AC centrifugal booster pump to avoid cavitation. The reservoir tank can either be exposed to the atmosphere or be pressurized with a sealed lid on top. The fluid temperature is controlled using a manually operated heater installed at the bottom of the tank. To ensure the uniform fluid temperature in the tank, a variable speed propeller has been used to stir the fluid when the tank is exposed to the atmosphere. When the tank is pressurized, a heavy metal lid with a rubber gasket is placed on the top of the tank. The lid is equipped with a small port for the placement of the Omegaette RTD temperature and the Rosemount 2051 pressure sensors.

The AC booster pump is placed underneath the reservoir tank and it only runs at 3600 rpm. The pump is then connected to the Endress+Hauser Promass 8°F Coriolis mass flowmeter, set up to show the volumetric flow rate in gallon per minute (gpm) and the temperature of the liquid or mixture in degrees Fahrenheit. Right after the flow meter, there is a gas injection port, connected to the nitrogen, supplying set-up of the facility. After the gas injection port, the liquid flows to the single-stage centrifugal pump, which is controlled by a variable frequency drive (VFD). The pressure generated by the pump is measured using a Rosemount 2051 differential pressure transmitter. The pipe viscometer installed downstream from the pump is used to monitor the rheology of the fluid exiting the centrifugal pump under the flowing condition. The pipe viscometer is a pipe section with two fitted ports, which are connected to another Rosemount 2051 differential pressure transmitter. A choke valve is placed after this setup to control the flow rate of the system. After the choke valve, the fluid stream returns to the tank.

Anton Paar MCR 302 rotational viscometer is shown in Figure 2. As expected, the viscosity of the fluid dropped non-linearly as the temperature increased from 21 to 66°C. In the lower temperature range, the reduction in viscosity has been much more significant. The reduction in viscosity illustrates that at lower temperature testing, extra effort would be needed to maintain the temperature constant to reduce fluctuations in viscosity. The density of the fluid was measured at the planned testing temperatures (from 43 to 49°C) using the mud balance, and was found to be constant at 865 kg/m^3.

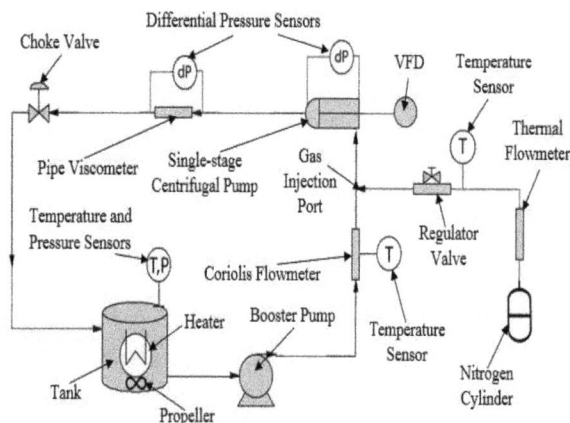

Figure 1: A schematic of the pump facility.

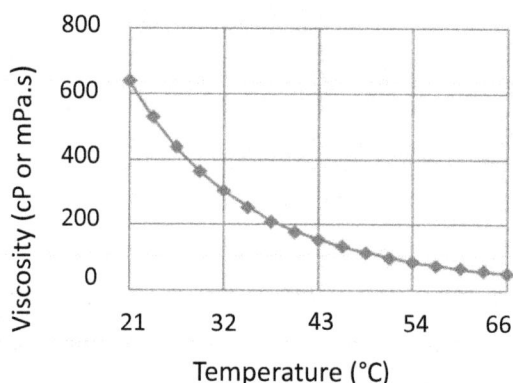

Figure 2: Oil viscosity-temperature curve; the nitrogen cylinder was pressurized to about 13.8 MPa.

The gas which is needed for the study is supplied by nitrogen cylinders. The flow rate of the gas is measured by the Endress+Hauser t-mass 65F thermal mass flow meter. To control the flow rate of nitrogen, an analog gas regulator valve is employed. The temperature of the gas is monitored and recorded by the Omegaette RTD temperature sensor.

Test Fluids

For this work, the Conoco R&O Multipurpose 220, which is a Newtonian fluid, has been used. This reduces effects from variations due to the fluid rheology during testing. A viscosity-temperature curve produced by the

Pipe Viscometer Rheology Measurements

Commonly used for research purpose and in-line viscosity measurements the pipe viscometer (Figure 3) can characterize the fluid under flowing conditions. A standard pipe viscometer system has a flow rate and pressure loss measuring instrumentations.

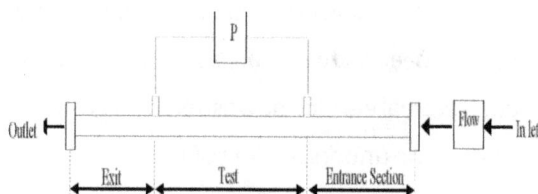

Figure 3: Pipe viscometer system.

The shear stress at the pipe wall can be calculated by using a given frictional pressure drop, the length of the pipe viscometer, and the radius of the pipe:

$$\tau_w = \frac{R}{2}\left(-\frac{\Delta P}{\Delta L}\right) \qquad (1)$$

On the other hand, the shear rate at pipe wall can be calculated with the mean velocity (obtained from the flow meter) and the pipe diameter [13]:

$$\dot{\gamma}_w = \left(\frac{3N+1}{4N}\right)\frac{8U}{D} \qquad (2)$$

where, N is the generalized flow behavior index, which can be determined as follows:

$$N = \frac{d\left(\ln \tau_w\right)}{d\left(\ln \frac{8U}{D}\right)} \qquad (3)$$

The apparent viscosity of the fluid is then calculated based on the shear stress vs. the shear rate curve. In this facility, the length of the viscometer pipe section is 0.984 m and the internal diameter of the pipe is 0.0213 m.

Single-phase Oil Test

In all the tests, three temperatures (viscosities) were chosen: 43°C (155 cP), 46°C (134 cP), and 49°C (115 cP). To ensure the temperature uniformity during a test run, a propeller was used to mix the fluid inside the tank. The propeller was driven by a variable speed motor. At a high speed, the risk of creating a mixing vortex, which might lead to pump cavitation, was high. A trial run at 49°C at the maximum pump speed determined that mixing vortices would not occur below 75% of the maximum propeller speed. To confirm the absence of entrained gas, a sample of oil was taken for visual check. Extra precaution was also taken that the propeller motor was often operated at a half speed. Each viscosity test was run at three different centrifugal pump rotational speeds, in other words 3600, 3300, and 3000 rpm. For each viscosity and rotational speed combination, the pump performance test was repeated three times.

Two-phase Foamy oil Test

A two-phase foamy-oil test presented the largest challenge in this study. Most of the papers studying the effects of two-phase fluids on a centrifugal pump or an ESP have used air and water. In these studies, gas was injected constantly into the liquid phase. A conventional separator is sufficient to separate the gas out of the liquid before the liquid returns back into the tank. The ratio between the volumetric rate of gas injection and the volumetric rate of the liquid in in-situ condition would be the gas-oil-ratio (GOR):

$$GOR = \frac{Q_g}{Q_o} = \frac{V_g}{V_o} \qquad (4)$$

However, this approach was not suitable for conducting our tests because of two reasons: (1) commercial separators are not able to separate gas out of high viscous fluid, and hence the GVF is undetermined; (2) continuously injecting gas into the liquid phase leads to difficulty in attaining the foamy state because it would require conditioning of the oil-gas mixture by a single-stage centrifugal pump to have a foamy mixture.

To solve this issue, the liquid tank was modified into a pressurized tank. A metal lid with a rubber gasket was put on top of the tank to trap any escaped gas, which would increase the pressure inside the tank. The total amount of gas injected was measured by the flow meter and the amount of free gas escaped from the oil during injection was monitored by the pressure increase inside the liquid tank. The volume of the remaining gas inside the oil could then be calculated based on the ideal gas law. With the known oil volume, the GOR of the mixture was calculated as the ratio between the remaining gas volume in the oil and the volume of the oil, as defined in Equation 4.

In order to check if the liquid tank held the testing pressures, two trial runs would be conducted to find the maximum holding pressure as illustrated in Figure 4. The first trial run was carried out for 110 minutes at 193 kPa. The second trial run was tested for almost four hours at 345 kPa. The results from the two runs revealed that the tank could hold pressure well for pressures up to 345 kPa. In this study, all the tests were carried out at pressures in the system of less than 276 kPa. Therefore, the seal should be able to maintain the pressure during the actual experiments.

For the data analysis, not only did the gas pressure inside the tank have to be determined, but the temperature of the gas also needed to be measured. A temperature probe was installed on top of the tank to measure the gas temperature inside the tank. This temperature, together with the gas pressure, was utilized to calculate the amount of free gas inside the tank.

For the two-phase pump performance test, the selected temperatures were 43, 46, and 49°C. At each temperature, the same three rotational speeds were used as in the single-phase, viscous test: 3600, 3300, and 3000 rpm. Finally, there were four GOR's tested during each test. The GOR was determined by the duration of gas injection, namely 1, 3, 6, and 9 minutes at a constant gas injection rate. For two-phase tests, a matrix of three temperatures, three rotational speeds, and four GOR's were planned. From the GOR, The GVF λg was then calculated as follows:

$$\lambda_g = \frac{GOR}{1+GOR} \qquad (5)$$

Equation 5 can be used only under homogenous foamy-oil flow. This foamy-oil is treated as a homogeneous two-phase mixture. Therefore, under flowing condition, it is assumed that there is no slip between the tiny bubbles and the liquid phase. It was then found out that 150°F was the temperature needed to remove entrained gas and possible solution gas. Therefore, after each test, the fluid was heated up to at least 65°C. Then, it was allowed to cool down to the set temperature for another test run. Optimally, at each temperature and speed combination, each injection time could be built upon the previous one.

RESULTS AND DISCUSSIONS

Transient Behavior of the Testing Loop under a Two-phase Oil-gas Flow

Figure 5 shows how the Coriolis flowmeter, the pump, and the pipe viscometer behaved in one of the two-phase test runs. A typical two-phase test can be divided into four different stages:

Stage 1: as the gas is injected into the loop, the pump does not develop any pressure (i.e. under

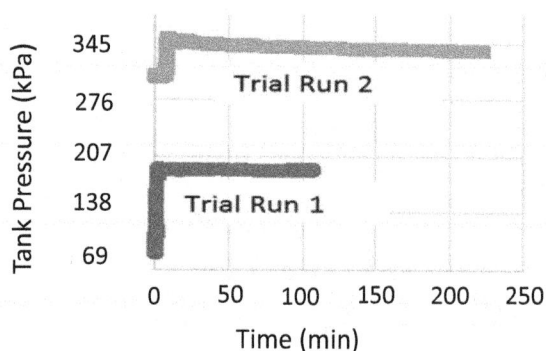

Figure 4: Transient Tank pressure; trial Run 1 confirms that the tank can hold the applied pressure for the experiment (193 kPa); trial Run 2 identifies the maximum pressure the tank can hold (about 345 kPa).

gas-lock condition). Trevisan and Prado (2010) characterized this condition in an intermittent gas pattern, under which the large bubbles are not broken up. At the same time, the liquid flow rate is greatly fluctuating. This is because the gas injection port, which is installed downstream of the liquid flowmeter, will exert a backpressure restricting the fluid flow.

Stage 2: when the gas injection is stopped and the liquid is kept circulating, the bubble size becomes smaller. As a result, the pump performance gradually recovers back to the steady state performance. During this initial period, the liquid flow rate drops significantly right after the gas shut-off. After a short phase of conditioning the fluid, the liquid flow rate rises and approaches a steady state flow rate. This behavior is because the oil-gas mixture re-circulates back into the loop. The initial large gas bubble size interferes with the flow of the mixture inside the Coriolis flowmeter.

Stage 3: as soon as the pump returns back to a steady state performance, the liquid flow rate and the pump reach a steady state indicating that the micro-sized bubbles have insignificant effects. During the recovery period after the gas injection shut-off and before reaching steady foamy flow, the pump intake pressure stays relatively constant. A 69-kPa increment in intake pressure right after steady state suggests that the booster pump should also reach a steady, higher performance state.

Stage 4: The different behaviors of the liquid rate and the intake pressure are obtained by adjusting the choke valve after the pipe viscometer. This step helps map the two-phase performance of the pump.

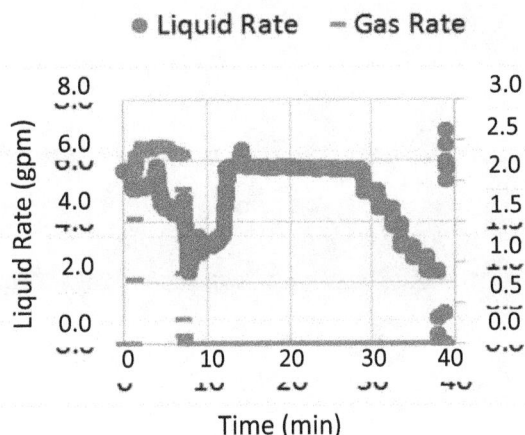

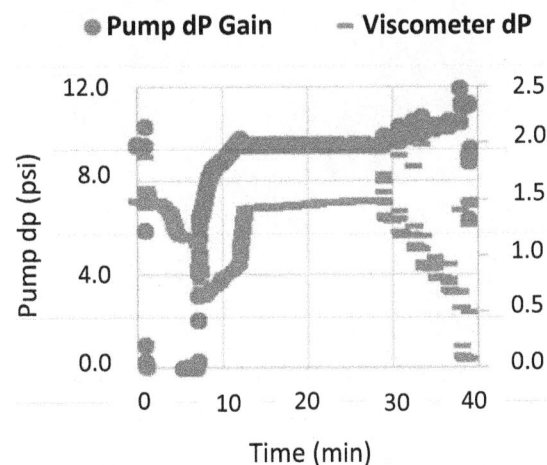

Figure 5: Transient behavior of the flowmeter, the pump, and the viscometer in one test run; Stage 1: gas injection, Stage 2: bubble breaking, Stage 3: homogeneous flow, and Stage 4: pump test.

Oil samples were collected before Stage 1 and after Stage 3 for visual demonstration (Figure 6). The original oil was translucent with an amber color. On the other hand, the foamy-oil mixture was cloudy and the color changed into yellow. The color change was due to tiny bubbles, which refracted light, not visible on the pictures. If the foamy oil were exposed to open air, its appearance would remain unchanged for hours, days, or even weeks depending on the ambient temperature.

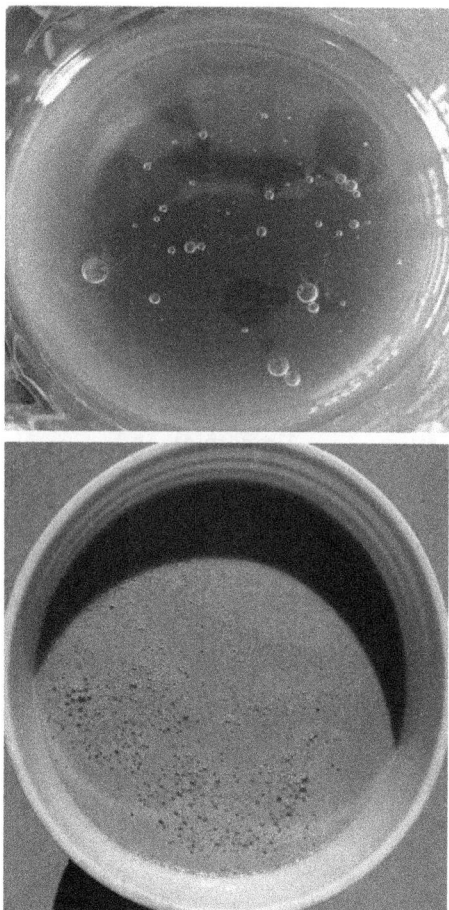

Figure 6: Single-Phase Oil (top) and Foamy-Oil (bottom)

Pipe Viscometer Measurements Analysis

Using Equations 1 to 3, the shear stress and shear rate under single-phase flow were calculated based on the pressure drop across the pipe viscometer and the flow rate of the oil at all three different rotational speeds (3600, 3300, and 3000 rpm). A linear trend line which was going through the origin was used to fit the analyzed data, and the slope of the line was the viscosity of the fluid; this trend line represented a Newtonian fluid. Table 1 shows the summary of the measured viscosities by the rotational viscometer and the calculated viscosities by the pipe viscometer under a single-phase flow. With the maximum error of 5.9%, there is strong agreement between the two viscometers.

Thus, the pipe viscometers can be used with confidence to calculate the viscosity of the foamy oils. The fact that the calculation did not consider the pipe roughness might explain the higher error at higher viscosities.

Table 1: Rotational viscometer vs. pipe viscometer results under a single-phase flow.

Rotational Viscometer Measured Viscosities (cP)	Pipe Viscometer Calculated Viscosities (cP)	Error from the Measured (%)
155	146	5.9%
134	129	3.9%
115	114	1.4%

Similar calculations of pipe viscometer data were then performed under a two-phase foamy-oil flow. Figure 7 shows shear stress as a function of shear rate under a single-phase and foamy-oil flow of different viscosities at three different rotational speeds. The slope of the linear trend line is the fluid viscosity in centipoise. With up to a GVF of about 37.8% and several cool-heat cycles of the mixtures, the foamy-oils have similar viscosity as single-phase oil. This is similar to the findings by Alshmakhy and Maini [9]. Furthermore, foamy oils behave like a Newtonian fluid. Higher GVF data would be needed to find the critical GVF value, where the foamy oil rheology would deviate from the single-phase oil.

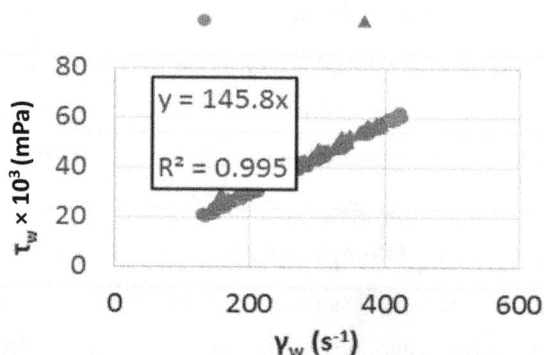

Figure 7: Shear stress vs. shear rate of 155 cP oil from 3000 to 3600 rpm for single-phase oil and foamy oils with a GVF of 11.4-37.8%.

Single-phase Oil vs. Two-phase Foamy Oils at 155 cP

To compare the single-phase and the two-phase foamy oil pump performance, the pump pressure increment was expressed as a function of a liquid-only flow rate. Under a foamy flow, the mixture flow rate was measured by the Coriolis mass flowmeter. This is because the foamy mixture can be treated as an incompressible homogeneous mixture, in which density only depends on the temperature of the mixture. With the measured mass flow rate and density, the flowmeter can calculate and display the volumetric flow rate of the foamy mixture. The liquid flow rate was then calculated as follows:

$$Q_l = Q_m \left(1 - \lambda_g\right) \tag{6}$$

Figure 8 presents the relationship between pump-developed pressure and the liquid flow rate under a single-phase and foamy oil flow at 155 cP and 3600 rpm. Single-phase oil is expressed as having a GVF of 0%. The solid curve of single-phase oil pump performance represents the average pump performance of the three test runs. At the same flow rate, the pump pressure under a two-phase foamy-oil, up to a GVF of 30%, are similar to that of the single-phase oil. This is atypical since given the same viscosity (as shown previously in the pipe viscometer data analysis), a centrifugal pump would give the same head performance regardless of the density of the fluid; as the fluid density decreases, the pump generates less ΔP to produce the same head (Equation 7). On the other hand, under a foamy-oil flow, there has been an additional pressure gain to compensate for the pressure loss due to the reduction in density. This pressure gain can be explained by less energy losses of foamy oil within the pump in comparison to that of the single-phase oil.

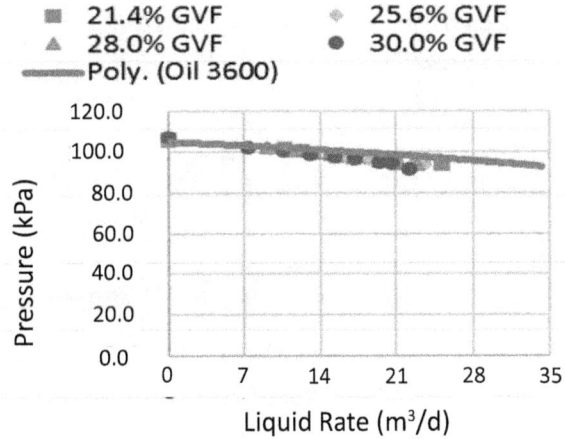

Figure 8: Single-phase oil vs. two-phase foamy oil pump performance at 155 cP and 3600 rpm.

The relationship between pump developed pressure and pump head is given by:

$$H = \frac{\Delta P}{0.052 \times \rho} \tag{7}$$

In the foamy oil case, the density of the mixture was calculated as follows:

$$\rho_m = \rho_{oil} \left(1 - \lambda_g\right) + \rho_{gas} \lambda_g \tag{8}$$

Figure 9 shows the pump head performance as a function of liquid flow rate under a single-phase oil (having a viscosity of 155 cP) and two-phase foamy oil flow at 3600 rpm with ASME (American Society of Mechanical Engineers) 95% uncertainty analysis of the head; a further explanation of the approach is presented in the Appendix. Since the pump pressure is similar between the single-phase oil and foamy oil (Figure 8), a higher-GVF, lower-density foamy mixture has a higher pump head performance. Despite having a similar viscosity as the single-phase oil and the appearance of a homogeneous mixture, it would not behave like a lower-density, single-phase oil. This behavior of foamy oil correlated well with the higher primary production under solution-gas-drive of the heavy-foamy-oil reservoir. Furthermore, the 95% confidence interval of the head shows highly precise measurement of each reading, less than 0.5% deviation from the mean.

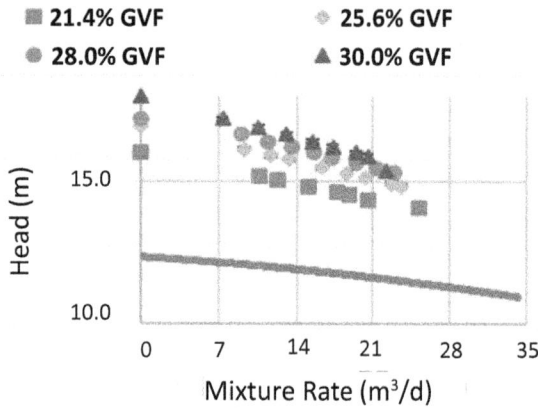

Figure 9: Single-phase oil vs. two-phase foamy oil head performance at 155 cP and 3600 rpm.

Homogeneous Model

Foamy oil appearance suggests its homogeneous nature. Thus, the experimental data was compared to the prediction of the homogeneous model (Hom.). Figure 10 shows that the homogeneous model under-predicts the pump pressure performance under a foamy flow up to 30% difference. Furthermore, the homogeneous model predicts that the higher the GVF is, the lower the pump pressure would become. This is because the homogeneous model assumes that the head performance will remain constant under a two-phase flow, while the two-phase pressure performance degrades due to the decrease in density. As a result, the homogeneous model is incapable of predicting the behavior of heavy-foamy oils.

Prediction Model

Based on the effect of foamy oil on the pump pressure and the head performance outlined in the single-phase oil vs. two-phase foamy-oils section, correction factors for head and capacity have been proposed:

$$C_{H_{foamy-oil}} = \left(\frac{\rho_{oil}}{\rho_{oil}\left(1-\lambda_g\right)+\rho_{gas}\lambda_g} \right) \qquad (9)$$

$$C_{Q_{foamy-oil}} = \left(1-\lambda_g\right) \qquad (10)$$

The single-phase oil head-capacity pump performance curve has measured at the facility and used as the base line. As seen in Figure 11, the correction head and capacity factors (in Equations 9 and 10) are then used to predict the performance of 155 cP foamy oils at four different GVF's: 16.4, 25.0, 34.2, and 37.8%. The difference between the modified model and experimental data is less than 3%. If only water head-capacity pump performance curve is available, other models which are predicting pump performance, such as the Hydraulic Institute chart, will be used to get the single-phase viscous pump performance curve first before applying the foamy correction factors in Equations 9 and 10.

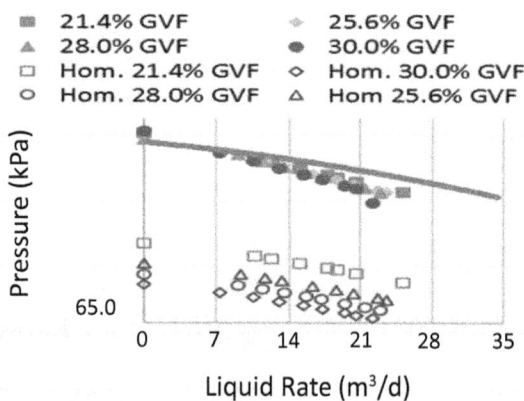

Figure 10: Oil pressure performance: experimental data vs. homogenous model at 155 cP and 3600 rpm.

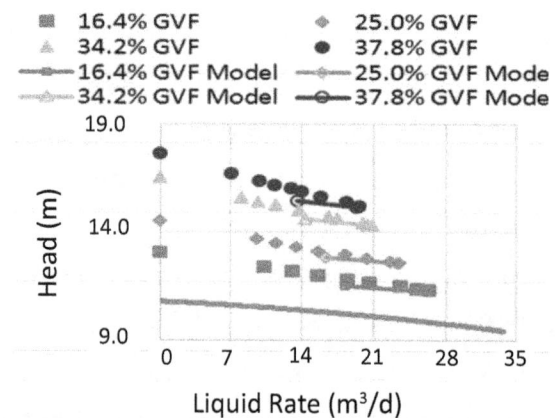

Figure 11: Head performance comparison: foamy correction model vs. experimental data.

APPENDIX

An experimental uncertainty assessment deals with analyzing the uncertainty in a measurement and determines how large an error might be in the result. During measurement in an experiment, the results will be affected by errors due to instrumentation, methodology, and other cofounding effects. The uncertainty analysis considered in this study is accepted by the American Society of Mechanical Engineering (ASME). An error consists of two main components: systematic error (bias) and random error (precision). Human error is assumed to be absent due to good engineering practice.

Systematic errors can come from various sources, but in this paper, only the instrumental errors as the major source are considered. Each instrument has its own elemental systematic uncertainty, bi. The combined effect of different instruments, Br¬, can be calculated using the root sum squared method:

$$B_r = \sqrt{\sum \left(b_i\right)^2} \tag{11}$$

For each measurement, the first step to determine the random uncertainty is to calculate the variance, Sx, of the sample data:

$$S_x = \sqrt{\frac{1}{n-1} \sum_{i=1}^{n} \left(x_i - \bar{X}\right)^2} \tag{12}$$

The mean variance, $S_{\bar{x}}$, is then estimated:

$$S_{\bar{x}} = \frac{S_x}{\sqrt{n}} \tag{13}$$

Then, the degrees of freedom, *df*, must be computed to get the 95% confidence interval *t*-statistic from statistics tables. This should combine all the degrees of freedom of both random and systematic uncertainties:

$$df = \frac{\left(\sum \left(S_{\bar{x},i}\right)^2 + \sum \left(\frac{b_i}{2}\right)^2\right)^2}{\left(\sum \left(\left(S_{\bar{x},i}\right)^4 / df_i\right) + \sum \left(\frac{b_i}{2}\right)^4 / df_i\right)} \tag{14}$$

From this *df*, the corresponding t_{95} value can be found in statistics table. The experimental uncertainty, U_{ASME}, can then be determined for each data point based on the ASME method:

$$U_{ASME} = t_{95} \left[\left(\frac{B_r}{2}\right)^2 + \left(S_{\bar{x}}\right)^2\right]^{0.5} \tag{15}$$

This number is presented as the error bars on the plot, as seen in Figure 9. Due to the highly precise measurement, the uncertainty is small, less than 0.5%, and does not show up well in Figure 9. Figure 12 shows a blown-up view of just one test run at a GVF of 21.4% to demonstrate the 95% confidence interval.

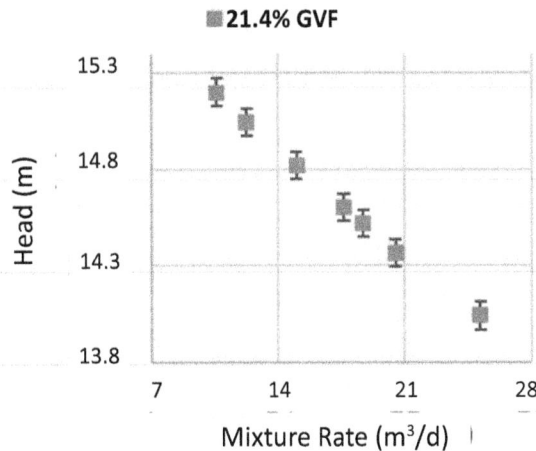

Figure 12: 95% confidence interval analysis.

CONCLUSIONS

This study experimentally investigated the effects of high viscosity oil and foamy oil flow on a single stage centrifugal pump performance. At the same time, the rheology between high viscosity single-phase oil and foamy oils was compared using the pipe viscometer. The correction factors had also been introduced to predict the performance of a centrifugal pump when handling foamy oils. Moreover, the following conclusions are drawn from the study:

• Up to a GVF of 37.8%, foamy oils had the same viscosity of dead oil and behaved as a Newtonian fluid.

• Up to GVF of 37.8%, foamy oils had similar pump increment pressure performance curves as single-phase oil with up to 8% difference in a comparable capacity. As a result, foamy oil induced a gain in pump head performance.

• Homogeneous model was unable to predict the pump pressure performance with acceptable accuracy.

• The new head and capacity correction factors for foamy flow were proposed. The prediction closely matched the experimental data, with an error less than 3% in both head and capacity.

ACKNOWLEDGEMENTS

The authors would like to acknowledge the support of the member companies of the Production and Drilling Research Project (PDRP) at New Mexico Institute of Mining and Technology.

NOMENCLATURE

ΔL	Differential Length (m)
ΔP	Differential Pressure (kPa)
cP	centipoise (cP)
D	Pipe Diameter (m)
ESP	Electrical Submersible Pump (N/A)
GOR	Gas-Oil-Ratio (N/A)
GVF	Gas Volume Fraction (N/A)
H	Head (m)
N	Rotational Speed (RPM)
Q	Volumetric Flow Rate (m³/d)
Q_g	Gas Volumetric Flow Rate (mscf/d)
Q_l	Liquid Volumetric Flow Rate (m³/d)
Q_m	Mixture Volumetric Flow Rate (m³/d)
Q_o	Oil Volumetric Flow Rate (m³/d)
R	Pipe Radius (m)
RPM	Rounds Per Minute (N/A)
U	Mean Velocity (m/s)
V	Volume (L)
VFD	Variable Frequency Drive (N/A)

Subscript

g	Gas
l	Liquid
m	Mixture
o	Oil

Greek

γ_w	Wall Shear Rate (Pa.s)	
γn	Nominal Shear Rate	(s⁻¹)
λ_g	Gas Void Fraction	(N/A)
ρ	Density (kg/m3)	
ρm	Mixture Density (kg/m³)	
τw	Wall Shear Stress (Pa)	

REFERENCES

1. Lea J. F. and Bearden J. L., "Effect of Gaseous Fluids on Submersible Pump Performance," *Journal of Petroleum Technology*, **1982**, *34*, 2922-2930.

2. Pessoa R. and Prado M., "Experimental Investigation of Two-phase Flow Performance of Electrical Submersible Pump Stages," in SPE Annual Technical Conference and Exhibition, New Orleans, **2000**.

3. Beltur R., Duran J., and Pessoa R., "Analysis of Experimental Data of ESP Performance under Two-phase Flow Conditions," in SPE Production and Operations Symposium, Oklahoma City, **2003**.

4. Duran J. and Prado M., "ESP Stages Air-water Two-phase Performance: Modeling and Experimental Data," Society of Petroleum Engineers, SPE-87627-MS, **2003**.

5. Fonseca Jr. R. d., Domingos P. V. S. R., and Reis D. C. d., "Experimental Evaluation of Behavior of Intermittent Flow in Scenario of Application of Electrical Submersible Pump," in SPE Artificial Lift Conference, Salvador, Bahia, **2015**.

6. Trevisan F. E. and Prado M. G., "Experimental Investigation on the Viscous Effect on Two-phase Flow Patterns and Hydraulic Performance

of Electrical Submersible Pumps," in Canadian Unconventional Resources and International Petroleum Conference, Calgary, **2010**.

7. Maini B., "Foamy Oil Flow in Heavy Oil Production," *Journal of Canadian Petroleum Technology*, 06, *35*, **1996**.

8. Albartamani N. S., Ali S. M., and Lepski B., "Investigation of Foamy Oil Phenomena in Heavy Oil Reservoirs," in International Thermal Operations/Heavy Oil Symposium, Bakersfield, **1999**.

9. Alshmakhy A. B. and Maini B. B., "Foamy-oil Viscosity Measurement," *Journal of Canadian Petroleum Technology, 51,* **2012** .

10. Solano E. A., "Viscous Effects on the Performance of Electro Submersible Pump (ESP's)," The University of Tulsa, Tulsa, **2009**.

11. J. Gamboa and M. Prado, "Experimental Study of Two-phase Performance of an Electric-Submersible-Pump Stage," SPE Production & Operations, **2012**.

12. Sun D. and Prado M., "Single-phase Model of ESP's Head Performance," in SPE Production and Operations Symposium, Oklahoma City, **2003**.

13. Aadnoy B. S., Cooper I., Miska S. Z., Mitchell R. F. et al., "Advanced Wellbore Hydraulics," in Advanced Drilling and Well Technology, Richardson, TX, Society of Petroleum Engineers, **2009**, 197-198.

7

Analysis of the Photo Conversion of Asphaltenes Using Laser Desorption Ionization Mass Spectrometry: Fragmentation, Ring Fusion, and Fullerene Formation

Socrates Acevedo [1]*, Henry Labrador [2], Luis Puerta [2], Brice Bouyssiere [3] and Hervé Carrier [4]

[1]Universidad Central de Venezuela, Facultad de Ciencias, Escuela de Química, Caracas 1053, Venezuela

[2]Universidad de Carabobo, FACYT. Departamento de Química, Lab. Petróleo, Hidrocarburos y Derivados(PHD), Valencia Edo. Carabobo.

[3]CNRS / UNIV Pau & Pays de l'Adour, Institut des Sciences Analytiques et de Physico-Chimie pour l'Environnement et les Matériaux, LCABIE, UMR 5254, 64053 Pau, France.

[4]CNRS/TOTAL/UNIV PAU & PAYS ADOUR, LFCR-IPRA UMR 5150, 64000, PAU, FRANCE, Av. de Université-BP1155-64013. Pau, Cedex France.

ABSTRACT

The conversion or photo conversion of asphaltenes to polycyclic aromatic hydrocarbons (PAH's) promoted by a laser source is analyzed using both experimental and theoretical methods. We propose that during measurements performed at an intermediate laser power, fragmentation to afford PAH's and ring fusion to yield fused PAH's (FPAH's) may occur either within molecular clusters (resin case) or within molecular aggregates (asphaltene case) which are vaporized or sublimed after ionization by the laser source. These events change the initial molecular mass distribution (MMD) of the sample to a continuous statistical MMD that can be fitted to a log-normal distribution. At a high laser power, the experimental MMD is converted to a sequence of C_n bands (n is an even number) which are separated by a 24-amu, the characteristic of a mixture of fullerene compounds.

Keywords: LDI, Asphaltenes, Photo Conversion, PAH's, Fullerenes

INTRODUCTION

The formation of photocompounds during laser desorption ionization mass spectrometry (LDI MS) examination of asphaltenes and other carbonaceous materials is becoming an important research area related to fullerene and carbon nanotube formation, extra-terrestrial bodies, and, of course, asphaltenes. Several research groups have presented spectra showing a 24-amu difference, corresponding to C_2. These spectra are produced when asphaltene or other carbonaceous materials are examined by LDI MS.

When asphaltenes were examined using LDI (+)-FT-ICR MS, asphaltenes showed a wide Gaussian band extended from 500 to 2000 Da with a 24-Da separation over the entire range. These bands, promoted by laser power, were attributed to fullerene-type materials [1] Similar results were reported by Rizzi et al. in 2006 [2] when Italian asphaltenes were analyzed using LDI and SALDI MS techniques. The abovementioned sequential band of 24 Da, which was found in a wide band

*Corresponding author

Socrates Acevedo

Email: socrates.acevedo@gmail.com

extending from 500 to 2000 Da and beyond, was promoted by the laser and attributed to the formation of polycyclic aromatic hydrocarbons (PAH's) [2]. Carbon clusters or fullerenes with a typical 24-amu separation were reported for a vacuum residue examined using LDI [3]. The examination of PAH's under LDI MS conditions similar to the one employed for asphaltenes led to the formation of multimers of these PAH's [4]. Authors have ascribed these multimers to the presence of non-covalent clusters during the desorption process [4]. Apicella and co-workers have also reported both multimer formation and fragmentation when PAH's were studied using LDI. The abovementioned 24-amu sequence was also observed when a soot sample was studied using LDI techniques [5]. Covalent clusters of PAH's were detected when they were examined using an LDI assisted by collision-cooling with nitrogen as the inert gas [6]. According to these authors [6], the collisions of radical ions with the gas molecules dissipate energy, thereby reducing unwanted reactions such as fragmentation. However, C-H and C-C fragmentations were observed as there were covalent clusters with a molecular mass of 2M-4H, in which M is the mass of the parent cation radical and the dimer lost 4 hydrogens during the fusion of two PAH's [6,7] . A statistical MMD comprising a sequence of bands separated by 24 amu was reported by Santos et al. after the examination of an asphaltene sample using LDI MS [8]. When a mixture of polycyclic aromatic hydrocarbon standards was examined under similar conditions, it afforded a similar MMD that allowed the authors to conclude that these MMD were the result of PAH fusion and the formation of fullerene-type homologous photocompounds. According

to these authors [8], the "C_2 sequence" has been observed previously [9,10], and was erroneously ascribed to the compounds present in the sample under analysis. The C_2 sequence was also found during the LDI treatment of Hassi-Messaoud asphaltenes [11] and Algerian asphaltenes [12]. Becker et al. in 2008 [13] used LDI MS combined with ion mobility to study asphaltenes and DAO, and observed fullerenes only in the case of asphaltenes. High resolution transmission electron microscopy (RTEM) was used for the detection of fullerenic structures in crude oils [14]. Traveling wave ion mobility (TWIM) was combined with LDI to separate fullerenes from other asphaltene components formed during LDI MS experiments. The technique employed gives a two-dimensional plot where fullerenes are clearly distinguished from other components [15].

Mass spectra showing the removal of hydrogen from cation radicals to afford multiradicals (diradicals, triradicals, and so on) under the energy environment provided by a laser (henceforth, laser conditions) were reported in 2014 [16]. Thus, all the hydrogens of the PAH cation radical $[C_{66}H_{26}]^+$ were removed as laser power steadily increased. MMD obtained using SEC with a UV detector leads to a log-normal distribution when octylated asphaltenes (OA) are used as the sample [17]. Presumably, the octyl groups inhibit the column-sample interaction responsible for the long tail and band deformation typical of the SEC of asphaltenes. Due to the employment of a UV detector, the observed MMD corresponds to the random distribution of PAH's comprising the core of the asphaltene molecules.

The reported LDI mass spectra of asphaltenes and octylated asphaltenes with up to five octyl groups per 100 carbons showed a similar MMD, indicating that these alkyl groups have no impact on the MMD obtained. As is now recognized [8], the obtained MMD corresponds to a log-normal distribution of fused PAH's (see below). A combination of LDI, SALDI, and L^{2}MS techniques led authors to propose the presence of asphaltene nanoaggregates in the vapor phase [18], and, according to other authors, the use of ion mobility techniques confirm aggregate formation in the gas phase for molecules with an MM > 3000 Da [13]. The partial fragmentation of alkylated PAH was observed during its analysis with desorption/ionization two laser μL^2-MS equipment [19], a technique employed to reduce or suppress fragmentation and ring fusion. When LDI MS was employed to measure the MMD of asphaltenes obtained after evaporation from toluene solutions, a large shift to a higher MM was observed. The authors ascribed this finding to asphaltene aggregates presented in the sample [20]. LDI MS applied to Algerian [11] and asphaltenes from another source [21] also confirmed the C$_2$ sequence referred to above. Curved PAH structures, containing five member rings such as corannulene and corannulene derivatives, were proposed by Pope et al. as the precursors of fullerene structures [22,23].

EXPERIMENTAL PROCEDURES
Materials and Methods

Asphaltenes and resins, both from Venezuela, were obtained by routine methods described elsewhere [24]. Briefly, corresponding crude oils mixed with toluene (1:1) were diluted (at a ratio of 1:60) with heptane, and the flocculated asphaltenes were filtered and dried. The resins were obtained after the extraction of asphaltenes in a Soxhlet using boiling n-heptane. Asphaltenes were obtained from Cerro Negro (CN) and Furrial crude oils, while the resins were obtained from Furrial crude oil.

Equipment

LDI MS TOF was determined using a laser desorption ionization system with a time-of-flight mass spectrometer (LDI TOF MS Voyager DE-STR Applied Biosystems Instrument). The laser system comprised a 337 nm-N$_2$ excimer laser at 20 Hz, which was used to analyze diluted (approximately 100 mg.L^{-1}) tetrahydrofuran solutions. These solutions were spread on the sample holder, and after solvent evaporation, these solutions were introduced into the LDI system under vacuum. The TOF detector was used in the linear mode, and the laser shots (LS) or laser powers were in a range of 1500 to 3000. No signal was detected below the LS equal to 1500.

Methods

The semi-empirical quantum mechanical method PM6, developed by Stewart et al. in 2007, was used for the required calculations [25].

The ionization and fragmentation of asphaltenes leading to the formation of PAH systems, along with ring fusion to form larger systems (FPAH's) and C$_2$ sequence fullerene are the processes considered herein. The route examined here is shown in the scheme below (Figure 1); all the stages are promoted by the laser source, where the power increases from one stage to the next in an LS range of 1500 to 3000.

In the first stage, asphaltene aggregates are ionized to a cation radical (CR) and sublimed from a condensed state that could be either the solid state or the plume.

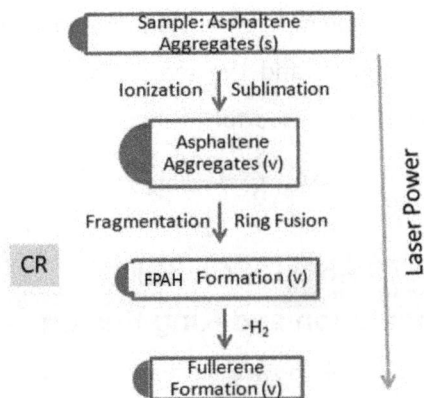

Figure 1: A schematic representation of the sequence of events promoted after laser contact with the sample. Ionization to a cation radicals (CR) and sublimation occurs in the first stage followed by fragmentation and ring fusion to afford FPAH ion radicals at a higher laser power. At even a higher laser power, the dehydrogenation of FPAH's is followed by fullerene formation. The letters "s" and "v" stand for solid and vapor respectively.

Once in the vapor phase as aggregates, the sample undergoes the fragmentation and ring fusion to afford FPAH's in the second stage. Finally, after H_2 is lost, fullerenes are formed in the third stage. Relevant fragmentation, multiradical formation, and the formation of FPHA's and fullerene-type materials were simulated and calculated with convenient models using PM6. In this work, multiradicals are defined as the structure resulting after the removal of all or some hydrogen from the corresponding PAH.

For the resin samples, the scheme is similar to the one presented in Figure 1 with the following differences; after ionization to a CR, a molecular cluster is evaporated. After fragmentation and fusion, the cluster is converted to FPAH's. No fullerene formation was observed in this case within the same LS range used for asphaltenes.

Thermodynamic Parameters

Thermodynamic parameters were calculated at different temperatures using the PM6 method.

Molar entropies were obtained at the required temperature from the frequency facility of the PM6 method. The calculation is illustrated by the reaction scheme given below:

$$A + B \xrightarrow{hv} C + 2H_2 \tag{1}$$

where, A is a CR, and B is a neutral reactant; C comes from the fusion of A and B. In this photoreaction, two moles of hydrogen are produced when two moles of reactant are consumed. For each component, the geometry was optimized, and the entropy S was calculated at the required temperature (T). The following changes were then computed:

$$\Delta S_R = \frac{2S_H + S_C - S_A - S_B}{2} \tag{2}$$

$$\Delta S_{PAH} = \frac{S_C - S_A - S_B}{2} \tag{3}$$

where, ΔS_R is the overall entropy change, and ΔS_{PAH} is the entropy change of the PAH species. These changes were evaluated in a temperature range of 500 to 1000 K.

For these calculations, the reaction between one pyrene molecule and its CR homolog was used to afford a dimer CR of a molecular mass of 2M-4, where M is the molecular mass of pyrene. This occurs according to the reaction given below (Figure 2):

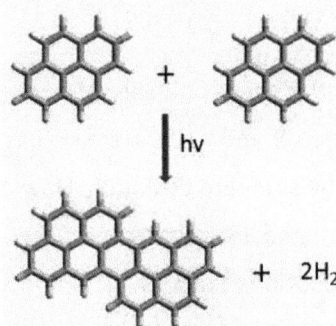

Figure 2: The reaction between one pyrene molecule and its CR homolog.
It should be noted that the MM of pyrene dimer is 2M-4. This dimer was detected when pyrene was examined with LDI MS under the conditions of collision cooling [16].

Closed or Fullerene-type Structures

The few fullerene-type and curved surface structures considered herein were constructed using the isolated pentane rule (IPR), in which no two pentagons share an edge [8]. When required, these structures were calculated using the PM6 method.

The number-(M_n) and weight-(M_w) average molecular masses were obtained using Equations 4 and 5.

$$M_n = \frac{\sum_i n_i M_i}{\sum_i n_i} = \frac{\sum_i I_i M_i}{\sum_i I_i} \qquad (4)$$

$$M_w = \frac{\sum_i n_i M_i^2}{\sum_i n_i M_i} = \frac{\sum_i I_i M_i^2}{\sum_i I_i M_i} \qquad (5)$$

The geometric mean (M_g) of these values was obtained using Equation 6.

$$M_g = \sqrt{M_n M_w} \qquad (6)$$

where, n_i, I_i and M_i are the corresponding number of molecules, intensity, and molecular mass respectively. The intensity and molecular mass were determined from the experimental results.

Log Normal MMD

For statistical MMD, the log-normal distribution was used, which is easily obtained as follows:
The error distribution could be defined using Equation 7:

$$dP = \frac{1}{\sigma'\sqrt{2\pi}} \exp\left[-\frac{1}{2}\frac{(x-x_0)^2}{\sigma'^2}\right] dx \qquad (7)$$

As usual, dP is the probability of finding the value of x between x and $x+dx$. Herein, x_0 is the mean, and σ' is the standard deviation. Now, the variable can be changed in a way that $x = \log M$, which leads to Equations 8 and 9.

$$dP = \frac{1}{\sigma\sqrt{2\pi}} \exp\left[-\frac{1}{2}\frac{(\log M - \log M_0)^2}{\sigma^2}\right] d\log M \qquad (8)$$

$$dP = \frac{1}{M\sigma\sqrt{2\pi}} \exp\left[-\frac{1}{2}\frac{(\log M - \log M_0)^2}{\sigma^2}\right] dM \qquad (9)$$

where σ, M, and M_0 are the standard deviation, the molecular mass, and the molecular mass mean. Equation 10 was used for fitting.

$$y = y_0 + \frac{A}{M\sigma\sqrt{2\pi}} \exp\left[-\frac{1}{2}\frac{(\log M - \log M_0)^2}{\sigma^2}\right] \qquad (10)$$

RESULTS AND DISCUSSION
Fragmentation and Ring Fusion

Figures 3 and 4 show a series of LDI MS spectra taken of Furrial and CN asphaltene samples at different laser powers (LS's).

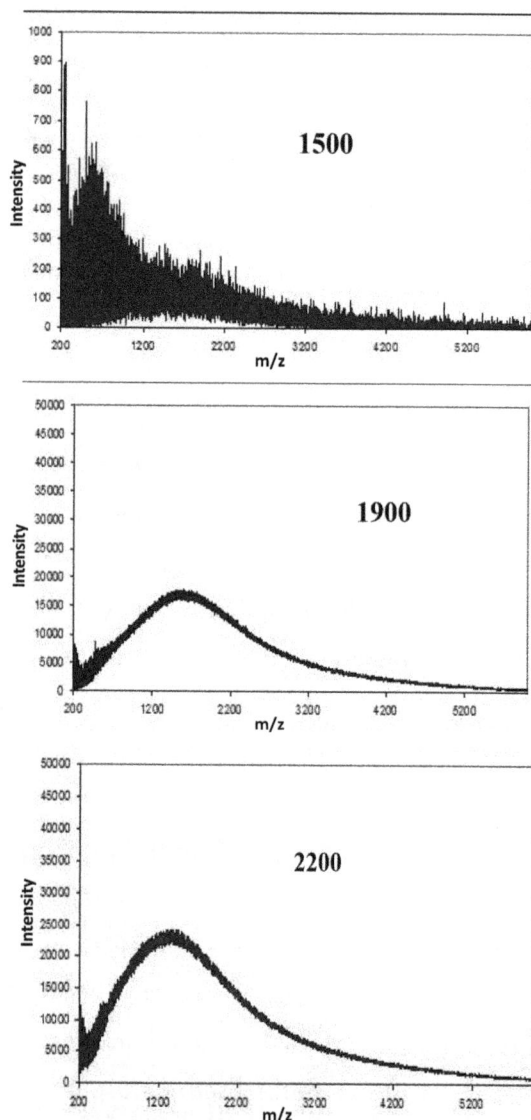

Figure 3: Sequences of LDIMS of asphaltenes (Furrial) measured at increasing LS power. The increase in LS apparently promotes both fragmentation and ring fusion, as suggested by the observed large MMD changes observed in this LS range (200-5200 amu range).

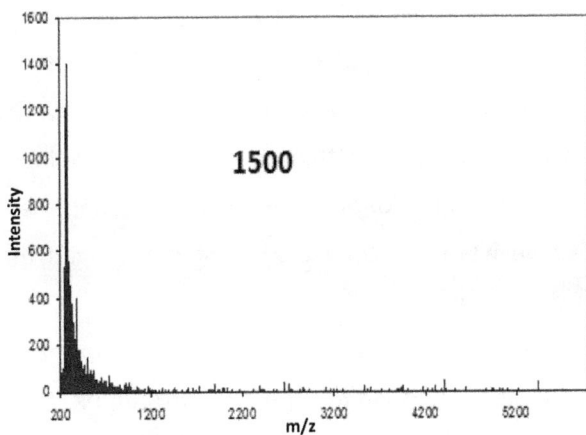

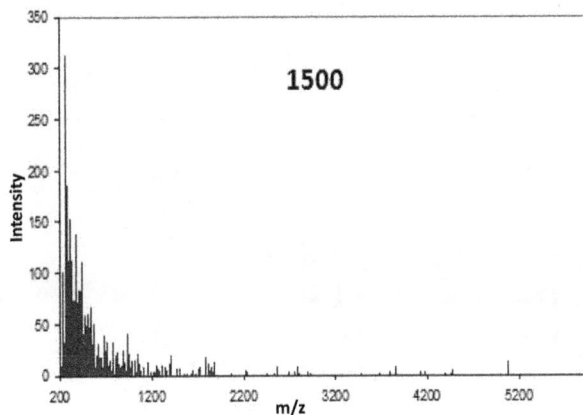

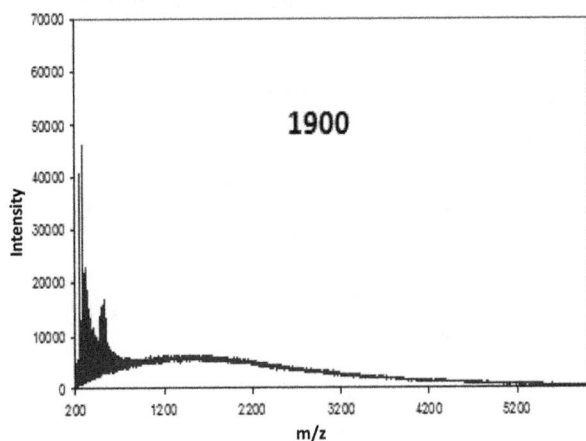

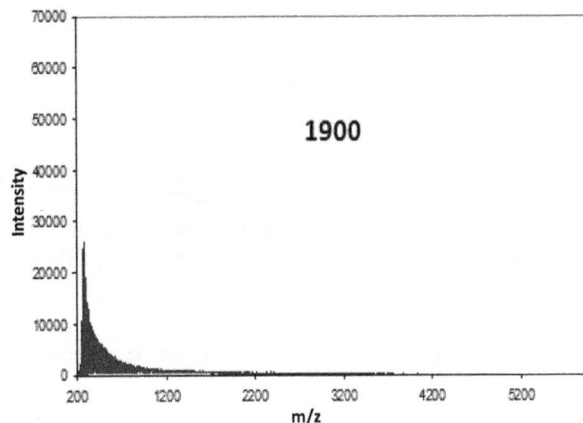

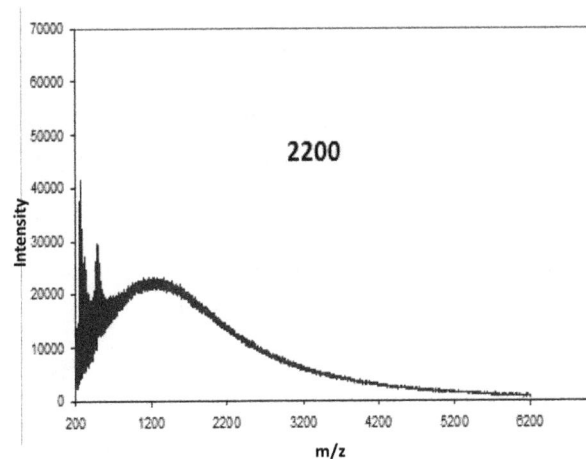

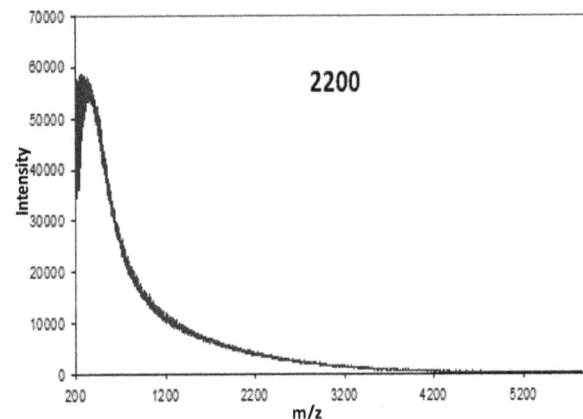

Figure 4: Sequences of LDIMS of CN asphaltenes measured at increasing LS power. Evidence for ring fusion is apparent from LS = 1900 onward in this LS range(200-5200　amu).

Apparent conversion to FPAH's was also observed for the resins (Figure 5).

Figure 5: Resin spectra measured at different relative laser powers (LSs) in the range of 200 to 6200 m/z. Apparently, as LS increases, the sample is steadily converted to a mixture of FPAH's.

The sequence shows how the initial sample changes when the LS increases to the point where fragmentation is readily apparent as the initial complex distribution is converted to a quasi-continuous MMD. This result is coherent with the simultaneous operation of both fragmentation and ring fusion and with the mass maxima shown by these samples in Figures 6 and 7.

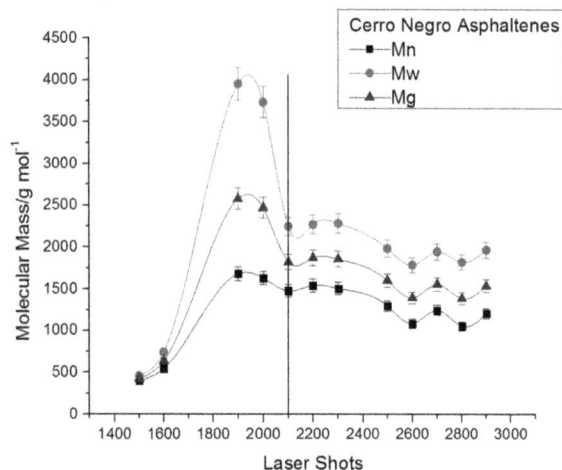

Figure 6: Dependence of molecular mass on LS for CN asphaltenes. The maximum below LS = 2100 is consistent with the simultaneous operation of ring fusion and fragmentation, which have opposite effects on the sample mass. The sector to the right of LS = 2100 corresponds to fullerene formation and the loss of hydrogen and other elements such as nitrogen, sulfur, and oxygen.

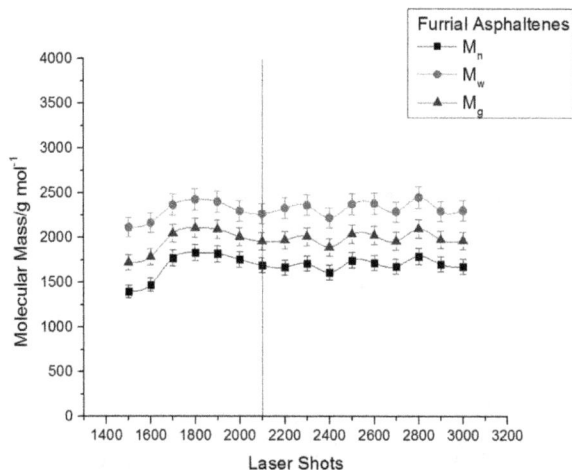

Figure 7: Dependence of the molecular mass on LS for Furrial asphaltenes. The maximum below LS = 2100 is consistent with the simultaneous operation of ring fusion and fragmentation, which have opposite effects on the sample mass. The sector to the right of LS = 2100 corresponds to fullerene formation and the loss of hydrogen and other elements.

The increase in the mass of ring fusion is opposed by fragmentation (mass decrease). As described above (see Introduction), fragmentation and ring

fusion have been observed using model PAH's [4,6,19]. At an LS value larger than about 2100, a series of maxima and minima are observed; however, it appears that most of these values are within the experimental error suggesting that sample has been converted to stable derivatives such as fullerenes.

It is likely that in this case, the sample is vaporized as molecular clusters, thereby promoting ring fusion. The molecular mass distribution of PAH coming from ring fusion (MMDRF) could be well fitted to a log-normal statistical distribution, as shown in Figure 7.

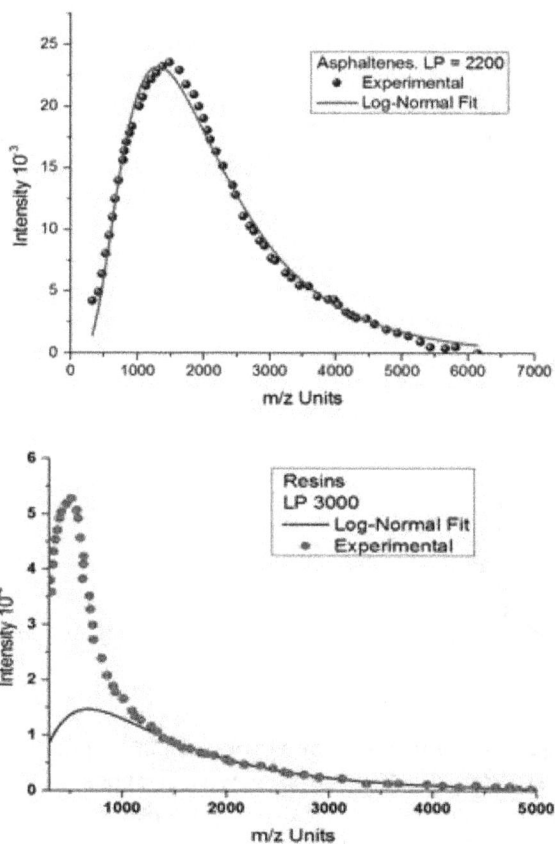

Figure 8: Fitting a log-normal distribution (Equation 10) to experimental data for the asphaltenes (top) and the resins (bottom) at the shown laser power. Apparent conversion to FPAH's is complete for the asphaltenes and partial (approximately 60%) in the case of the resins.

Log-normal distribution was well fitted to the experimental data only at the tail of the mass spectra of the resins; for asphaltenes, conversion to ring fusion was complete, whereas for the resins it was approximately 60%. The parameters corresponding to these fits are collected in Table 1.

Table 1: Log-normal fitting parameters.

Sample	M_0	σ	y_0	A	Adjusted R^2
Asphaltenes	1863	0.596	-151.2	5.47E+07	0.98
Resins	676	0.8	-289	3.00E+04	0.99

See Equation 10 for parameter definition.

In comparison with the asphaltenes, the resins have higher and faster volatility, which could account for their partial conversion; they are also consistent with the higher LS value required for ring fusion.

Previously, MMD's obtained from LDI-MS were also fitted by a log-normal distribution [4,27].

Asphaltene molecules are thought of as aromatic cores with a high substitution of hydrogen in the form of aliphatic chains [28]; in other words, they contain a very low percentage (approximately 5%) of aromatic hydrogens [24]. This means that for sufficient laser power (see Figures 2 and 3), the fragmentation of alkyl chains would result in the formation of very reactive multiradicals, which, if close enough, will react with the nearest neighbor homologs, leading to ring fusion and to a random MMD.

Figure 7 is a simple depiction of fragmentation using molecular models. Because multiradicals are extremely reactive, they could react to form new 6- or 5-member aromatic rings.

The presence of diradicals is possible only in the presence of the high energy provided by the laser. Fragmentation could occur either by direct energy coming from the laser or by contact with highly excited CR or homologs (see Figure 9 for an example of this process).

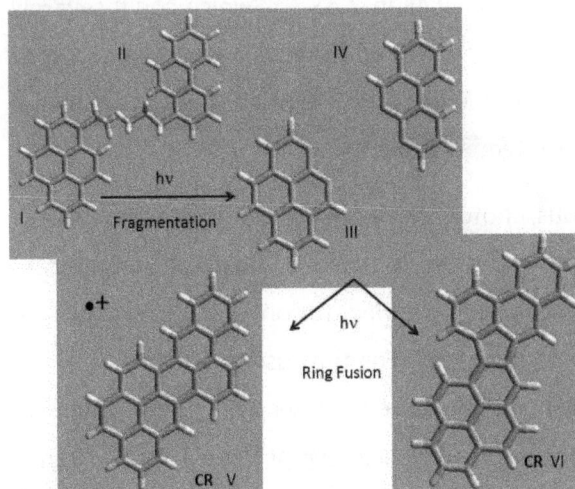

Figure 9: An example of ring fusion considered herein, in which the alkyl aromatics, represented by 1-ethyl-pyrene (I) and 4-ethyl-anthracene (II), are fragmented by the laser to afford diradicals III and IV, which leads to fused system V or VI cation radicals (CR).

Ring fusion in the open vapor phase is unlikely to occur. The gathering of many molecules in the vapor state would have a high entropic cost, as suggested by the calculations shown in Table 2.

Table 2: Entropy changes computed for the fusion of pyrene and the pyrene cation radical[a].

T/K	$T\Delta S$ /kCal.mol^{-1}	
	ΔS_R [b]	ΔS_{PAH} [c]
500	3.8	-26.0
600	4.9	-32.0
700	6.1	-38.2
800	7.1	-44.7
900	8.2	-51.4
1000	9.2	-58.4

The PM6 method; b: overall change; c: change between PAH species.

Although the overall $T\Delta S_R$ is positive, due to the formation of hydrogen, bringing the reactants together for the reaction to occur involves a large negative change in $T\Delta S_{PAH}$, which could severely impair or prevent all contacts among reactants. As shown in Table 2, this hindering effect increases with temperature.

Thus, photoreactions in the vapor phase are unlikely and expected to have a small or insignificant contribution to ring fusion and other possible photoreactions. For the fusion of three or more PAH's, larger, more negative $T\Delta S_{PAH}$ are obtained. Moreover, very high temperatures, in the range of 4000-8000 K, are expected in ns pulsed laser plasma commonly used in LDI [23].

On the other hand, due to close contact within an aggregate or cluster that has been evaporated or sublimed from the corresponding condensed state, ring fusion will more be favored for an associated mixture such as the asphaltenes. Of course, in this case, little contribution of ΔS_{PAH} to the reaction is expected. These arguments are congruent with the method employed in the L^2 MS technique described above, in which, to avoid ring fusion, the two lasers are tuned so that the first sublimes the sample and the second (the one used to ionize the sample) is shot at the vaporized sample a few μs later [19,27].

In view of the above comments and results, it is concluded that under the present conditions and after ion radical formation, the samples are either sublimed (the asphaltenes) or vaporized (the resins) as aggregates or molecular clusters respectively. Subsequently, they undergo fragmentation and ring fusion before being converted to fullerenes at higher LS values. As mentioned in Introduction, the

presence of asphaltene nanoaggregates has been suggested in the vapor phase using a combination of LD techniques [4,18].

Fullerene Formation

Figure 10 shows the LDI MS spectrum of Furrial asphaltenes measured at an LS value of 2900.

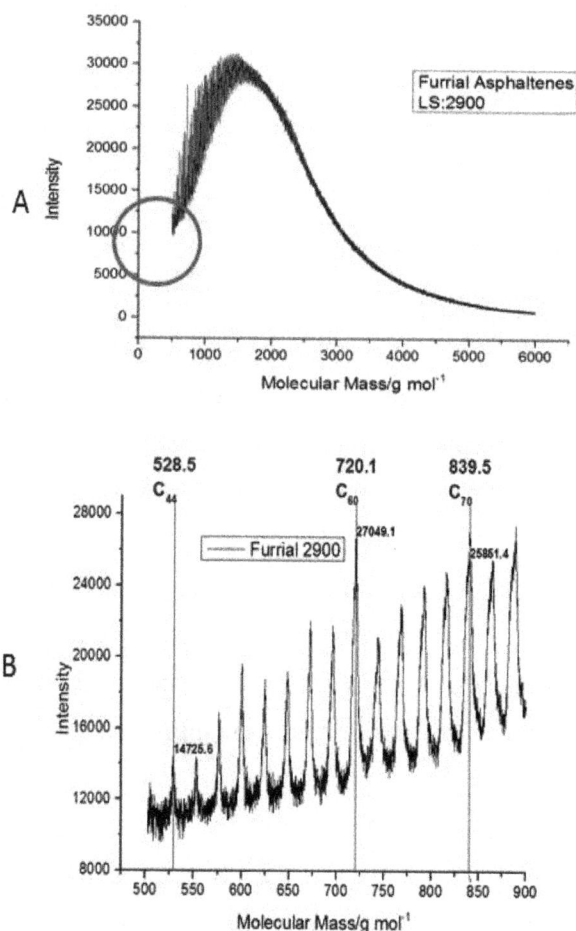

Figure 10: LDI spectra of Furrial asphaltenes measured at LS = 2900 laser power. A) full range; B: expansion of the 500-900 amu, corresponding to the inset range showing the C_2 = 12 amu band sequence. No fullerene bands were detected below 500 g mol^{-1} or above 2300 g mol^{-1}. Bands and intensities corresponding to C_{44}, C_{60}, and C_{70} are distinguished by vertical lines.

At this high energy, the sequence of C_2 bands shown in the expansion is very clear. We could detect fullerenes from C_{42} (MM = 720) to approximately C_{214} (MM = 2570). As suggested elsewhere [30,31], C_{60} and C_{70} are relatively stable fullerenes, and this is consistent with the intensities of the corresponding bands (Figure 9-B). Similar results were obtained for CN asphaltenes (not shown). Thus, in view of similar findings regarding the C_2 sequence described in Introduction, this behavior appears to be a characteristic of asphaltenes under

high laser energy conditions.

At energies high enough to promote the removal of hydrogens from C-H corresponding to the formed FPAH cation, multiradicals will form. These multiradicals are very reactive species; a hint of this is shown in Figure 11, where a very large and negative change in the enthalpy of formation could be estimated.

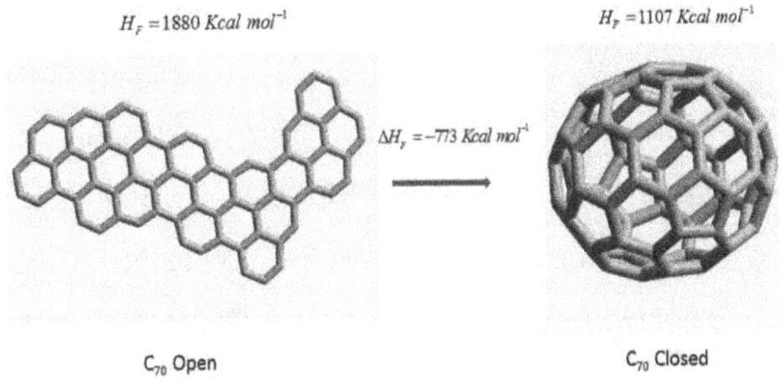

Figure 11: Depiction of the conversion of a multiradical cation C_{70} to a closed C_{70} cation radical fullerene type structure

As it is well known [22,23,30], the PAH's avoid the formation of multiradicals by forming curved surfaces in which the rupture of C-H would occur simultaneously with C_2 additions and C-C bond formation. These curved surfaces are formed after intramolecular rearrangements of the planar hexagonal structures of FPAH's, which become curved after the inclusion of pentane rings.

A plausible evolution of some stages leading to C_{72} is shown in Figure 12.

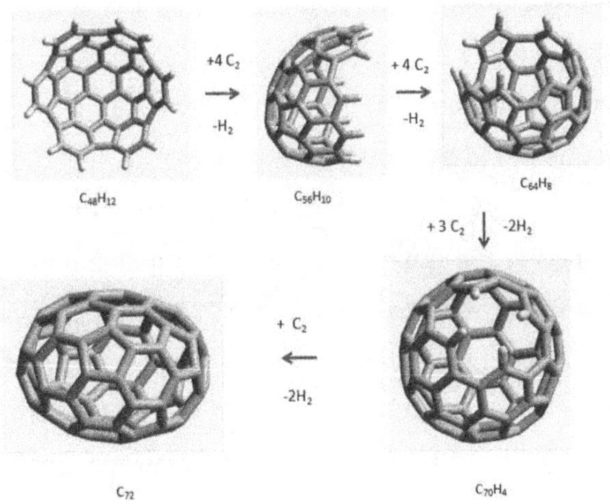

Figure 12: A schematic representation of the formation of the fullerene-type structure C_{72} by means of progressive stages in which the growth of the spheroid surface, after C_2 transfer, allows for hydrogen loss without the formation of multiradicals. Transfer of C_2 is intramolecular and comes from somewhere else within the molecule (not specified for simplicity).

It should be mentioned that, by using appropriate computation algorithms, it is possible to build fullerenes to the required size and molecular mass [32]. In this way, the observed C_2 sequence can be possible in terms of the above arguments irrespective of the fullerene size.

It should also be mentioned that heteroatoms initially present in the asphaltene sample should be removed at some stage during laser contact; otherwise, the observed C_2 band sequence will not exist. It is also interesting that under the same LS used for the asphaltenes, the resins do not yield fullerenes. Finally, it is very important to remark that, as it is the case for fullerenes, the statistical molecular mass distribution obtained at low and medium values of laser power for the asphaltenes is an artifact produced by laser contact with the sample and is not at all the proper molecular mass distribution of the asphaltenes. Thus, some previous results regarding this issue need revising [27].

CONCLUSIONS

The sublimation of aggregates (the asphaltenes) and the vaporization of clusters (the resins) were found out to be consistent with both fragmentation and ring fusion. The fragmentation promoted either by the laser or by contact with high energy homologs leads to very reactive PAH multiradicals, which, in contact with each other in the aggregates or clusters, result in ring fusion and in the formation of FPAH's having a log-normal MMD. At a high laser power, hydrogen is removed from FPAH's, leading to a typical fullerene MMD in which each band is separated from the next by C_2 amu. It could be summarized that to avoid the formation of multiradicals, the mainly planar FPAH's with hexagonal rings undergo rearrangement to include both hexagonal and pentagonal rings, which results in the closed curved 3-D structures typical of fullerenes.

ACKNOWLEDGMENTS

We thank the Post Graduated Cooperation Program between France and Venezuela for the partial financial support of this work. Additionally, the financial support from the Conseil Régional d'Aquitaine (20071303002PFM) and FEDER (31486/08011464) is acknowledged.

NOMENCLATURES

CDA   : Coal derived asphaltenes
CR   : Cation radical
CN   : Cerro Negro
FPAHs : Fused polycyclic aromatic hydrocarbons
IPR   : Isolated pentane rule
LS   : Laser shots
LDI-TOF : Laser desorption/ionization-time of flight
MM   : Molecular mass
MMD   : Molecular mass distribution
MS   : Mass spectrometry
PAHs   : Polycyclic aromatic hydrocarbons
RTEM   : Resolution transmission electron microscopy
TWIM   : Traveling wave ion mobility

REFERENCES

1. Pereira T. M. C., Vanini G., Tose L.V., Cardoso F. M. R., and et al., "FT-ICR MS Analysis of Asphaltenes: Asphaltenes Go in, Fullerenes Come Out," *Journal of Fuel*, **2014**, *131*, 49-58.

2. Rizzi A., Cosmina P. Flego C., Montanari L., Seraglia R., and et al., "Laser Desorption/Ionization Techniques in the Characterization of High Molecular Weight Oil Fractions. Part 1: Asphaltenes," *Journal of Mass Spectrometry*, **2006**, *41*, 1232-1241.

3. Palacio L., Orrego-Ruiz J. A., Barrow M. P., Hernandez R. C., and et al., "Analysis of the Molecular Weight Distribution of Vacuum Residues and their Molecular Distillation Fractions by Laser Desorption Ionization Mass Spectrometry," *Journal of Fuel*, **2016**, *171*, 247-252.

4. Hortal A. R., Hurtado P., Martínez-Haya B., and Mullins O. C., "Molecular-weight Distributions of Coal and Petroleum Asphaltenes from Laser Desorption/ionization Experiments," *Journal of Energy and Fuels*, *21*, 2863-2868.

5. Apicella B., Alfè M., Amoresano A., Galano E., and Ciajolo A., "Advantages and Limitations of Laser Desorption/ionization Mass Spectrometric Techniques in the Chemical Characterization of Complex Carbonaceous Materials," *International Journal of Mass Spectrometry*, **2010**, *295*, 98-102.

6. Gámez F., Hortal A. R., Martínez-Haya B., Soltwisch J., and et al., "Ultraviolet Laser

Desorption/ionization Mass Spectrometry of Single-core and Multi-core Polyaromatic Hydrocarbons under Variable Conditions of Collisional Cooling: Insights into the Generation of Molecular Ions, Fragments and Oligomers," *Journal of Mass Spectrometry*, **2014**, *49*, 1127-1138.

7. Daaou M., Modarressi A., Bendedouch D., Bouhadda Y., Krier G., and et al., "Characterization of the Nonstable Fraction of Hassi–Messaoud Asphaltenes," *Journal of Energy and Fuels*, **2008**, *22*, 3134-3142.

8. Santos V. G., Fasciotti M., Pudenzi M. A., Klitzke C. F., and et al., "Fullerenes in Asphaltenes and other Carbonaceous Materials: Natural Constituents or Laser Artifacts," *Journal of Analyst*, **2016**, *141*, 2767-2773.

9. Buseck P. R., Tsipursky S. J., and Hettich R., "Fullerenes from the Geological Environment," *Journal of Science*, **1992**, *257*, 215-217.

10. Becker L., Bada J. L., Winans R. E., Hunt J. E., and et al., "Fullerenes in the 1.85-billion-year-old Sudbury Impact Structure," *Journal of Science*, **1994**, *265*, 642-645.

11. Daaou M., Larbi A., Martínez-Haya B., and Rogalski M., "A Comparative Study of the Chemical Structure of Asphaltenes from Algerian Petroleum Collected at Different Stages of Extraction and Processing," *Journal of Petroleum Science and Engineering*, **2016**, *138*, 50–56.

12. Fergoug T., Boukratem C., Bounaceur B, and Bouhadda Y., "Laser Desorption/Ionization-Time of Flight (LDI-TOF) and Matrix-Assisted Laser Desorption/Ionization-Time of Flight (MALDI – TOF) Mass Spectrometry of an Algerian Asphaltene," *Egyptian Journal of Petroleum*, **2016**, *26*(3), 803-810.

13. Becker C., Qian K., and Russell D. H., "Molecular Weight Distributions of Asphaltenes and Deasphaltened Oils Studied by Laser Desorption Ionization and Ion Mobility Mass Spectrometry," *Journal of Analytical Chemistry*, **2008**, *80*, 8592–8597.

14. Camacho-Bragado G. A., Espinosa M. M., Romero E. T., Murgich J., and et al., "Fullerenic Structures derived from Oil Asphaltenes," *Journal of Carbon*, **2002**, *40*, 2761–2766.

15. Koolen H. H. F., Klitzke C. F., Cardoso F. M. R., Rosa P. T. V., and et al., "Fullerene Separation and Identification by Traveling Wave ion Mobility Mass Spectrometry in Laser Desorption Processes during Asphaltene Analysis," *Journal of Mass Spectrom*, **2005**, *51*, 254–256.

16. Zhen J., Castellanos P., Paardekooper D., Linnartz H., and et al., "Laboratory Formation of Fullerenes from PAHs: Top-Down Interstellar Chemistry," *The Astrophysical J. Letters*, **2014**, *797*, 30.

17. Acevedo S., Escobar G., Ranaudo M. A., and Rizzo A., "Molecular Weight Properties of Asphaltenes Calculated from GPC Data for Octylated Asphaltenes," *Journal of Fuel*, **1998**, *77*(8), 853-858.

18. Wu Q., Pomerantz, A. E. Mullins O. C., and Zare R. N., "Laser-based Mass Spectrometric Determination of Aggregation Numbers for Petroleum and Coal-derived Asphaltenes," *Journal of Energy and Fuels*, **2014**, *28*, 475-482.

19. Mahajan T. B., Elsila J. E., Deamer D. W., and Zare R. N., "Formation of Carbon-carbon Bonds in the Photochemical Alkylation of Polycyclic

Aromatic Hydrocarbons," *Journal of Origins of Life & Evolution of the Biosphere*, **2003**, *33*, 17-35.

20. Lobato M. D., Pedrosa J. M., Hortal A. R., Martínez-Haya B., and et al., "Characterization and Langmuir Film Properties of Asphaltenes Extracted from Arabian Light Crude Oil," *Colloids and Surfaces A: Physicochemical and Engineering Aspects*, **2007**, *298*, 72–79.

21. Araujo P., Mendes M., and Oller N., "Contribution of Mass Spectrometry for Assessing Quality of Petroleum Fractions," *Journal of Petroleum Science & Engineering,* **2013**, *109*, 198-205.

22. Pope C. and Howard J., "Thermochemical Properties of Curved PAH and Fullerenes- a Group Additivity Method Compared with $mm^3(92)$ and Mopac Predictions," *Journal of Physical Chemistry,*" **1995**, *99*, 4306-4316.

23. Pope C. J., Marr J. A., and Howard J. B., "Chemistry of Fullerenes C_{60} and C_{70} Formation in Flames," *Journal of Physical Chemistry*, **1993**, *97*, 11001-11013.

24. Acevedo S., Mendez B., Rojas A., Layrisse I., and et al., "Asphaltenes and Resins from the Orinoco Basin," *Journal of Fuel*, **1985**, *64*, 1741-1747.

25. Stewart J. J. P., "Optimization of Parameters for Semi-empirical Methods V: Modification of NDDO Approximations and Application to 70 Elements," *Journal of Molecular Modeling*, **2007**, *13*, 1173-1213.

26. Brinkmann G., Goedgebeur J., and McKay B. D., "The Generation of Fullerenes," *Journal of Chemical Information and Modeling*, **2012**, *52*, 2910-2918.

27. Acevedo S., Gutiérrez L., Negrin G., Pereira J., and et al., "Molecular Weight of Petroleum Asphaltenes: A Comparison between Mass Spectrometry and Vapor Pressure Osmometry," *Journal of Energy & Fuels*, **2005**, *19,* 1548-1560.

28. Yen, T. F., "Structural Differences between Asphaltenes Isolated from Petroleum and from Coal Liquid," *Journal of Chemistry of Asphaltenes*, **1981**, *195*, 39-51.

29. Phipps C., "Laser Ablation and its Applications Laser Ablation of Energetic Polymer Solutions: Effect of Viscosity and Fluence on the Splashing Behavior," *Journal of Applied Physics A, Springer*, **2009**, *94*(3), 657-665.

30. Cash G. G., "Heats of Formation of Curved PAH's and C_{60}: Graph Theoretical vs MM3(92) Predictions," *Journal of Physical Chemistry A*, **1997**, *101*, 8094-8097.

31. Yen T. F., "Structural Differences between Asphaltenes Isolated from Petroleum and from Coal Liquid," *Journal of Chemistry of Asphaltenes*, **1981**, *195*, 39-51.

32. Brinkmann G., Goedgebeur J., and McKay B. D., "The Generation of Fullerenes," *Journal of Chemical Information and Modeling*, **2012**, *52*, 2910-2918.

8

A New Discovery about Inflow Control Devices in Controlling Water and Increasing Oil Recovery

Li Mingjun*[1,2], Ma Yongxin[2], and Yi Xiangyi[1]

[1] Chengdu University of Technology; Chengdu, 610059, Chengdu, China
[2] CNOOC -Ltd Zhanjiang. Zhanjiang, 524057, Zhanjiang, China

ABSTRACT

Inflow control devices (ICD), which prevent water breakthrough by controlling the inflow profile of a well, have been used successfully in many oilfields. This paper will introduce a new discovery and an unsuccessful example. Moreover, this paper investigates meticulously and thoroughly to find the application conditions of the new discovery. Based on permeability rush coefficient and permeability differential, a series of plans are carried out to study ICD application conditions. Finally, a new discovery is developed. There are inflow-control device applications for ICD's, which can work well in heterogeneous reservoirs for controlling water and increasing oil production, but they cannot work well in a homogeneous formation. The effect of ICD on controlling water and increasing oil is very sensitive to the degree of reservoir heterogeneity. The cumulative oil production increases with increasing permeability rush coefficient and permeability differential.

Keywords: Inflow Control Devices, Controlling Water, Increasing Oil Recovery, Application Conditions

INTRODUCTION

ICD, as a sort of completion tools, have been used in many oilfields, in many countries. From these papers, we can understand that the effect of ICD is very good, which can prevent successfully water breakthrough and improve the oil production. In Kristian Brekke's paper [1], ICD's was used to enhance the performance of long horizontal wells producing from high-permeability formations; moreover, ICD increased the plateau rate of a frictionless well by 50% in comparison with a standard completion well. In Augustine's paper [2], the use of ICD can result in increased NPV and increased ultimate recovery. However, neither thick pay zones nor short horizontal wells are good candidates for horizontal wells with ICD. In Michael Lorenz's paper [3], the inflow control technology can balance production inflow and prevent gas or oil coning; also, the inflow control technology can minimize the risk of erosion and achieve sand control without gravel packing. In Henriksen's paper [4], the use of ICD yield the higher volumetric recovery of oil from each well compared to more conventional sand control completion methods in the Troll oil subsea field. In McIntyre's paper [5], ICD are used successfully in the UK sector of

*Corresponding author
Li Mingjun
Email: 529145486@qq.com

the North Sea, which can prevent early water and gas breakthrough, and increase oil recovery. In Henriksen's paper [6], the use of ICD can increase recovery and keep oil rates at higher levels compared to conventional methods. In Cudahy's paper [7], ICD are very effective in improving sustainable well productivity compared to past horizontal completion methods. The combination of ICD design geo-steering calibration with real-time log data, installation procedures, and fluid historical production results through time are also described elsewhere [8]. ICD's show good performance with increased oil production and lower water production in all wells in Ecuador. It is also reported that horizontal producers with inflow controls and NCBF's greatly improve the recovery of oil and reduce the production of water under a water flood scenario [9]. In Zhuoyi Li's paper [10], they investigated how and when an ICD should be used. ICD can be used to improve well performance and increase recovery, but it is not a universal solution for production problems. The application requires a thorough understanding of long-term reservoir behavior and upfront reservoir characterization for implementation. Gualdron reported that ICD's are used in three horizontal wells in Rubiales field, and the performance of the ICD was found to reach the highest cumulative oil production in comparison to neighboring wells [11]. Shad's stated the advantages of ICD-equipped wells over the conventional dual-tubing and slotted liner completions for SAGD operation [12]. The improved well performance, increased bitumen production, and longevity of the wells will compensate for the additional cost of ICD installation in a short period of time.

On the other hand, the oil-field service companies

owning ICD technology also publicize that ICD's have an amazing effect on controlling water breakthrough and increasing oil recovery. However, in our work and study, we found a surprising problem that ICD cannot work well in preventing water breakthrough. Thus, a series of deep studies about ICD application conditions have been carried out, and as a result, we have found some new discoveries.

EXPERIMENTAL PROCEDURES
Problem Finding

This is a thin bottom water reservoir with sufficient bottom water energy in South China Sea. The reservoir thickness is about 5 m. The type of reservoir is single porosity, and reservoir physical properties are very good; the porosity is 30-36%, and the permeability is 655-5242 mD; the reservoir pressure is about 8.9 MPa, and the reservoir temperature is about 67.1 °C. The ground oil is characterized by "three highs and two lows", namely high density (0.9366-0.944 g/cm^3), high viscosity (178.5-237.6 mPa.s), high colloid content (7.96-15.81%), low freezing point (-l8-8.9°C) and low wax content. The underground oil density is 0.8975 g/cm^3, featured by high oil viscosity (70 mPa.s), low solution gas/oil ratio, low saturation pressure (0.8 MPa) and great difference between reservoir pressure and saturation pressure.

In order to improve oil recovery, horizontal wells are used in this oilfield. The simulation results show that all horizontal wells with conventional completion techniques cannot prevent water rising, and water rises fast. Single well cumulative oil production is very low, and the average is less than 5×10^4 m^3; the minimum is about 2×10^4 m^3 with a small pressure drop as shown in Figure 1.

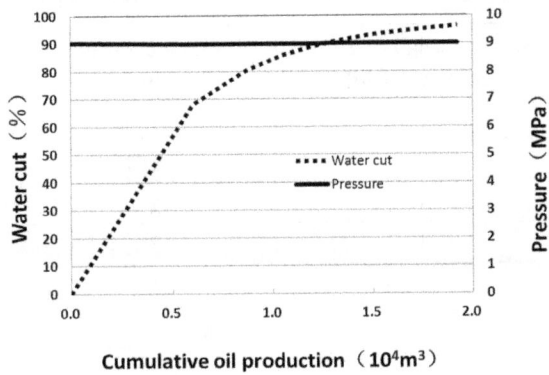

Figure 1: Single well production curve.

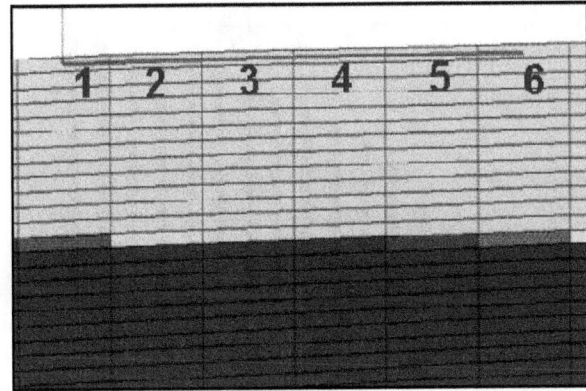

Figure 3: Well (A1H) trajectories.

As the water rises fast and single well production is low, inflow control devices are considered to control bottom water coning and improve oil production. A low-yielding well, A1H was selected as an experimental well to carry out ICD numerical simulation research. The model plane grid is 50 m×50 m; in addition, vertical grid is 0.6 m. The number of grids in X, Y, and Z direction is 312, 200 and 51 respectively. The main parameters of A1H well are as follows: the wellbore length is 300 m, and the production rate is 500 m³/d. Because it is a heavy oil reservoir, underground oil viscosity is 70 mPa.s, and Nozzle-based ICD are used (Figure 2), which are independent of oil viscosity. Eclipse simulator is used to carry out the ICD research.

Figure 4: Water cut versus position along horizontal segment.

Finally, ICD's are placed in the heel to control water. The ICD place, the dimensionless flow coefficient for the valve, the cross-section area for flow in the constriction etc. are carefully optimized. The simulation results are shown in Figures 5 and 6.

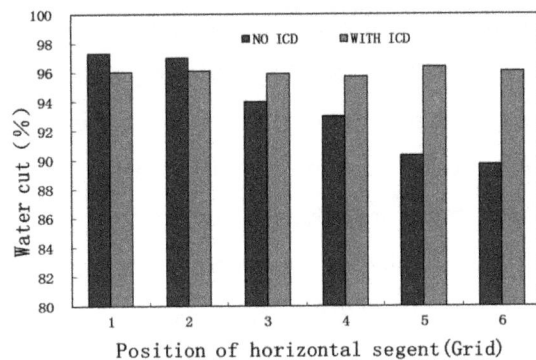

Figure 2: Nozzle-based ICD.

First, every grid is numbered, through which the horizontal section passes from 1 to 6 in the model (Figure 3). By analyzing the PRT file, the part of high water cut is found, which is the heel of the horizontal section (Figure 4).

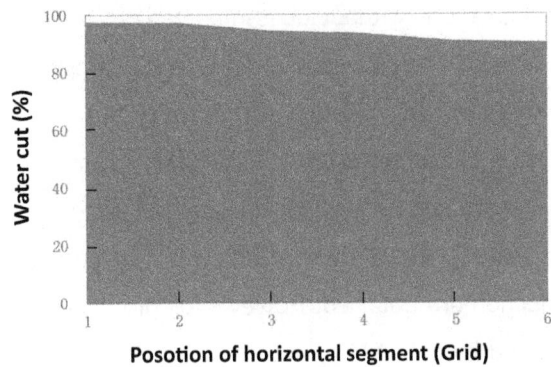

Figure 5: Water cut versus position along horizontal segment for two simulations: No ICD and With ICD.

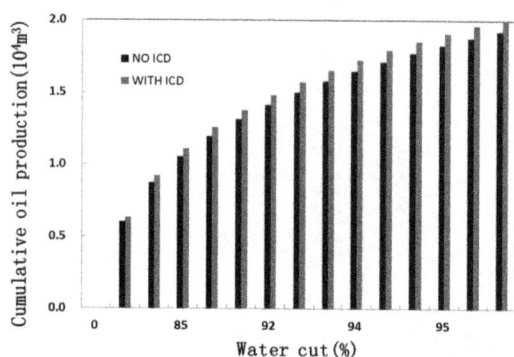

Figure 6: Water cut versus cumulative oil production for two simulations: No ICD and with ICD.

Two simulations, without ICD (symbolized as "No ICD") and with ICD are shown in Figures 5 and 6. Compared the cases of with ICD and without ICD, the water cut along horizontal segment with ICD is adjusted to equilibrium, and heel water is controlled; toe water is increased, and a lower water cut and more oil production are obtained. However, oil production is increased by only 4.6%, which is about 900 m³. The effect of increasing oil is unsatisfied for ICD.

The Analysis of s Causes of Reservoir Heterogeneity

Why could ICD not prevent water rising and increase oil production in well A1H? Maybe, our simulation method with ICD was wrong; that was our first thought, because all the news about ICD is positive. For this reason, on the one hand, two professional ICD Companies were invited to make an ICD program in well A1H. Unfortunately, the two professional ICD Companies obtained the same result as ours. On the other hand, we thought, perhaps, there are suitable conditions for ICD, so the reservoir physical properties along well A1H were studied deeply. Finally, it was found out that the average permeability along the horizontal segment was close and high (Table 1).

Table 1: Average permeability along the horizontal segment.

Position along horizontal segment (Grid)	Average permeability (mD)
1	1618
2	1575
3	1573
4	1504
5	1476
6	1453

From Table 1, it is clear to see that the reservoir along well A1H is homogeneous. In order to investigate the reservoir heterogeneity degree, the reservoir heterogeneity parameter standards are introduced into the model as shown in Table 2. There are three parameters characterizing reservoir heterogeneity: variation coefficient, rush coefficient, and differential.

Table 2: Reservoir heterogeneity evaluation standard.

Heterogeneity standard	Variation coefficient	Rush coefficient	Differential
weak	< 0.5	< 2	< 20
medium	0.5 ~ 0.9	2 ~ 3	20 ~ 30
strength	> 0.9	> 3	> 30

The permeability variation coefficient is the ratio of standard deviation to average given below:

$$K = \sqrt{\sum_{i=1}^{n}(K_i - K_{ave})^2/(n-1)}\Big/K_{ave} \qquad (1)$$

The permeability rush coefficient is the ratio of maximum to average, as reads:

$$K = K_{max}/K_{ave} \qquad (2)$$

The permeability differential is the ratio of maximum to minimum defined by:

$$K = K_{max}/K_{min} \qquad (3)$$

Based on the permeability rush coefficient and differential, the heterogeneity evaluation of A1H was carried out. The results show that it is a homogeneous formation, and permeability rush coefficient is 1.05, and permeability differential is 1.1. So far, the reason leading to water-control failure is still uncertain, but we speculate that the failure is related to homogeneous formation in well A1H. ICD have application conditions, which cannot work well in homogeneous formation. With the well-founded suspicion, a more in-depth study has been carried out.

Deep Study about Inflow Control Devices Application Situations

Based on permeability rush coefficient and permeability differential, the permeability along horizontal segment are changed to make the formation heterogeneous. The major changes of permeability along horizontal segment are heel and toe. From weak heterogeneity to strength heterogeneity, there are three plans; the brief parameters are shown in Table 3.

Table 3: The brief parameters of the three plans.

Position along horizontal segment (Grid)	Average permeability (mD)		
	Plan 1	Plan 2	Plan 3
1	809.0	64.7	16.2
2	787.5	63.0	15.8
3	1573.0	1573.0	1573.0
4	1504.0	1504.0	1504.0
5	738.0	59.0	14.8
6	726.5	58.1	14.5
Rush coefficient	1.5	2.8	3.1
Differential	2.2	27.1	108.3
Heterogeneity standard	weak	medium	strength

Firstly, the single well production without ICD has been simulated in three plans with Eclipse simulator. Based on the water cut along horizontal segment without ICD, high water cut positions are found. ICD are placed in high water cut positions, and the parameters of ICD are optimized carefully and reasonably in every plan. Finally, the results of two situations, with ICD and without ICD are compared with every plan. The increased proportion of oil in the situation with ICD are compared in three plans to study further ICD application condition and find the relation between the increasing oil effect and heterogeneity degree.

Plan 1

The water cut versus position along horizontal

segment and the water cut versus cumulative oil production in plan 1 for the two simulation cases without ICD and with ICD are shown in Figures 7 and 8.

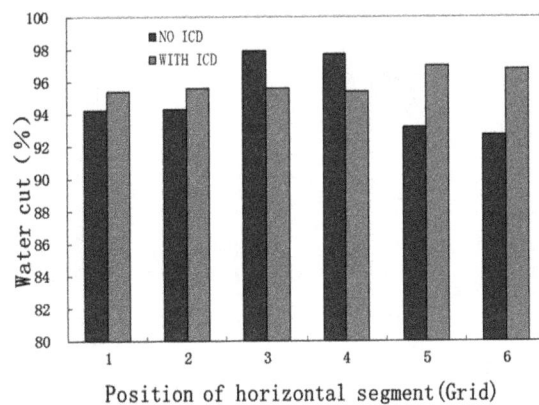

Figure 7: Water cut versus position along horizontal segment (plan 1).

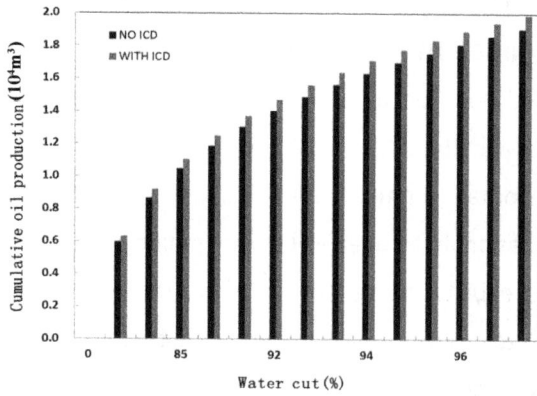

Figure 8: Water cut versus cumulative oil production (plan 1).

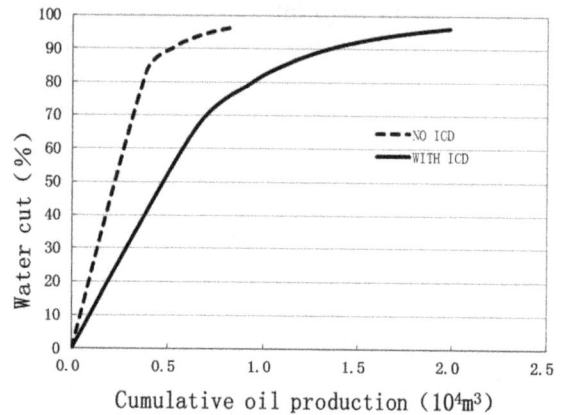

Figure 9: Water cut versus position along horizontal segment (plan 2).

In plan 1, the reservoir has weak heterogeneity; moreover, permeability rush coefficient and permeability differential are 1.5 and 2.0 respectively. As shown in Figure 7, there is a little change for water cut along horizontal segment, the water cut in center is slightly greater than the water cut in heel and toe for the situation that no ICD's are used (the permeability of center is bigger than the permeability of hell and toe). With ICD, the water cut along horizontal segment is adjusted to a balanced state. As can be seen from Figure 8, the water rising with ICD is slower than that of without ICD, and the oil production with ICD is higher than that of without ICD. However, oil production is increased by about 1000 m³ and the increased proportion is only 5%.

Plan 2

The water cut versus position along horizontal segment and the water cut versus cumulative oil production in plan 2 for two simulation cases of without ICD and with ICD are shown in Figures 9 and 10.

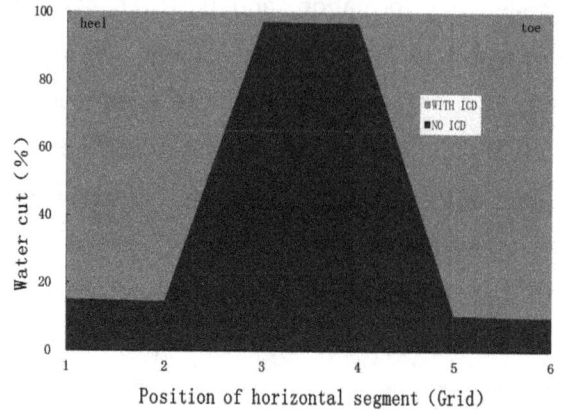

Figure 10: Water cut versus cumulative oil production (plan 2).

In plan 2, the reservoir has medium heterogeneity, and permeability rush coefficient and permeability differential is 2.8 and 27.0 respectively. As shown in Figure 9, there is a great change for water cut along horizontal segment; the water cut in center is greater than the water cut in heel and toe for the situation of without ICD. With ICD, the water cut along horizontal segment is adjusted to a balanced state. As can be seen from Figure 10, with ICD, the water rising is slower than that of without ICD, and the oil production with ICD is higher than that of without ICD. The increased oil production is about 1.15×10^4 m³ and the increased proportion is 139%.

Plan 3

The water cut versus position along horizontal segment and the water cut versus cumulative oil production in plan 3 for two simulation cases without ICD and with ICD are shown in Figures 11 and 12.

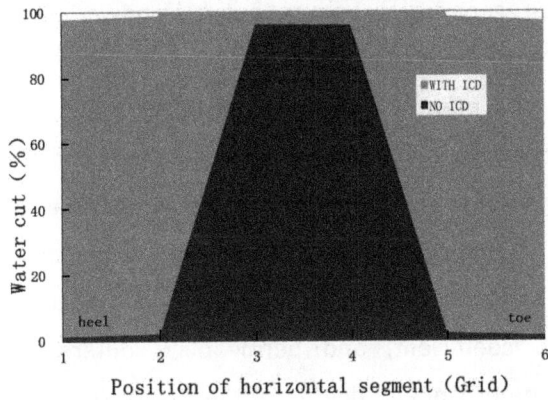

Figure 11: Water cut versus position along horizontal segment (plan 3).

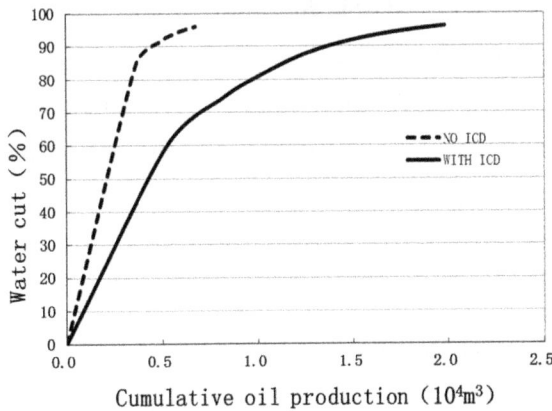

Figure 12: Water cut versus cumulative oil production (plan 3).

In plan 3, the reservoir has strong heterogeneity; additionally, permeability rush coefficient and permeability differential is 3.1 and 108.0 respectively. As shown in Figure 11, there is also a great change for water cut along horizontal segment; moreover, the water cut in center is greater than the water cut in heel and toe for the situation of without ICD. With ICD, the water cut

along horizontal segment is adjusted to a balanced state. As can be seen from Figure 11, with ICD, the water rising is slower than that of without ICD, and the oil production with ICD is higher than that of without ICD. The increased oil production is about 1.31×10^4 m^3 and the increased proportion is 203%. As can be seen from the three plans described above, there are application situations for ICD, which can work well in heterogeneous reservoirs, controlling water and increasing oil production. The failure reason for ICD in well A1H is that the reservoir is homogeneous. As can also be seen from the three plans, the water cut along horizontal segment is related to the reservoir heterogeneity degree; the more the heterogeneous is a reservoir, the more different the water cut along horizontal segment becomes.

Sensitivity Research

Based on the three plans, a further study about ICD has been carried out to find out the relation between the effect of controlling water, increasing oil and heterogeneity degree. The result is shown in Figures 13 and 14.

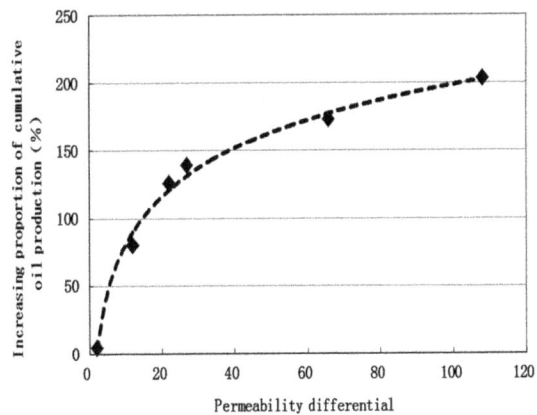

Figure 13: Increased proportion of cumulative oil production versus permeability rush coefficient.

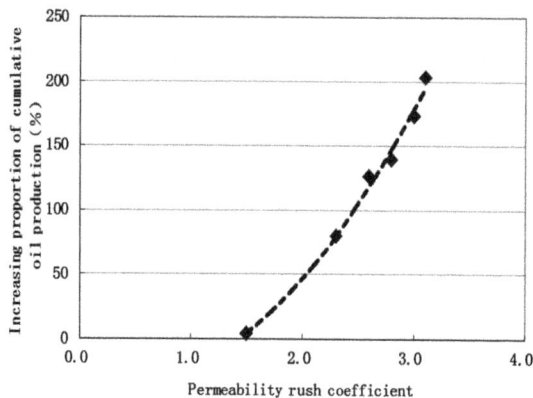

Figure 14: The increased proportion of cumulative oil production versus permeability differential.

As can be seen from the two above figures, the effect of ICD on controlling water and increasing oil is very sensitive to the heterogeneity degree of reservoirs. The cumulative oil production increases with an increase in permeability rush coefficient and permeability differential. The relation between the increasing proportions of cumulative oil production which increases with the permeability differential is a logarithmic relationship. The relational expression is as follows; its correlation coefficient is about 0.992.

$$Y = 51.281\ln(X) - 37.86 \qquad (4)$$

where, Y (%) is the increasing proportion of cumulative oil production, and X is permeability differential.

It has a multinomial relationship between the increasing proportion of cumulative oil production and increasing permeability rush coefficient. The relational expression is as follows; its correlation coefficient is about 0.991.

$$Y = 31.62X^2 - 25.94X - 27.79 \qquad (5)$$

where, Y (%) is the increasing proportion of cumulative oil production, and X is permeability rush coefficient.

CONCLUSIONS

(1) ICD can adjust the water cut along the horizontal section, but they cannot solve all problems. Based on a series of studies, the application situation of ICD is found, which is suitable for heterogeneous reservoirs; in other words, ICD can work well in heterogeneous reservoirs, but it is powerless in homogeneous reservoirs.

(2) The effect of ICD on controlling water and increasing oil is very sensitive to the heterogeneity degree of reservoirs, which increases with increasing the heterogeneity degree of reservoirs. Using permeability variation coefficient, permeability rush coefficient, and permeability differential, the heterogeneity degree of the reservoir can be determined to preliminary forecast the result of ICD on controlling water and increasing oil.

ACKNOWLEDGEMENTS

The authors would like to thank anonymous reviewers and the editors for constructive comments and suggestions for improving this paper. The authors also wish to thank their colleagues for their help and suggestions.

REFERENCES

1. Brekke K. and Lien S. C., "New and Simple Completion Methods for Horizontal Wells Improve the Production Performance in High-Permeability," *Thin Oil Zones, SPE Drilling & Completion*, **1994**, *9*(3), 205-209.

2. Augustine J. R., "An Investigation of the Economic Benefit of Inflow Control Devices on Horizontal Well Completions Using a Reservoir-Wellbore Coupled Model," *European Petroleum Conference*, Aberdeen, United Kingdom, SPE 78293, **2002**.

3. Lorenz M. D., Ratterman E. E. , and Augustine J.

R., "Uniform Inflow Completion System Extends Economic Field Life: A Field Case Study and Technology Overview," *SPE Annual Technical Conference and Exhibition*, San Antonio, SPE 101895, **2006**.

4. Henriksen K. H., Gule E. I. , and Augustine J. R., "The Application of Inflow Control Devices in the Troll Field," *SPE Europe/EAGE Annual Conference and Exhibition*, Vienna, SPE 100308, **2006**.

5. Augustine J. R., McIntyre A., Adam R. J., and et al., "Increasing Oil Recovery by Preventing Early Water and Gas Breakthrough in a West Brae Horizontal Well: A Case History," *SPE/DOE Symposium on Improved Oil Recovery*, Tulsa, Oklahoma, SPE 99718, **2006**.

6. Henriksen K. H., Augustine J. R., and Wood E. T., "Integration of New Open Hole Zonal Isolation Technology Contributes to Improved Reserve Recovery and Revision in Industry Best Practices," *SPE International Improved Oil Recovery Conference in Asia Pacific*, SPE 97614, **2005**.

7. Al-Qudaihy D. S., Al-Qahtani H. B., Sunbul A. H., and et al., "The Evolution of Advanced Well Completions to Enhance Well Productivity and Recovery in Saudi Aramco's Offshore Fields," *IADC/SPE Asia Pacific Drilling Technology Conference and Exhibition*, SPE 103621, **2006**.

8. Vela I., Viloria-gomez L. A., Caicedo R., and et al., "Well Production Enhancement Results with Inflow Control Device (ICD) Completions in Horizontal Wells in Ecuador," *SPE EUROPEC/ EAGE Annual Conference and Exhibition*, May, Vienna, SPE 143431, **2011**.

9. Sierra L., East L., Soliman M. Y., and et al., "New Completion Methodology To Improve Oil Recovery and Minimize Water Intrusion in Reservoirs Subject to Water Injection," *SPE*

Journal, **2011**, *16*(3), 648-661.

10. Li Z., Fernandes P. X., and Zhu D., "Understanding the Roles of Inflow-Control Devices in Optimizing Horizontal-Well Performance," *SPE Drilling & Completion*, **2011**, *26*(3), 376-385.

11. Gomez M., Anaya A. F., Araujo Y. E., and et al., "Passive Inflow Control Device (ICDs) Application in a Horizontal Wells Completions in the Rubiales Area Heavy-Oil Reservoir," *SPE Heavy and Extra Heavy Oil Conference,* Latin America, SPE 171040, **2014**.

12. Shad S. and Yazdi M. M., "Wellbore Modeling and Design of Nozzle-Based Inflow Control Device (ICD) for SAGD Wells," *SPE Heavy Oil Conference-Canada*, Alberta, Canada, SPE 170145, **2014**.

13. Sadeghnejad S. and Masihi M., "Water Flooding Performance Evaluation Using Percolation Theory," *Journal of Petroleum Science and Technology*, **2011**, *1*(2), 19-23.

14. Taheri Nakhost A. and Shadizadeh R., "A Simulation of Managed Pressure Drilling in Iranian Darquain Oil Field," *Journal of Petroleum Science and Technology*, **2013**, *3*(2), 45-56.

9

Assessing the Biological Inhibitors Effect on Crude Oil Wax Appearance Temperature Reduction

Ali Reza Solaimany Nazar[1]*, Navvab Salehi[1], Yavar Karimi[1], Roha Kasra Kermanshahi[2], and Masoud Beheshti[1]

[1]Chemical Engineering Department, University of Isfahan, Isfahan, Iran

[2] Department of Biology, Alzahra University, Tehran, Iran

ABSTRACT

To assess the effect of micro-organisms on the reduction of wax appearance temperature (WAT) of waxy crude oil, some appropriate strains were obtained from contaminated samples exposed to hydrocarbon compounds for a long time. By conducting some screening tests, four strains were chosen and aerated in a bioreactor; they were then grown in some hydrocarbon environments in order to produce biological inhibitors. The ability of the biological inhibitors in wax deposition prevention or reduction of WAT is assessed. The WAT is determined by means of the optical absorption spectroscopy method. The absorption plots show that biological compounds are highly effective in reducing WAT; however, different strains are not of the same efficiency. In some cases, the efficiency of biological inhibitors is more than chemical inhibitors. The optimization experiments were run with the objective of achieving the maximum WAT reduction through the Taguchi design method, and the optimum cultivation condition was identified. According to the analysis of variance, pH with a contribution of 49.63% is the most influential factor on the cultivation of the most efficient micro-organisms. The factors of temperature, the inoculation fluid, and nitrogen concentrations are ranked after pH with the contributions of 32.39, 7.92, and 1.39% respectively.

Keywords: Waxy Crude Oil, Biological Inhibitors, Wax Appearance Temperature

INTRODUCTION

Wax deposition phenomenon always threats oil industry companies by serious problems such as damaging the oil reservoir, chocking the pipelines, causing accumulation of oil sludge at the bottom of oil tanks [1], increasing viscosity, necessitating to pump power increment, and forming waxy gel leading to serious situations after long shutdowns periods [2]. More frequent pipeline cleaning operations and implication of various wax inhibitor methods impose more costs on crude oil transportation companies [3]. Since wax deposition removal operations, especially in deep-water pipelines, are technically very difficult and costly, oil companies always pursue for new methods to prevent wax deposition or its removal. There exists various methods, including pipelines thermal insulation [4], coating the pipelines inner surfaces [5], mechanical methods [6,7], and chemical methods [8-10] in order to control and

*Corresponding author

Ali Reza Solaimany Nazari

Email: asolaimany@eng.ui.ac.ir

reduce wax deposition or remediate wax deposit. Very detailed schedules should be utilized to minimize the cost and frequency of mechanical pipeline cleaning operations like pigging [6] or a combination of several chemical and mechanical methods which are implemented to increase the wax removal efficiency [11]. Normally, efficient and economic strategies are adopted with respect to determining deposition rate, the characteristics of the studied location, available methods, and previous experiences [12].

The wax appearance temperature (WAT) is one of the important stability characteristics of waxy crude oil and is referred to the highest temperature where the first wax crystals are formed during cooling condition [13]. The pipeline design depends on the WAT [14], and the efficiency of wax inhibitors can be assessed by measuring the influence of their operational factors. Several laboratory methods are suggested to measure this characteristic, but the accuracy of these methods depend on the techniques and materials applied [15]. These techniques include the ASTM D2500, light transmission measurement, viscosity and density measurement [16-18], microscopic techniques based on cross polarized light [15], differential calorimetry, Fourier transform infrared spectroscopy (FTIR) [19], and near-infrared spectrum (NIR) [20].

Biological methods are considered as the manners against heavy deposits in addition to the mechanical, thermal, and chemical methods since the early 90s [1, 21-25], and the number of academic researchers and oil companies R&D's interested in adopting these methods are rising. Although the strains identifying techniques capable of destructing hydrocarbons have improved, conditions and methods of their development [26], diagnosis of metabolic pathways of degradation, preparation of databases of biochemical metabolic, genetic engineering tools, and many of the principals involved in this issue still remain unexplored [27]. Several methods such as applying bacteria which is able to consume the sole carbon source [28], using extracellular enzymes capable of gradual destruction of chemical bonds within or between wax deposit molecules [29], and producing metabolic compounds to increase the dissolution and availability of heavy wax deposit [21] are proposed to explain the micro-organism activities in preventing, dispersing, and remediating wax deposits. In general, micro-organisms able to degrade the organic compounds which create deposits or produce biological solvents and bio-surfactants can be used to prevent or remove deposits. Bio-surfactants with a variety of structures and chemical compositions are produced by many types of micro-organisms. According to many studies, consuming inhibitors in order to prevent wax deposit is one of the most rewarding and economical methods [30-33]. Considering the wide range of inhibitors properties and many difficulties involved in chemical inhibitor synthesis process, it is clear that biological inhibitors are a good choice to encounter wax deposition problem [25].

The first pilot project of microbial treatment was run successfully in the 1980's [34]. In a study, some strains isolated from several hydrocarbon contaminated samples were used to biodegrade paraffin deposits successfully [29]. Some emulsifier compounds and commercial products were employed in enhancing the heavy crude oil flow in Venezuela, and the heavy oil viscosity was reduced from 105 to 70 mPa.s, which allowed the transport of heavy oil for 600 km from the reservoir. In a similar attempt, crude oil viscosity was reduced from an initial value of 20000 to 1 mPa.s by applying a combination of bio-surfactants leading to an easy transportation in a long pipeline [35]. In another study, the viscosity, surface tension, and pour point of a crude oil sample were reduced by 34, 31, and 31% respectively [36]. One of the most successful results was obtained in Kuwait oil industry; the researchers used two tons of microbial solution containing bio-surfactants for remediating the formed deposits in the bottom of a large storage tank and retrieved a significant portion of the accumulated petroleum hydrocarbons worth 150 thousand dollars per tank [37]. The results of a study on the effect of three groups of bacteria on the removal of paraffin wax deposits in four wells in China indicated that microbial coalition contributed to reducing these deposits, which led to an increase in the production perfectly [38]. A similar study showed the high ability of micro-organisms in the degradation of paraffin wax and deposits removal within the wells and pipelines for the easier transportation of waxy crude oil [39]. According to the results obtained in another work [40], some biological surfactants could reduce the adhesion of wax crystals on steel pipeline wall and cause the deposits removal. The thermophilic Bacillus subtilis could reduce the crude oil viscosity efficiently [41]. In addition, the high efficiency of Z25 strain in the removal of wax deposits, the degradation of alkanes, pour point reduction, and crude oil flow properties improvement is verified elsewhere [23]. Adding microbial strain to two crude oil types in China led to a destruction rate of 64% and the deposits removal of 55% [42]. The advantage of this method compared to chemical treatment methods is less harmful to human and the environment, thereby attracting the attention of the involved in this field [24].

In order to compare WAT reduction performance of some biological inhibitors extracted from various strains, Iranian waxy crude oil sample and the optical spectroscopy technique were used and implemented. The best microorganisms' cultivation state, which leads to the highest reduction of wax appearance temperature, is identified by adopting the Taguchi experimental design method. The objective here is to emphasis the importance of properly adjusting the cultivation conditions in order to achieve the most effective strains and to provide the practical comparison between the biological and chemical inhibitor efficiencies.

EXPERIMENTAL PROCEDURES
Material and Method

Dehloran (SW, Iran) crude oil specimen is used for the experiments in this study. The subject sample is dead oil. The kerosene with low sulfur content, glycerol, methanol, a chemical surfactant Tween 80, agar, nutrient broth and nutrient agar were purchased from Merck Company (Germany) and the solid paraffin was provided by Sepahan Oil Company (Isfahan, Iran). The related physical properties of Dehloran crude oil and solid paraffin are tabulated in Tables 1 and 2 respectively.

Table 1: Physical properties of Dehloran crude oil.

Properties	Values
Density at 15°C (g/m³)	0.8659
Viscosity at 45 °C (mm²/s)	0.5123
Wax content (wt.%)	12.66
Asphaltene content (wt.%)	0.86

Table 2: Physical properties of solid wax.

Property	Density at 15 °C (kg/m³)	Oil content (wt.%)	Flash point (°C)	Melting point (°C)
Method	ASTM D 4052/96	ASTM D 721	ASTM D 92	ASTM D 127
Value	850-884	22-25	245-262	54-62

Resources and Mediums for Strain Cultivation

In order to isolate the strains which are able to digest the petroleum hydrocarbons and produce suitable bio-surfactants, the soil samples contaminated with crude oil and some sludge samples were collected from six national wide refineries. Since the most successful method for the isolation of desired strains is a mineral salt reinforced medium containing the hydrocarbon source, three types of the following mineral media are consumed:

Mineral medium (g/L): K_2HPO_4 (0.775), KH_2PO_4 (0.35), $(NH4)_2SO_4$ (0.1), $FeSO_4.7H_2O$ (0.001), and $CaCl_2$ (0.04).

Bushnell-Haas medium (g/L): K_2HPO_4 (1), KH_2PO_4 (1), NH_4NO_3 (1), $MgSO_4.7H_2O$ (0.2), $FeCl_3.6H_2O$ (0.05), and $CaCl_2$ (0.02).

Oil Medium (g/L): K_2HPO_4 (1), KH_2PO_4 (1), $(NH_4)_2SO_4$ (1), $MgSO_4.7H_2O$ (0.04), and $FeCl_3.6H_2O$ (0.004).

The effect of iron as a multi-electrovalence metal on the production of biological inhibitors is assessed by adding $FeCl_3.6H_2O$ to the Bushnell-Haas medium. It should be noted that $FeCl_3.6H_2O$ is not added to the media when the crude oil or oil sludge are used as the carbon source in the Bushnell-Haas medium.

Apparatuses

Jasco UV-VIS-NIR V570 spectrometer device (Japan) equipped with quartz cuvette and 1 cm-light pathlength was utilized to obtain the absorption plots of the waxy samples. A laboratory bioreactor was made with the working volume of 5.1 L and a height to diameter ratio of 0.6. The sterile air is blown into the medium by two spargers after passing through some 0.45-micron filters.

Microorganism Isolation

The microorganisms' isolation is performed in two direct and indirect procedures. In the direct procedure, some cultivation plates containing Bushnell-Haas agar and mineral salts medium agar were prepared, and the oil contaminated soil and oil sludge were placed on each solid medium by loop . In the indirect approach, a 100-mL Erlenmeyer flask containing mineral salts, Bushnell-Haas [43], and nutrient broth was prepared. In order to isolate the appropriate strains from the contaminated soil samples, 0.5 grams of soil was screened on each of the media, and to isolate the possible strains available in the sludge, 1 mL of the same sludge was added to the minerals and Bushnell-Haas mediums. After 72 hours, 2 mL of each Erlenmeyer flasks were inoculated to the mediums containing sterile hydrocarbons.

Strain Production and Purification

After one week, the strains produced by the direct procedure were purified. For this purpose, three Erlenmeyer flasks with different carbon sources at a concentration of 1% were prepared [44], and the contents of each plate were inoculated to each of the Erlenmeyer flasks. The strains grew during 4 to 10 days of incubation at 30 °C while stirring at 120 rpm. The growth of the strains was confirmed after the mediums turbidity and microscopic observation. In order to re-purify the strains, they were cultivated on a solid medium with a similar composition. It should be noted that due to the very low solubility of hydrocarbons in this medium, Kiyohara standard method [45] was used to prepare the plates containing hydrocarbons [46]. The purification process in the indirect procedure was performed in the same manner. Thus, after the growth of strains, the proper plates were inoculated by the loop .

Screening Experiments

After successive stages of separation and purification through the direct and indirect procedures, a total of 44 types of bacteria capable of consuming hydrocarbons, as the sole carbon source, are isolated from the collected samples. In order to select the most appropriate strains, several screening experiments such as the determination of growth potential of strain in different hydrocarbon sources, the emulsification index, the ability to disperse or remove solid deposits, the growth rate of microorganisms, and the amount of bioactive

compounds production were run. Provided that adding an organic solvent to an aqueous solution leads to emulsion or the appearance of two phases of foam and liquid, it could be concluded that there exist bioactive compounds in the solution [47]. By definition, mediums containing bioactive compounds form flat drops, while a medium without bioactive compounds forms round drops [48, 49], indicating the lack of the two phases emergence. The distribution of medium on the oil surface is an indication of the presence of bioactive compounds [50]. The emulsification index is a quantitative measure of the strain ability to form the emulsion of hydrocarbon, while surface tension is the best criterion to confirm the presence of bioactive compounds in the culture medium. Running these tests according to the procedures mentioned in [51], the four strains named S_0, S_2, S_4, and S_8 are selected for further experiments; these belong to a gram-positive strain group and their microscopic images are shown with a magnification of 1000x in Figure 1.

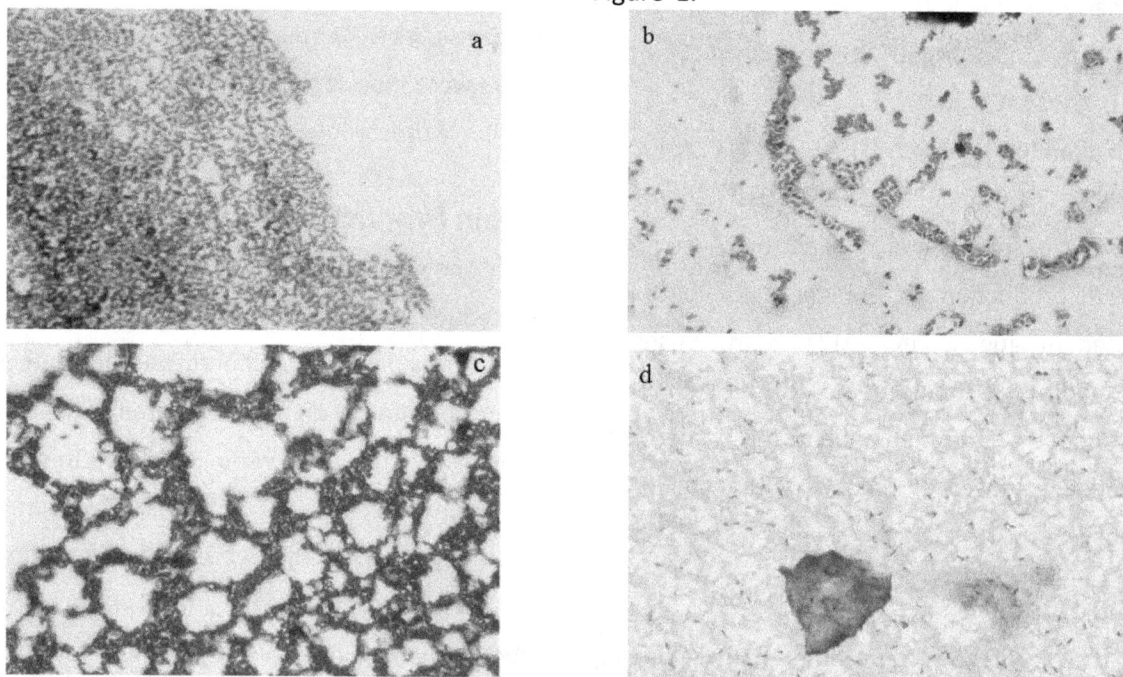

Figure 1: Microscopic images of the strains; a) S_0; b) S_2; c) S_4; and d) S_8.

Extraction of the Biological Inhibitors

The extraction of bioactive compounds was performed through centrifugation after the cells growth in the aqueous phase (inorganic medium). The centrifugation took place at a constant temperature of 4 °C since a higher temperature may inactivate the compounds. The pH of cell-free cultivation medium was adjusted at 2 by adding the concentrated hydrochloric acid, and the medium was kept at 4 °C overnight. The solution containing the sediment was centrifuged for 10 min at 8000 rpm at 4 °C. The obtained sediment was isolated by a lower volume of distilled water, and the pH was adjusted at 7 by adding required amount of soda [44, 51-54]. In order to dry the compounds, they were poured on the pre-weighted plates and put inside a desiccator. After 4 to 6 hours, the plates were dried in the oven for 4 hours at 35 °C.

WAT Determination through Spectroscopy Technique

The near-infrared spectroscopy (NIR) uses the NIR spectra in the range of 780 to 2500 nm [20]. The WAT determination through this method is based on the strong dependency of waxy crude oil absorption on temperature in a wide range of NIR spectrum. One appropriate wave length must be selected to perform the further NIR tests. Since the appropriate wave length depends on the oil sample, at first the absorption curve of the waxy crude oil is determined within a reasonable NIR range. The appropriate wave length is the wavelength at which the minimum absorption occurs. Next, the sample absorption value can be measured at different temperatures, at appropriate wavelength through which the absorption plot is obtained. The temperature at which the slope of the plot changes

in a sudden manner is reported as the WAT of the waxy crude oil [55].

Taguchi Design Method

The experimental design methods, including Taguchi design is a statistical method that distinguishes the important factors affecting the response from the unimportant factors and determines the conditions to achieve the optimal performance of system. This method reduces the number of required experiments in achieving the desired results [56]. The first and most important step in this approach is selecting the factors and their levels. To assess the effect of various factors on the strain cultivation condition in order to achieve the maximum efficiency in WAT reduction, the four factors of temperature, inoculums concentration, pH, and concentration of nitrogen source are of concern. The factors and their levels are tabulated in Table 3.

Table 3: Factors and their levels.

Factor	Level 1	Level 2	Level 3
Temperature (°C)	30	40	45
Coalition liquid (%)	25	50	75
pH	5	6	7
Nitrogen source concentration (g/L)	0.8	1	1.2

The coalition liquid is made of four strains S_0, S_2, S_4, and S_8. According to the results obtained from the screening tests, the concentration of S_2 is found to be more than the rest of the strains, and in the Taguchi table, inoculation concentration is indicative of the concentration of S_2 in the coalition liquid. Finally, by applying the L_9 Taguchi table (Table 4) the experiments are conducted.

Table 4: Design of experiments: an L_9 Taguchi table.

Run	Temperature (°C)	Coalition liquid (%)	pH	Nitrogen source concentration (g/L)
1	30	25	5	0.8
2	30	50	6	1
3	30	75	7	1.2
4	40	25	6	1.2
5	40	50	7	0.8
6	40	75	5	1
7	45	25	7	1
8	45	50	5	1.2
9	45	75	6	0.8

RESULTS AND DISCUSSION
WAT Determination

The minimum wavelength at which the oil medium has the minimum absorption must be determined because at this wavelength oil absorption is very low, and it helps us to detect the solid phase as soon as it appears. When the medium is one liquid phase, the absorption is minimum, but once the first solid crystal appears, it absorbs the spectra, and the absorption value rises; hence, we can detect the solid crystal presence, and as a result WAT will be determined exactly.

The absorption plot of Dehloran crude oil sample in the range of 1200 to 1900 nm is shown in Figure 2, where the minimum absorption is observed in the vicinity of 1608 nm. After determining the appropriate wavelength (1608 nm), three grams of solid paraffin is added to 60 mL of the sample oil to determine WAT.

The results of measuring the absorption at different temperatures and at the wavelength of 1608 nm are expressed in Figure 3. Each of the 9 samples had the same thermal history.

The results indicate that a reduction in temperature will increase absorption. The significant absorption of wax crystals leads to a change in the slop of the absorption plot, which is very different in the single phase system (right side) compared to the two-phase system (left side). Since the technique applied in this study is of high accuracy, it is accepted that the measured temperature is a good estimation of the actual WAT. Based on the mentioned approach and using Figure 3, the WAT of Dehloran crude oil is 36.6 °C.

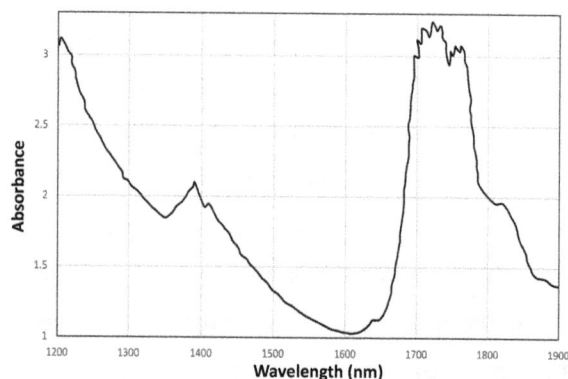

Figure 2: Absorbance data of Dehloran crude oil.

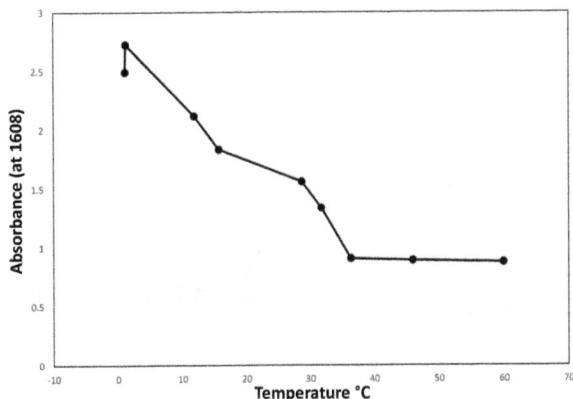

Figure 3: Absorbance-temperature plot of Dehloran crude oil.

The results of measuring the absorption at different temperatures and at the wavelength of 1608 nm are expressed in Figure 3. Each of the 9 samples had the same thermal history.

The results indicate that a reduction in temperature will increase absorption. The significant absorption of wax crystals leads to a change in the slop of the absorption plot, which is very different in the single phase system (right side) compared to the two-phase system (left side). Since the technique applied in this study is of high accuracy, it is accepted that the measured temperature is a good estimation of the actual WAT. Based on the mentioned approach and using Figure 3, the WAT of Dehloran crude oil is 36.6 °C.

Effect of Chemicals Inhibitor on WAT

In order to provide a standard for comparing the performance of biological inhibitors in WAT reduction, the effect of a mixture of some well-known chemical inhibitors is studied. For this purpose, 350 mL of xylene and methyl-ethyl-ketone mixture (at a ratio of 6 to 1) is added to 20 mL of the oil sample. The absorption values of this treated oil are measured at various temperatures (Figure 4). The trend of absorption-temperature plot reveals that as the temperature decreases to

25.3 °C, the absorption rate increases suddenly, indicating first detectable wax crystals formation; hence, this temperature is reported as the WAT of crude oil containing chemical inhibitor. The chemical inhibitor can reduce the WAT from 36.6 to 25.3 °C.

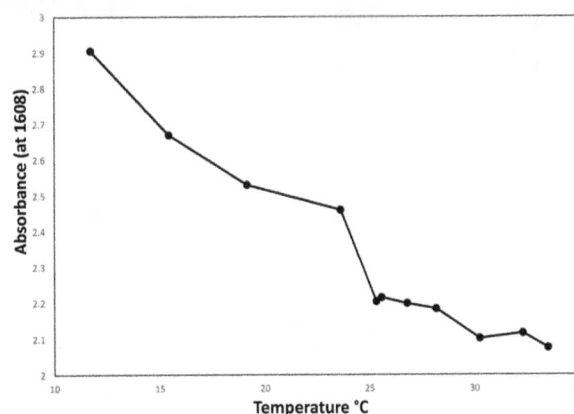

Figure 4: Absorbance-temperature plot of Dehloran crude oil containing methyl-ethyl-ketone at 1608 nm.

Effect of Biological Inhibitors on WAT

In order to produce the significant quantities of biological inhibitors in a short period of time, microbial strains are cultivated as pure and as coalition in the bioreactor. Four strains of S_0, S_2, S_4, and S_8 are selected as biological inhibitors by screening the obtained strains.

Each of the strains S_0, S_2, and S_4 is added to the sample at the concentration of 3.0 g/L. The corresponding absorption-temperature plots are shown in Figure 5. Based on these plots, the WAT of the oil samples containing S_0, S_2, and S_4 are 27.8, 22.5, and 26.8 °C respectively. In other words, the bio-inhibitors of strains S_0, S_2, and S_4 cause a drop of 24, 38.5, and 26.77% in WAT values respectively. Usually, the use of a coalition of bacteria has some advantages in a medium with some growth restrictions. In order to evaluate this claim, a coalition of strains S_0, S_2, S_4, and S_8 are cultivated together in the bioreactor. This coalition consists of 25% of S_0, 50% of S_2, 12.5% of S_4, 12.5% of S_8. According to Figure 5-d, the WAT of the oil sample containing microbial coalition is 19.5 °C, indicating a decrease of about 46.72% in WAT.

(a)

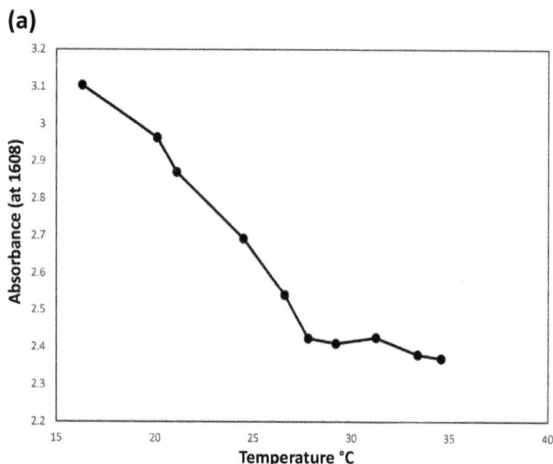

(b)

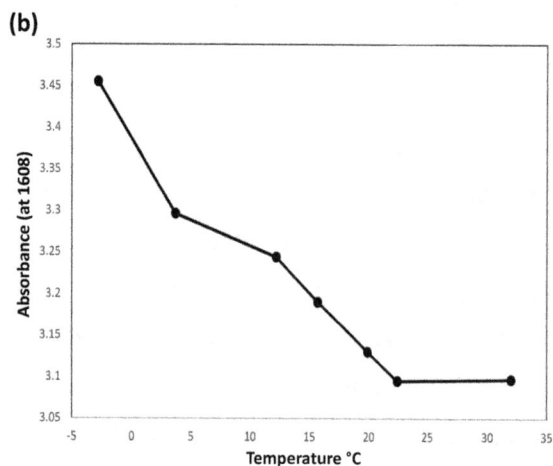

(c)

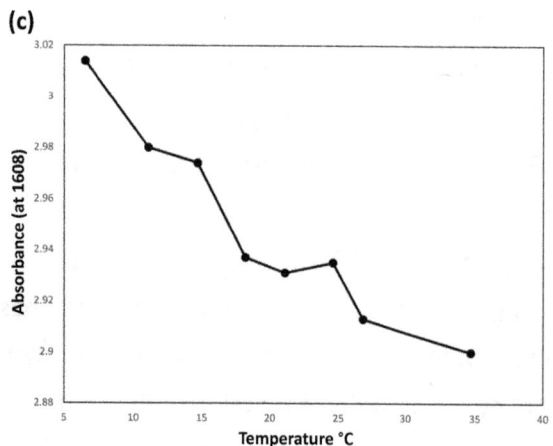

(d)

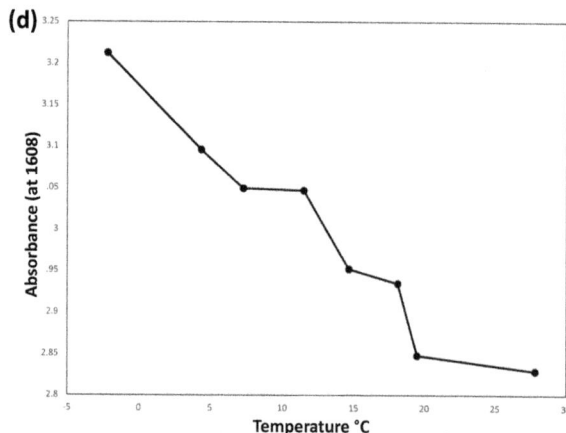

Figure 5: Absorbance data of Dehloran crude oil containing a) S_0, b) S_2, c) S_4, and d) microbial coalition.

Comparison of Performance of Different Strains in WAT Reduction

The treatment of the crude oil by the chemical inhibitor and biological inhibitors of S_0, S_2, S_4, and microbial coalition respectively leads to an 11.3, 8.8, 14.1, 9.8, and 17.1 °C drop in the WAT of untreated oil sample. Therefore, biological inhibitors of microbial coalition is the most influential WAT reducer, followed by bio-inhibitor of S_2, xylene, and methyl-ethyl-ketone chemical mixture, bio-inhibitors of strain S_4, and S_0 respectively.

As observed, the slope of the absorption plot of S_4 (Figure 5-b) is more than that of the S_0 plot slope (Figure 5-a), indicating that in the presence of S_4 the wax crystal formation decreases significantly. In addition to the fact that S_4 is a better WAT reducer than S_0, it can reduce the wax crystals concentration below the WAT in a significant manner.

Although the use of microbial coalition leads to such a desired result in this study, the detailed analysis of this mechanism is complicated. In a hypothetic sense, the dual nature compounds (hydrophilic-hydrophobic) can be the justification of these observation and findings. The result of a test where $FeCl_3 \cdot 6H_2O$ is added to coalition strains medium indicates that the use of this agent only causes an improvement of about 0.6 °C in the WAT reduction. It is found that this agent has no significant effect on reducing the wax crystals production rate below WAT. Therefore, due to the difficulties in experimental measurement, consuming $FeCl_3$. $6H_2O$ is not reasonable under normal conditions.

Parameters Optimization

After the cultivation of the strains under the condition adjusted in accordance with the experiments design table (Table 4), the obtained biological compounds are consumed to reduce the WAT of the crude oil samples. Each test of table L_9 Taguchi is repeated once and the findings are tabulated in Table 5.

Since the purpose of determining the optimum conditions is achieving the maximum reduction of WAT, the "small is better" option in the Taguchi method is chosen to analyze the results through the Qualitek-4 software. The results of the analysis of variance are tabulated in Table 6, where the pH with an influence of 49.63% is the most influential factor. The temperature, inoculum concentration, and the concentration of nitrogen source are ranked next by 32.39, 7.92, and 1.39% respectively.

Table 5: Results of wax appearance determination.

Temperature (°C)	Coalition liquid (%)	pH	Nitrogen source concentration (g/L)	Wax appearance temperature (°C)		Average (°C)
				First run	Second run	
30	25	5	0.8	25	26	25.5
30	50	6	1	16	17	16.5
30	75	7	1.2	23	20	21.5
40	25	6	1.2	15	16	15.5
40	50	7	0.8	18	20	19
40	75	5	1	21	21	21
45	25	7	1	26	27	26.5
45	50	5	1.2	23	25	24
45	75	6	0.8	21	20	20.5

Table 6: ANOVA table.

Factor	DF	Sum of Square	Variance	F-ratio	Influence Percentage
Temperature	2	80.11	40.05	32.77	32.39
Coalition liquid	2	21.44	10.72	8.77	7.92
pH	2	121.44	60.72	49.68	49.63
Nitrogen Source Concentration	2	5.77	2.88	2.36	1.39
Error	9	10.99	1.22	-	8.667

Effect of Temperature

The performance of biological inhibitors produced by the strains grown at 40 °C (medium level) is more than the others at other temperatures (Figure 6). Temperature affects the solubility of the hydrocarbons in aqueous phase same as the physiological activity of micro-organisms. The micro-organisms produce the bioactive compounds in order to increase the availability of carbon source, and an increase in temperature increases the solubility and availability of hydrocarbons; therefore, the micro-organisms have no tendency to produce bioactive compounds at high temperatures. Also, the biological inhibitors are ineffective at low temperatures, because these bacteria are isolated in high temperature mediums and cannot grow at temperatures lower than 40 °C, which may have a negative effect on their bio-inhibitor production capacity.

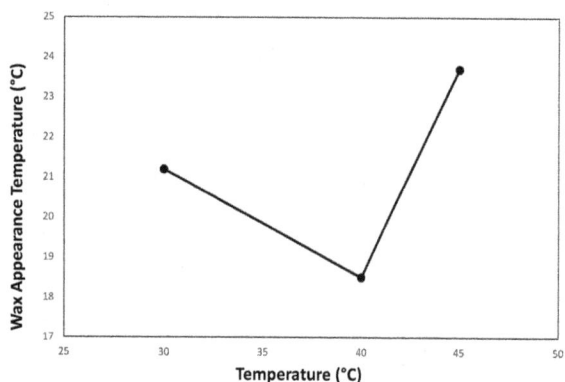

Figure 6: Effect of temperature on WAT.

Inoculums Concentration

The effect of the inoculum concentration on the performance of the strains on WAT reduction is shown in Figure 7. According to the previous results, S_2 strain outperforms the other strains; however, inoculation strain is better than S_2 in WAT reduction, which might be a result of an increase in desirable co-metabolic effects. There exists an optimum concentration at which the strain performance is at the highest level. By increasing the amount of inoculum concentration to more than 50%, WAT is increased as well, because here S_2 is predominant. The inoculums concentrations of other strains are neither low so that their effects are subtle nor high so that they would diminish the S_2 effect.

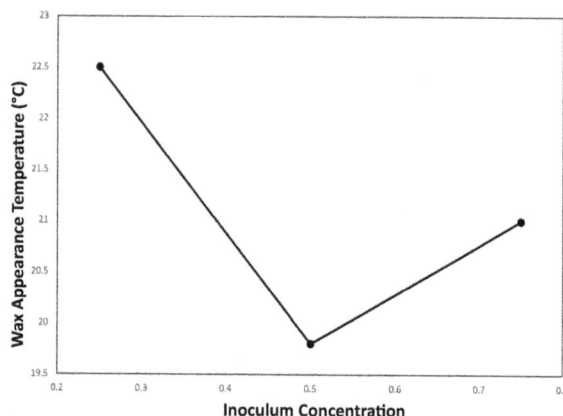

Figure 7: Effect of inoculums concentration on WAT

Effect of pH

The assessment of pH effect on the ability of strains in WAT reduction is conducted at three levels. pH is one of the factors influencing the growth and activity of micro-organisms. The strains' growth is very difficult and restricted in basic mediums. Provided that pH is adjusted at 6 or is slightly acidic, the strain performance will be appropriate which would lead to a higher WAT reduction (Figure 8).

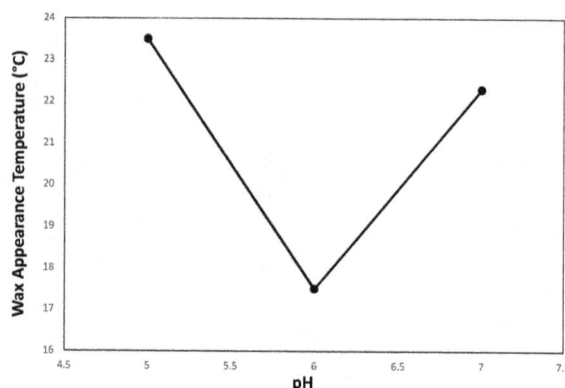

Figure 8: Effect of pH on WAT.

The Effect of the Concentration of Nitrogen Source

The results of published studies indicate that the concentration of nitrogen source is one of the important factors influencing the type and amount of produced bioactive compounds. According to Figure 9, an increase in the concentration of nitrogen source leads to an increase in the performance of the produced bioactive compounds. However, this parameter with an influence percentage of 1.39% and an F-ratio of 2.36 has no statistical significance effect on the response of the system (WAT) in the confidence level of 95%.

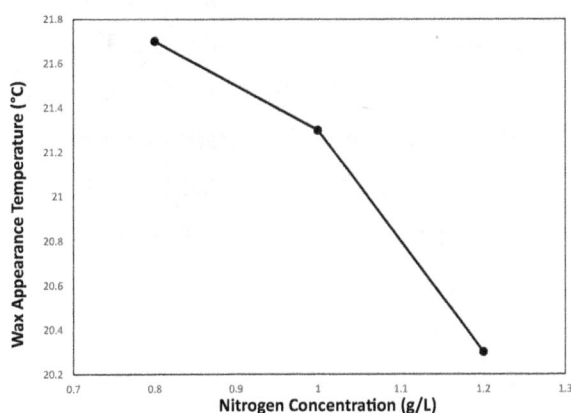

Figure 9: Effect of concentration of nitrogen source on WAT.

Optimum Condition

The optimum condition is determined by the Qualitek-4 software (Table 7). The best result or the highest WAT reduction is obtained when the temperature, inoculum concentration of S_2, pH, and the concentration of nitrogen source are set on 40 °C, 50 vol.%, 6, and 1.2 g/L respectively. The absorption plot of Dehloran crude oil in the presence of strain coalition which has grown under optimum conditions is shown in Figure 10, and the corresponding WAT value is 15 °C. It means that when the cultivation medium condition of the

micro-organisms is adjusted under the optimum condition, WAT reduction is 21.6 °C or 60%, which is a good achievement in this context.

Table 7: Optimum conditions.

Factor	Level	Value
Temperature	2	40
Concentration of S_2 in coalition liquid (%v.)	2	50
pH	2	6
Nitrogen source concentration	3	1.2

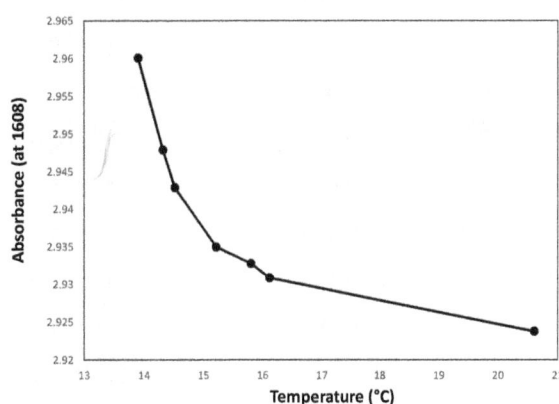

Figure 10: Absorbance data of Dehloran crude oil treated in the optimum condition.

CONCLUSIONS

In this study, the effects of biological inhibitors on the WAT of waxy crude oil are studied. Some strains are cultivated in a medium containing waxy crude oil (as the sole carbon source), isolated, and then screened in order to achieve the most effective strains for the WAT reduction of an Iranian waxy crude oil sample. The WAT determination is performed by means of the near-Infrared spectroscopy (NIR) technique, which obtains very accurate results. The biological inhibitors produced by the strains lead to a considerable decrease in WAT value. The analysis of variance method is implemented to assess the factors influencing the strains cultivation condition and the determination of the desirable condition suitable for the growth of most the powerful strains. The Taguchi design results indicate that the influence percentage of pH and temperature, as the most influential factors among the others, are equivalent to 49.63 and 32.39% respectively. The most powerful inhibitors in the WAT reduction are obtained when the temperature, the inoculum concentration of S_2, pH, and the concentration of nitrogen source are set on 40 °C, 50 vol.%, 6, and 1.2 g/L respectively. As a result, the WAT of Dehloran waxy crude oil reduces from 36.6 to 15 °C when the producer strains grow in this optimum condition. These results are highly desirable since, even in practical applications, the control of these parameters is very easy. Furthermore, an empirical comparison between the performance of chemical and biological inhibitors shows that the biological inhibitors are more successful than their counter parts. Despite the high efficiency of biological inhibitors, many comprehensive studies are necessary in order to improve the strain cultivation methods and medium condition optimization.

REFERENCES

1. Towler B. F. and Rebbapragada S., "Mitigation of Paraffin Wax Deposition in Cretaceous Crude Oils of Wyoming," *Journal of Petroleum Science and Engineering*, **2004**, *45*(1), 11-9.

2. Hénaut I., Vincké O., and Brucy F., "Waxy Crude Oil Restart: Mechanical Properties of Gelled Oils," *SPE Annual Technical Conference and Exhibition*, SPE, **1999**, 1-12.

3. Wardhaugh L. and Boger D, "Flow Characteristics of Waxy Crude Oils: Application to Pipeline Design," *AIChE Journal*, **1991**, *37*(6), 871-885.

4. Quenelle A. and Gunaltun M., "Comparison between Thermal Insulation Coatings for underwater Pipelines," *Offshore Technology Conference*, Offshore Technology Conference, **1987**, 1-11.

5. Patton C. C. and Casad B. M., "Paraffin deposition from Refined Wax-solvent Systems," *Society of Petroleum Engineers*, **1970**, *10*(01), 17-24.

6. Wang W. and Huang Q., "Prediction for Wax deposition in Oil Pipelines Validated by Field Pigging," *Journal of the Energy Institute*, **2014**, *87*(3), 196-207.

7. Tiratsoo J., "Pipeline Pigging Technology," (1st ed.), *Gulf Professional Publishing*, **1992**, 1-249.

8. Xu J., Xing S., Qian H., Chen S., and et al., "Effect of Polar/nonpolar Groups in Comb-type Copolymers on Cold Flowability and Paraffin Crystallization of Waxy Oils," *Fuel*, **2013**, *103*, 600-605.

9. Behbahani T. J., "Experimental Investigation of the Polymeric Flow Improver on Waxy Oils," *Petroleum and Coal*, **2014**, *56*(2), 139-142.

10. Al-Sabagh A. M., El-Hamouly S. H., Khidr T. T., El-Ghazawy R. A., and et al., "Synthesis of Phthalimide and Succinimide Copolymers and their Evaluation as Flow Improvers for an Egyptian Waxy Crude Oil," *Egyptian Journal of Petroleum*, **2013**, *22*(3), 381-393.

11. Ford P., Ells J., and Russell R., "Pipelining High Pour Point Crude Pt. 3. Frequent Pigging Helps Move Waxy Crude below its Pour Point," *Oil Gas Journal*, **1965**, *63*(19), 1-8.

12. McClaflin G. and Whitfill D., "Control of Paraffin deposition in Production Operations," *Journal of Petroleum Technology,* **1984,** *36*(11), 1965-

13. Visintin R. F. G., Lockhart T. P., Lapasin R., and D'Antona P., "Structure of waxy crude oil emulsion gels," *Journal of Non-Newtonian Fluid Mechanics*, **2008**, *149*(1-3), 34-39.

14. Sadeghazad A. and Christiansen R. L., "The Effect of Cloud Point Temperature on Wax Deposition," *Abu Dhabi International Petroleum Exhibition and Conference*, SPE, **1998**, 1-17.

15. Yen T. F. and Chilingarian G. V., "Asphaltenes and Asphalts," (2nd ed.), *Elsevier*, **2000**, 1-324.

16. Neto A. D., Gomes E., Neto E. B., Dantas T., and et al., "Determination of Wax Appearance Temperature (WAT) in Paraffin/solvent Systems by Photoelectric Signal and Viscosimetry," *Brazilian Journal of Petroleum and Gas*, **2010**, *3*(4), 149-57.

17. Coutinho J. A. and Daridon J-L., "The Limitations of the Cloud Point Measurement Techniques and the Influence of the Oil Composition on its Detection," *Petroleum Science and Technology*, **2005**, *23*(9-10), 1113-1128.

18. García d. C. M., Carbognani L., Orea M., and Urbina A., "The Influence of Alkane Class-types on Crude Oil Wax Crystallization and Inhibitors Efficiency," *Journal of Petroleum Science and Engineering*, **2000**, *25*(3-4), 99-105.

19. Roehner R. and Henson F. V., "Measurement of Wax Precipitation Temperature and Precipitated Solid Weight Percent Versus Temperature by Infrared Spectroscopy," U. S. Patent, **2005**.

20. Alex R. F., Fuhr B. J., and Klein L. L., "Determination of Cloud Point for Waxy Crudes Using a Near-infrared/fiber Optic Technique," *Energy & fuels*, **1991**, *5*(6), 866-868.

21. Sadeghazad A. and Ghaemi N., "Microbial Prevention of Wax Precipitation in Crude Oil by Biodegradation Mechanism," *SPE Asia Pacific Oil and Gas Conference and Exhibition*, **2003**.

22. Giangiacomo L. A., "Stripper Field Performance Comparison of Chemical and Cicrobial Paraffin Control Systems," SPE Rocky Mountain Regional meet, Richardson TX, **1999**, 1-11.

23. Zheng C., Yu L., Huang L., Xiu J., and Huang Z., "Investigation of a Hydrocarbon-degrading strain, Rhodococcus ruber Z25, for the Potential of Microbial Enhanced Oil Recovery," *Journal of Petroleum Science and Engineering*, **2012**, *81*, 49-56.

24. Patel J., Borgohain S., Kumar M., Rangarajan V., and et al., "Recent Developments in Microbial Enhanced Oil Recovery," *Renewable and Sustainable Energy Reviews*, **2015**, *52*, 1539-1558.

25. Sakthipriya N., Doble M., and Sangwai J. S., "Action of Biosurfactant Producing Thermophilic Bacillus Subtilis on Waxy Crude Oil and Long Chain Paraffins," *International Biodeterioration & Biodegradation*, **2015**, *105*, 168-177.

26. Etoumi A., "Microbial Treatment of Waxy Crude Oils for Mitigation of Wax Precipitation," *Journal of Petroleum Science and Engineering*, **2007**, *55*(1-2), 111-121.

27. Le Borgne S. and Quintero R., "Biotechnological Processes for the Refining of Petroleum," *Fuel Processing Technology*, **2003**, *81*(2), 155-169.

28. Bushnell L. and Haas H., "The Utilization of Certain Hydrocarbons by Microorganisms," *Journal of Bacteriology*, **1941**, *41*(5), 653-657.

29. Lazar I., Voicu A., Nicolescu C., Mucenica D., and et al., "The Use of Naturally Occurring Selectively Isolated Bacteria for Inhibiting

Paraffin Deposition," *Journal of Petroleum Science and Engineering*, **1999**, *22*(1), 161-169.

30. Denis J. and Durand J., "Modification of Wax Crystallization in Petroleum Products," Revue de l'Institut Français du Pétrole, **1991**, *46*(5), 637-649.

31. Hennessy A., Neville A., and Roberts K. J., "In-situ SAXS/WAXS and Turbidity Studies of the Structure and Composition of Multihomologous N-alkane Waxes Crystallized in the Absence and Presence of Flow Improving Additive Species," *Crystal Growth & Design,* **2004**, *4*(5), 1069-1078.

32. Hennessy A., Neville A., and Roberts K., "An Examination of Additive-mediated Wax Nucleation in Oil Pipeline Environments," *Journal of Crystal Growth*, **1999**, *198*, 830-837.

33. Tung N. P., Van Vuong N., Long B. Q. K., Vinh N. Q., and et al., "Studying the Mechanism of Magnetic Field Influence on Paraffin Crude Oil Viscosity and Wax Deposition Reductions," *SPE Asia Pacific Oil and Gas Conference and Exhibition*, SPE, **2001**, 1-8.

34. Bognolo G., "Biosurfactants as Emulsifying Agents for Hydrocarbons," *Colloids and Surfaces A: Physicochemical and Engineering Aspects*, **1999**, *152*(1), 41-52.

35. Banat I. M., Makkar R. S., and Cameotra S., "Potential Commercial Applications of Microbial Surfactants," *Applied Microbiology and Biotechnology,* **2000**, *53*(5), 495-508.

36. Feng Q., Zhou J., Ni F., Chen Z., and et al., "Application of Thermophilic Microbes In Waxy Oil Reservoirs at Elevated Temperature," *SPE Annual Technical Conference and Exhibition,* SPE, **2001**, 1-9.

37. Banat I. M., "Biosurfactants Production and Possible Uses in Microbial Enhanced Oil Recovery and Oil Pollution Remediation: a Review," *Bioresource Technology*, **1995**, *51*(1), 1-12.

38. He Z., Mei B., Wang W., Sheng J., and et al., "A Pilot Test Using Microbial Paraffin-removal Technology in Liaohe Oilfield," *Petroleum Science and Technology*, **2003**, *21*(1-2), 201-210.

39. Zhang J., Xue Q., Gao H., and Wang P., "Biodegradation of Paraffin Wax by Crude Aspergillus Enzyme Preparations for Potential Use in Removing Paraffin Deposits," *Journal of Basic Microbiology*, **2015**, *55*(11), 1326-1335.

40. Xiao M., Li W-H., Lu M., Zhang Z. Z., and et al., "Effect of Microbial Treatment on the Prevention and Removal of Paraffin Deposits on Stainless Steel Surfaces," *Journal of Bioresource Technology*, **2012**, *124*, 227-232.

41. Sakthipriya N., Doble M., Sangwai J. S., "Fast Degradation and Viscosity Reduction of Waxy Crude Oil and Model Waxy Crude Oil Using Bacillus Subtilis," *Journal of Petroleum Science and Engineering*, **2015**, *134*, 158-166.

42. Liu J. H., Jia Y. P., Chen Y. T., and Xu R. D., "Microbial Treatment for Prevention and Removal of Paraffin Deposition on the Walls of Crude Pipelines," *Indian Journal of Microbiology*, **2013**, *53*(4), 482-484.

43. Batista S., Mounteer A., Amorim F., and Totola M., "Isolation and Characterization of Biosurfactant/bioemulsifier-producing Bacteria from Petroleum Contaminated Sites," *Bioresource Technology*, **2006**, *97*(6), 868-675.

44. Rahman K., Rahman T., Lakshmanaperumalsamy P., Marchant R., and et al., "The Potential of Bacterial Isolates for Emulsification with a

Range of Hydrocarbons," *Acta Biotechnologica*, **2003**, *23*(4), 335-345.

45. Kiyohara H., Nagao K., and Yana K., "Rapid Screen for Bacteria Degrading Water-insoluble, Solid Hydrocarbons on Agar Pates," *Applied and Environmental Microbiology*, **1982**, *43*(2), 454-457.

46. Liu Y, Zhang J, and Zhang Z, "Isolation and Characterization of Polycyclic Aromatic Hydrocarbons-degrading Sphingomonas Sp. Strain ZL5," *Biodegradation*, **2004**, *15*(3), 205-212.

47. Schulte R., KKG Group Paraffin Removal, "Rocky Mountain Oilfield Testing Center," *Casper*, WY (US), **2001**.

48. Deziel E., Paquette G., Villemur R., Lepine F., and et al., "Biosurfactant Production by a Soil Pseudomonas Strain Growing on Polycyclic Aromatic Hydrocarbons," *Applied and Environmental Microbiology*, **1996**, *62*(6), 1908-1912.

49. Bodour A. A. and Miller-Maier R. M., "Application of a Modified Drop-collapse Technique for Surfactant Quantitation and Screening of Biosurfactant-producing Microorganisms," *Journal of Microbiological Methods*, **1998**, *32*(3), 273-280.

50. Youssef N. H., Duncan K. E., Nagle D. P., Savage K. N., and et al., "Comparison of Methods to Detect Biosurfactant Production by Diverse Microorganisms," *Journal of Microbiological Methods*, **2004**, *56*(3), 339-347.

51. Cooper D. G. and Goldenberg B. G., "Surface-active Agents from Two Bacillus Species," *Journal of Applied and Environmental Microbiology*, **1987**, *53*(2), 224-229.

52. Dehghan-Noudeh G., Housaindokht M., and Bazzaz B. S. F., "Isolation, Characterization, and Investigation of Surface and Hemolytic Activities of a Lipopeptide Biosurfactant Produced by Bacillus Subtilis ATCC 6633," *Journal of Microbiology Seoul*, **2005**, *43*(3), 272-276.

53. Grandison A. S., "Separation Processes in the Food and Biotechnology Industries," CRC Press (1st ed.), **1996**, 1-287.

54. Makkar R. and Cameotra S. S., "Production of Biosurfactant at Mesophilic and Thermophilic Conditions by a Strain of Bacillus Subtilis," *Journal of Industrial Microbiology and Biotechnology*, **1998**, *20*(1), 48-52.

55. Leontaritis K. J. and Leontaritis J. D., "Cloud Point and Wax Deposition Measurement Techniques," International Symposium on Oilfield Chemistry, Houston, Texas, SPE, **2003**.

56. Roy R. K., "Design of Experiments Using the Taguchi approach: 16 Steps to Product and Process Improvement (1st ed.)," *John Wiley and Sons*, **2001**, 1-531.

10

Association of the Flow Units with Facies Distribution, Depositional Sequences, and Diagenetic Features: Asmari Formation of the Cheshmeh-Khush Oil Field, SW Iran

Javad Honarmand[1]* and Abdolhossein Amini[2]

[1]Petroleum Geology Department, Research Institute of Petroleum Industry (RIPI), Tehran, Iran
[2]School of Geology, College of Science, University of Tehran, Iran

ABSTRACT

The Oligo-miocene Asmari formation is one of the most important hydrocarbon reservoirs in the southwest of Iran. In order to evaluate reservoir quality and the factors controlling reservoir properties, detailed geological and petrophysical studies were carried out on 242 m of core samples from Asmari reservoir. This study is a part of a larger project that examines depositional history and reservoir properties of the Asmari formation in the Cheshmeh-Khush field. Macroscopic and microscopic studies resulted in the determination of 5 shallow marine carbonate facies (from proximal open marine to tidal setting) and also 5 silisiclastic lithofacies (including channel, barrier, tidal, and shoreface sandstones). Based on the integrated results from sedimentological and paleontological studies, Sr isotopes dating, gamma-ray logs, and seismic data analysis, 5 depositional sequences with constituent system tracts were distinguished. In this research, the reservoir characterization of the Asmari reservoir were carried out through the integration of geological and petrophysical properties. In the first step, 21 hydraulic flow units (HFU or FU) were identified and then, to achieve better lateral correlation and modeling, HFU's were merged to 17. The results from this study showed different behaviors of the silisiclastic and carbonate facies next to the fluid flow. The findings of this study indicate that the lateral and vertical distribution of channel-filled sandstones (such as units 2 and 8) are strongly controlled by the geometry of depositional facies. Thus, the correlation and modeling of flow units, solely on the basis of lithology and thickness, and regardless of facies and its geometry, will cause different facies (such as coastal and channel-filled sandstones) with different geometry, and reservoir quality are placed incorrectly in a single flow unit.

In the carbonate parts of Asmari formation, the effect of diagenetic processes on reservoir quality is much higher than the facies. Hence, the LST limestones of unit 17, as a result of calcite cementation, were changed to a thick, distinct, correlatable, and barrier unit. On the other hand, dolomitic intervals that have not been affected by anhydrite cementation have formed porous and permeable carbonate reservoir units (such as units 18 and 21).

Keywords: Asmari Formation, Reservoir Quality, Flow Units, Sequence Stratigraphy, Facies Modeling

*Corresponding author

Javad Honarmand
Email: honarmandj@ripi.ir

INTRODUCTION

Categorization of a reservoir into permeable and non-permeable units is the most important step in the 3D modeling of reservoirs. These units, known as flow units, are widely used in reservoir characterization [1, 2, 3, 4, 5]. The term flow unit is applied to a distinct horizon in the reservoir with definite petrophysical characteristics [6]. Such units, which are discriminated from neighboring units based on their petrophysical properties, are mappable and correlatable in the field/regional scale.

The parameters that influence fluid flow are thought to be primarily related to the geometrical attributes of the pore-throat distribution. The pore geometry is controlled by mineralogy and texture. Therefore, an HFU can include several facies types, depending on their depositional characteristics.

Detailed analysis of facies and sequences of the hydrocarbon reservoirs, especially the carbonate types, and their correlation with reservoir properties show that the porosity and permeability of the reservoirs are not fully dependent to the facies characteristics. Although there are some correlation between textural and petrophysical properties of the facies, their reservoir characteristics are independent of facies type [7]. In this regard similar facies may show various petrophysical properties or various facies may show similar petrophysical characteristics. This is why most facies models do not demonstrate the distribution of reservoir parameters. Due to the significant role of diagenesis on the reservoir properties, an efficient reservoir model should include discrepancy of both facies characteristics and diagenetic features. A reservoir model that provides the pattern of petrophysical properties and correlates the properties with depositional setting provides great help in combination of geological descriptions and engineering calculations. Such a model provides superior view from the distribution of reservoir parameters.

The Oligo-Miocene Asmari formation in the Cheshmeh-Khush (CK) field is comprised of carbonate and siliciclastic facies with a variety of diagenetic features. The discrepancy of facies characteristics and diagenetic features through time (vertically) and space (laterally) have resulted in the heterogeneity of reservoir, especially in the carbonate part. This study aims to determine the flow units of the Asmari formation, using the porosity and permeability values from petrophysical logs and the stratigraphic modified Lorenz (SML) plot and to correlate them with the distribution of the facies, sequences, and diagenetic features. The results related to the facies analysis, sequence stratigraphy, and diagenetic studies are mainly taken from the first author's Ph.D. thesis.

EXPERIMENTAL PROCEDURES
Geological Setting

The Cheshmeh-Khush (CK) oil field is located in the northwest edge of Dezful embayment (DE), one of the major geological zones of Zagros range, about 180 km in the northwest of the Ahwaz city (Figure 1A). It is adjacent to Danan (the south of Lurestan) and Paydar (the north of DE) fields in the north and south respectively. The CK field has been the center of attention for petroleum geologist since 1966, when the first oil well was drilled in the field by National Iranian Oil Company [8].

The Oligo-Miocene succession in the DE consists of a relatively thick sedimentary system with a large variation in lithologies and depositional settings [9]. In different parts of the DE, lithologies differ from carbonate/evaporate to siliciclastic; moreover, depositional environments change from shallow/ deep marine carbonate platform to a siliciclastic shelf. These mixed carbonate-siliciclastic system is the most important hydrocarbon reservoir, producing oil and gas since the early twentieth century [9].

In the studied field, the Asmari formation is dominated by carbonate facies inter-fingering with sandstone intervals (Ahwaz member), so it is known as a mixed siliciclastic-carbonate reservoir (Figure 1B). The thickness of the formation is about 320 m in the field. In the CK field, the Oligo-Miocene Asmari formation is laid over the shales and argillaceous limestones of the Pabdeh formation concordantly and is overlaid concordantly by the evaporates of the Gachsaran formation (Figure 1B) [9].

On the basis of seismic data, the length and width of the CK anticline in the Asmari horizon is about 28.5 km and 4.5 km, respectively [8].

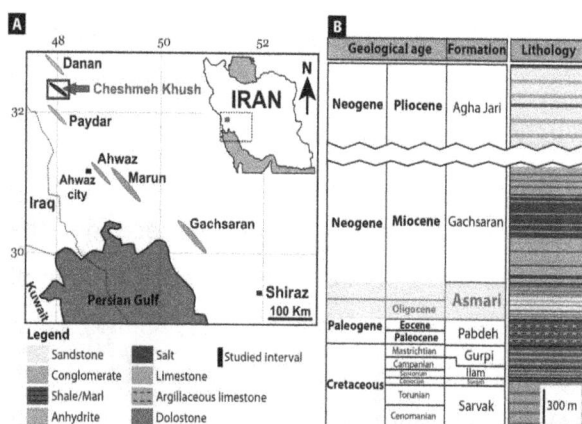

Figure 1: (A) the location map of the CK field and (B) its stratigraphic column [9].

Materials and Methods

Porosity and permeability, as the main controls of fluid flow in the hydrocarbon reservoirs, were used for the flow unit determination here. In this study, 242 m core samples from three key wells (wells: 5, 7, and 9) were examined. 720 core plugs were taken from the cored intervals of the Oligo-Miocene Asmari formation. The samples were studied for sedimentological analysis and were measured for its porosity and permeability. In

order to analyze visual pore spaces, thin section samples were impregnated with blue dyed epoxy. Wire line logs (including density, neutron, and sonic logs) were used to determine the porosity of entire (cored and uncored) intervals. In order to predict permeability, a quantitative integration of porosity/permeability measurements and well log data from the major reservoir intervals is carried out using an artificial neural network (ANN) [10, 11].

Since core measurements are usually taken at irregular spacing, comparisons to regularly spaced log data require some scheme for infilling. In this study, this was carried out through the interpolation of porosity, density, and permeability data.

The flow units are normally determined on the basis of the flow zone indicator (FZI) introduced by Ebanks et al. [6], or stratigraphic modified Lorenz (SML) plot introduced by Gunter et al. [12]. Both methods are tried here, but the latter is preferred due to the better correlation of its results with depositional facies and sequences especially in siliciclastic parts.

Facies Analysis and Sequence Stratigraphy

The Asmari formation in the CK field is comprised of two distinct parts. The lower part (Oligocene in age) is dominated by sandstone facies, whereas the upper part (Miocene in age) is dominated by carbonate (limestone and dolomite) facies. While compared with those in literature [14, 15, 16, 17, 18, 19, 20, 21, 9, 22, 13], the combination of the results from core description and petrographic studies led to the determination of 12 carbonate and 5 siliciclastic facies the characteristics of which are summarized in Table 1 and illustrated in Figure 2. The carbonate facies are classified into 5 facies association (facies belt) using Buxton and Pedley (1989) [23] and Flugel (2010) [24] criteria.

Table 1: Major characteristics of the carbonate and siliciclastic facies of the Asmari formation in the CK field [13].

Facies			Description	Depositional Environment
Carbonate facies		A	Planktonic foraminera bioclast wackestone	Distal open marine
	B	B1	Large benthic foraminifera bioclast packstone	Proximal open marine
		B2	Red algae, echinoderm, and Neorotalia bioclast packstone	
	C	C1	Ooid grainstone	Shoal/Lagoonal shoal margin
		C2	Bioclast, Faverina ooid grainstone	
		C3	Miliolid, Denderitina grainstone	
	D	D1	Miliolid bioclast packstone	Lagoon
		D2	Miliolid, Denderitina wackestone/packstone	
		D3	Coral boundstone	
	E	E1	Massive dolomitized mudstone	Intertidal zone
		E2	Dolomitized mudstone along with anhydrite interlayers	
		E3	Peyssonnelia* packstone	
Siliciclastic facies		F	Sandstone with intercalations of open marine carbonate, Planolithes and Paleophycus ichno-fossils	Lower shoreface to offshore
		G	Erosional based conglomerate, channel shape ((log correlation), chaotic nature (seismic data	Channel/incised valley fills
		H	Massive, fossil barren shale	Upper shoreface
		I	Unconsolidated, dolomite cemented, well sorted sandstone, with intercalations of lagoonal carbonates	Barrier
		J	Flasser bedded siltstone, sandstone and shale	Intertidal zone

Peyssonnelia is a genus of thalloid red alga*

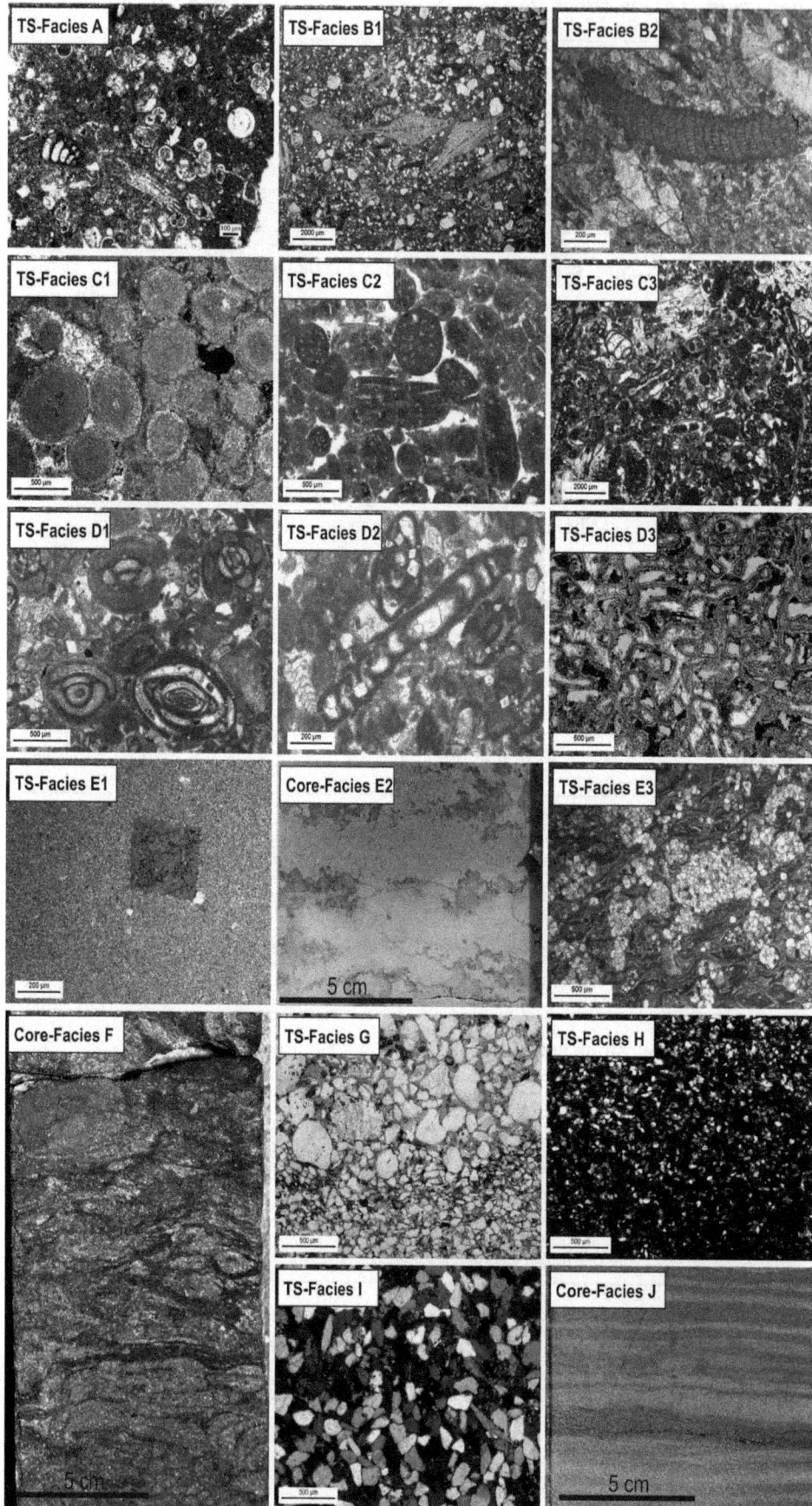

Figure 2: Macro and microscopic photographs from depositional facies of the Asmari formation in the studied field.

Sequence stratigraphy of the formation was carried out on the basis of the integrated results from sedimentological and paleontological studies, Sr isotopes dating, gamma-ray logs, and seismic data analysis [13]. Among sedimentological evidence like karstification (at the top of sequences 3 and 4), the presence of lowstand sandstone intervals (at the basal part of sequences 2 and 3) and vertical depositional pattern play the main role in the identification of sequence boundaries (Figure 3). Several index fossils such as Archaias sp., Elphidium sp.14, *Bolelis melocurdica*, and the association of large benthic foraminifera were used to determine the age of each sequence. On this basis, 5 depositional sequences with constituent systems tracts were distinguished (Figure 4).

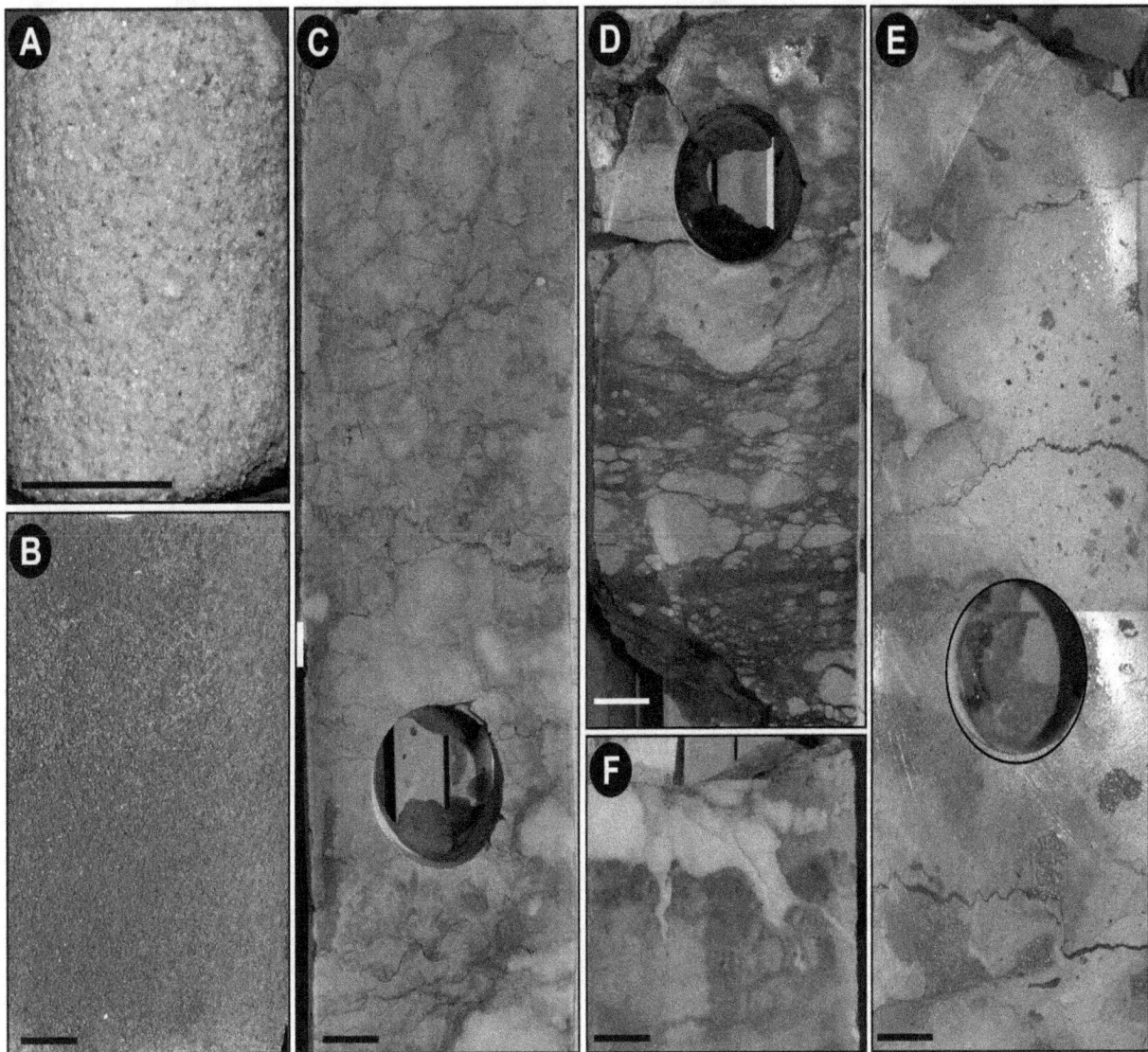

Figure 3: Core photographs from lowstand sandstones (A and B) and karstified surfaces (C-F) as two main pieces of evidence for identifying sequence boundaries in the Asmari formation in the studied field.

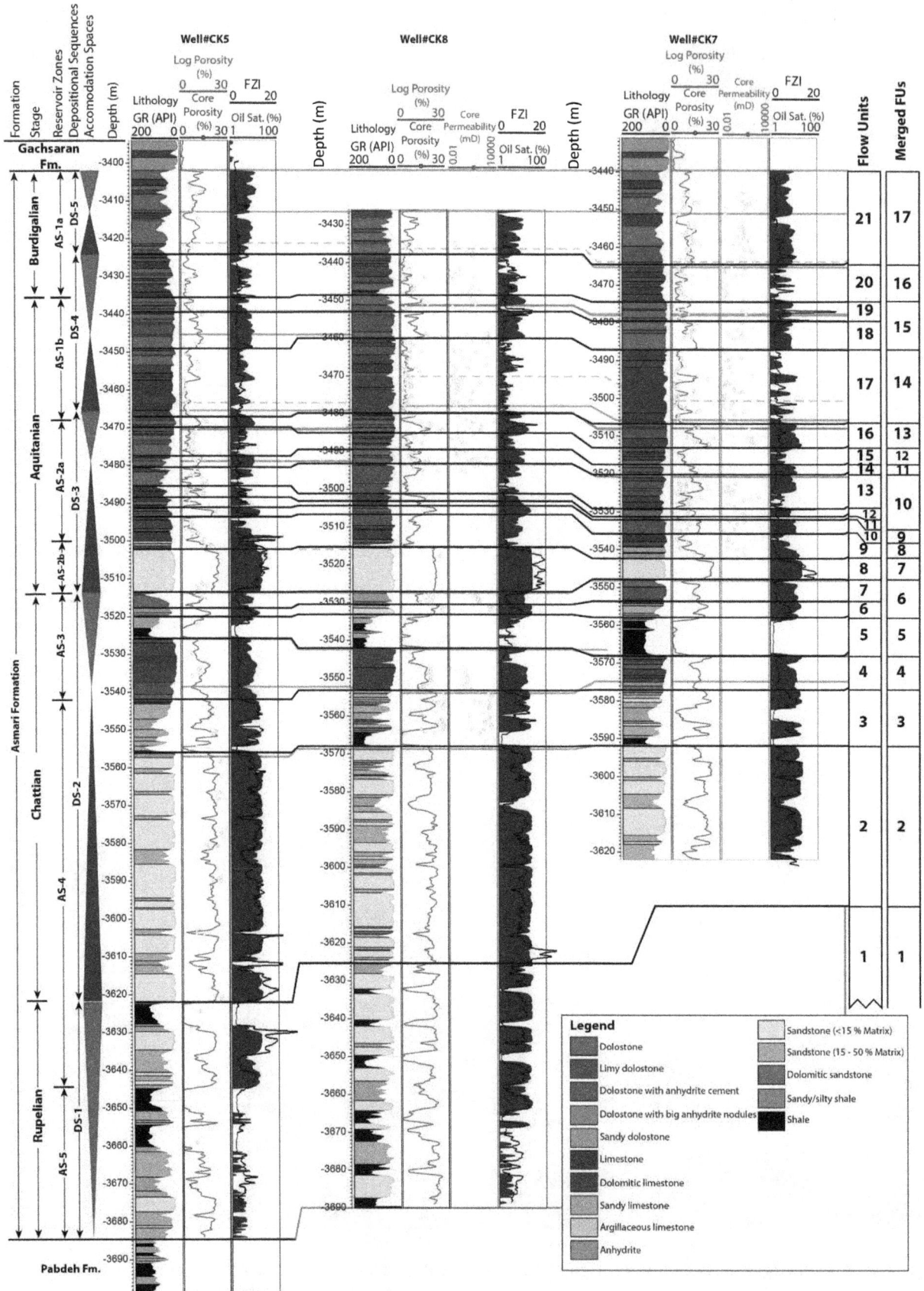

Figure 4: Comparison of flow units determined by SML and FZI methods in 3 key wells of the studied field.

Initially, depositional sequences were identified based on the presence of karstified surfaces (as sequence boundaries in carbonate sequences 4 and 5), channel-filled sandstones (as lowstand systems tracts in sequences 2 and 3), and vertical facies changes. Accordingly, five third order depositional sequences were introduced. Limestones with large-benthic foraminifera (in sequence 2) and ooid/skeletal grainstones (in sequences 3, 4 and 5), as the deepest facies, were attributed to maximum flooding surface. This study showed that due to the shallow carbonate facies and diagenetic effects, the relationship between petrophysical logs and sequence surfaces (SB & MFS) is not reliable. Hence, after identifying sequences based on sedimentological characteristics, the gamma log and Sr-dating results were used solely for the purpose of the correlation of time lines.

Ordering of the sequences was based on the results from the relative age determination of the formation [13] and data from Sr isotope dating [25]. Four systems tracts (i.e. LST, TST, HST, and FSST) were distinguished in most sequences [26, 27]. In the oldest sequence (seq. -1), only the falling stage systems tract is observed. In other words, the lower boundary of the formation is marked as a regressive surface here. The 4 major systems tracts (i.e. LST, TST, HST, and FSST) are distinguishable in the sequences 2, 3, and 4. The youngest sequence (seq. -5) is composed of the LST, TST, and HST. The falling stage of this sequence is recorded within the overlying Gachsaran formation, so the upper boundary of the formation is a regressive surface too (Figure 5).

Unlike the carbonate samples, in siliciclastic samples of the Asmari formation, there is a good linear relationship between the core porosity and permeability. On the other hand, the samples of the whole studied interval were not available. Therefore, in this study, after calibrating the core and well log porosities, log derived effective porosity were used.

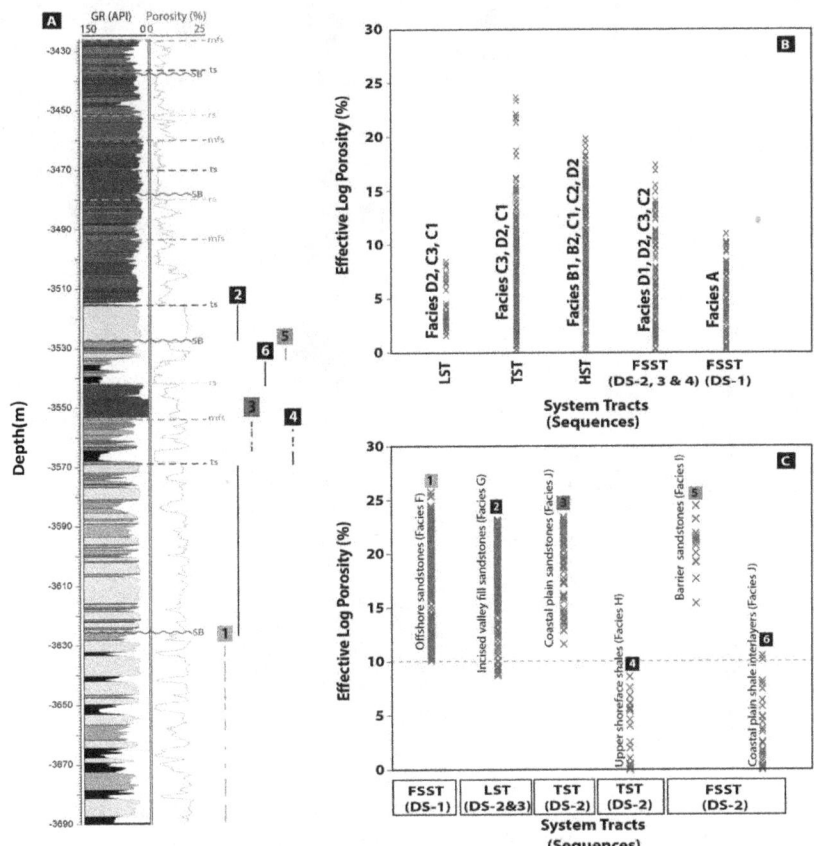

Figure 5: (A) position of the sequences and their constituent systems tracts and facies in well CK-8 along with log-derived porosity values. Correlation between sequences, systems tracts and porosity values in the (B) carbonated dominated and (C) siliciclastic dominated parts of the formation.

The detailed study of carbonate facies on macro and micro scales indicates the significant effects of diagenesis on the facies after deposition. A variety of post depositional changes in a single facies during burial seems responsible for its different petrophysical characteristics. In this study, diagenetic processes and the products of the formation are comprehensively investigated by core description, petrographic studies, SEM and CL analyses, and the stable isotopes (C & O) investigation of samples from 3 key wells [22, 13]. The results show that dolomitization, dissolution, and compaction are the main controls of reservoir quality in the carbonate-dominated part of the formation.

The role of compaction (Figure 6A) and cementation (Figure 6B) in the decrease and that of dolomitization (Figures 6G and 6H) and dissolution (Figure 6F) in the increase of reservoir quality is evident in the petrographic studies. In terms of composition, the cement occurs in calcite (Figures 6B and 6C) and anhydrite/celestine (Figures 6D and 6E) forms. Intercrystalline porosity, detectable in the SEM studies (Figure 6H) has created high porosity and permeability in the dolomitized facies. Limited cementation (early diagenetic) seems responsible for the minor compaction of some facies, the preservation of their inter-particle porosity, and finally high reservoir quality.

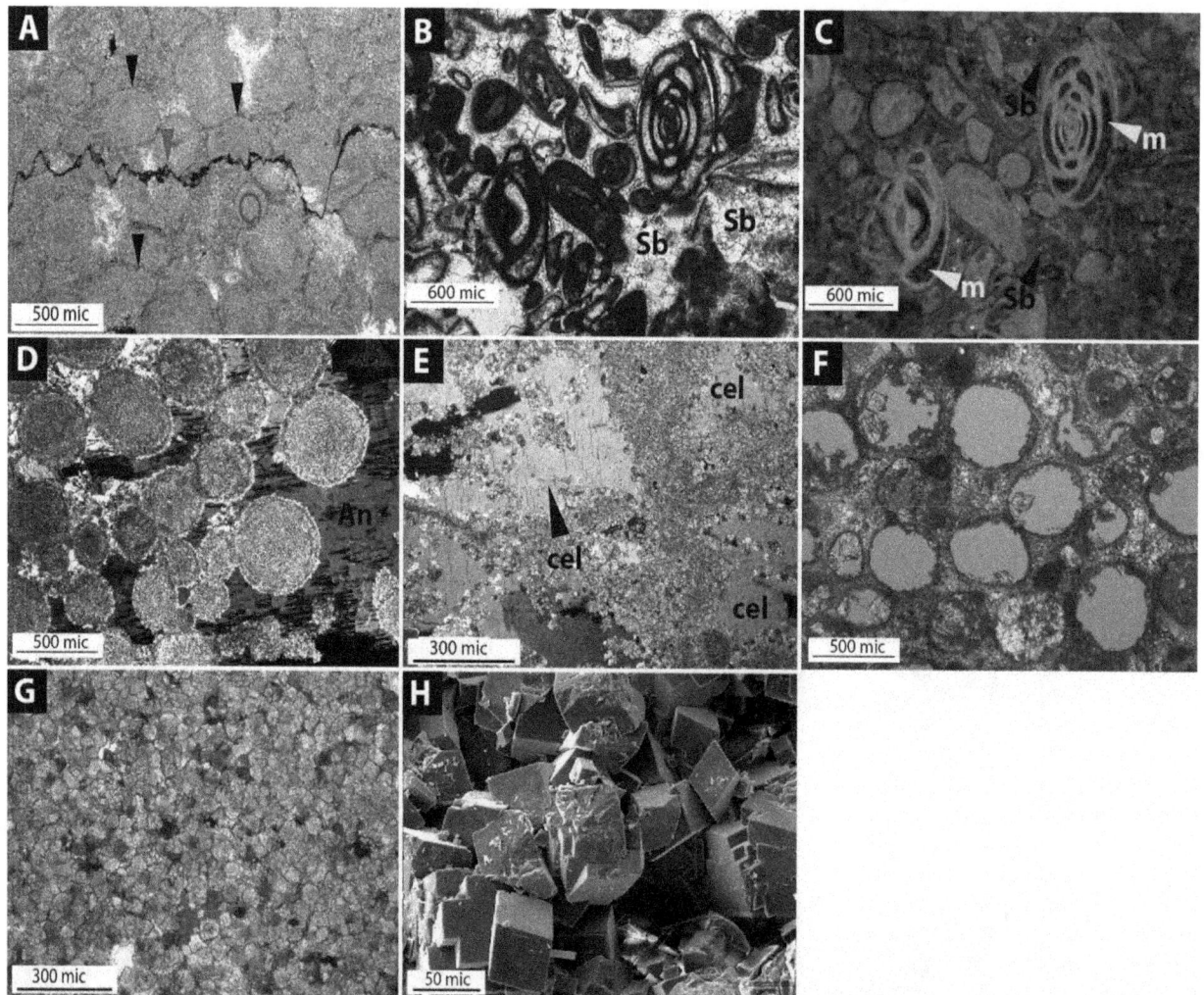

Figure 6: Photomicrographs representing major diagenetic features responsible for reservoir quality of the carbonate facies: (A) mechanical compaction; (B) shallow burial calcite cement (sb); (C) the shallow burial (sb) and marine (m) calcite cement differentiated by CL; (D) intergranular anhydrite cement (An); (E) celestine (cel) plugging; (F) dissolution of ooid grains producing moldic porosity (blue-died); (G) dolomitization responsible for the development of intercrystalline porosity; (H) the SEM photomicrograph of (G).

The frequency of the main diagenetic features in the carbonate facies indicates the extensive effect of nearly all facies by these features (Figure 7). Moldic and vuggy pore spaces are the main dissolution secondary porosities in the formation. Dissolution, as the main cause of porosity and permeability raise, is more frequent in the allochems (skeletal and non-skeletal), leaving biomoldic, oomoldic and vuggy porosity. The summation of more frequent dissolution porosities

(moldic and vuggy) in different facies is considered as a dissolution index in the carbonate facies (Figure 7D). Apart from petrographic studies, the density of the facies is used as a parameter for their dolomitization rate. It is based on the difference of the calcite, dolomite, and anhydrite densities, 2.71, 2.84, and 2.95 g/cm^3 respectively. In this regard, except for the facies of open marine (A and B), other facies are widely affected by dolomitization (Figure 7E).

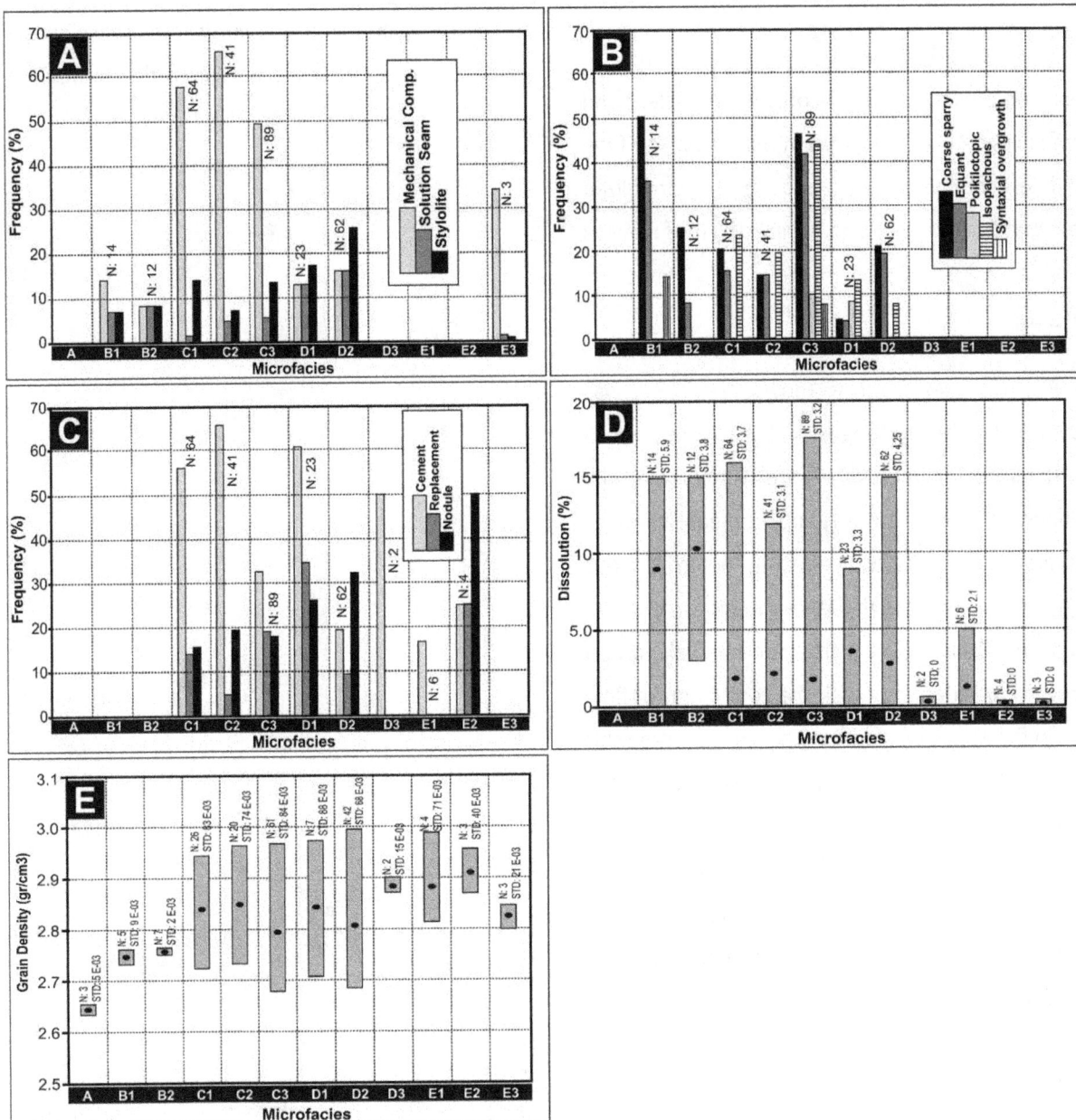

Figure 7: Frequency of major diagenetic features in the carbonate facies of the studied formation.

Unlike carbonate intervals, in siliciclastic parts of the Asmari formation, diagenetic process was very limited, so it had no significant effect on reservoir quality.

Flow Units Determination

In the FZI method [6] the flow units are determined on the basis of reservoir quality and normalized porosity indices. The reservoir quality index (RQI) is determined by permeability to porosity ratio using the fallowing equation [1]:

$$RQI = 0.0314\sqrt{\frac{K}{\phi}}$$

where the porosity value is expressed in percent and permeability (K) value is given in millidarcy. The flow unit determination by the RQI requires the normalized porosity index (NPI) that is determined by the fallowing equation:

$$NPI = \frac{\phi}{1-\phi}$$

The RQI/NPI ratio, which is known as the FZI, was calculated for the studying rocks by which the main flow units are discriminated (Figure 4). In this method, the units with various porosity and permeability values showed similar RQI. Moreover, in heterogeneous parts of the reservoir with a large variation in porosity and permeability values (carbonate intervals) and a high vertical variation of RQI, the correlation of flow units in field-scale was challenging. To overcome this problem, the SML plot method was also tried. In this method, the porosity, permeability, and thickness of units are used for the determination of the flow units. On the SML plot the cumulative flow capacity (CFC) is drown against the cumulative storage capacity (CSC), (Equations 1 and 2).

$$\left(K_h\right)_{cum} = \frac{K_1\left(h_1-h_0\right)+K_2\left(h_2-h_1\right)+\ldots+K_i\left(h_i-h_{i-1}\right)}{\sum K_i\left(h_i-h_{i-1}\right)} \quad (1)$$

$$\left(\phi_h\right)_{cum} = \frac{\phi_1\left(h_1-h_0\right)+\phi_2\left(h_2-h_1\right)+\ldots+\phi_i\left(h_i-h_{i-1}\right)}{\sum \phi_i\left(h_i-h_{i-1}\right)} \quad (2)$$

The porosity (φ) value is expressed in percent, while the permeability (K) value is given in millidarcy; h represents the sampling intervals in feet.

Using the log-derived porosity and permeability values and the sampling interval of 0.15 m, the CSC versus CFS diagrams were plotted based on the Gunter et al. [12]. The results led to the determination of 21 flow units in 3 key wells of the studied field (Figures 8 and 9). A slope discrepancy in the diagrams represents the variation of reservoir quality (porosity and permeability). In other words, the breaks on the diagram demonstrate the border of flow units.

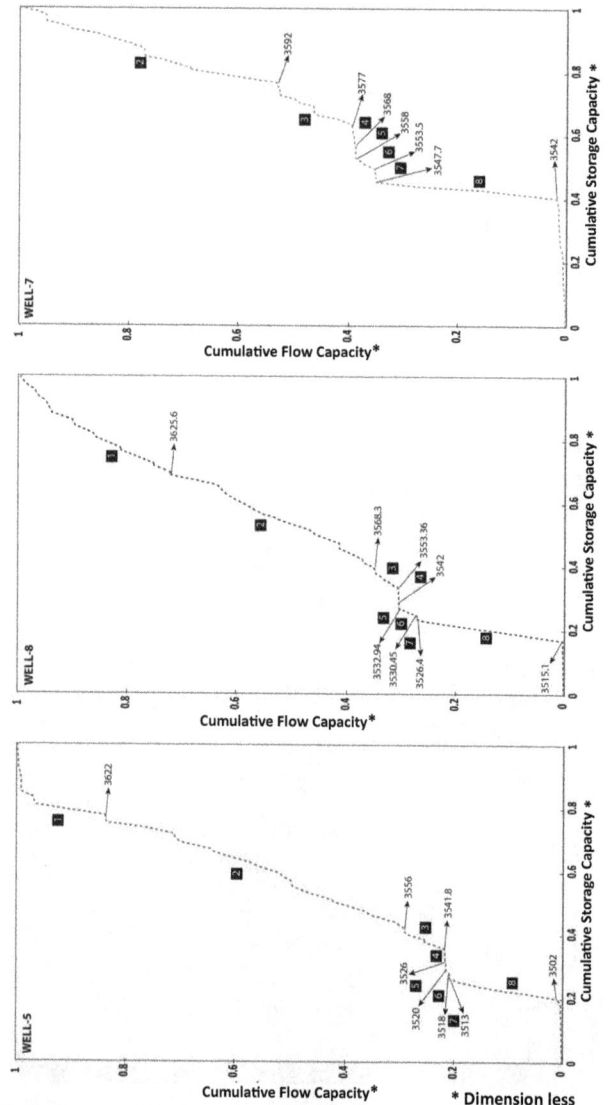

Figure 8: SML plot for siliciclas-tic parts of the Asmari formation in wells 5, 7, and 8 of the CK field. The lowest flow unit (FU-1) has not been penetrated in well 7.

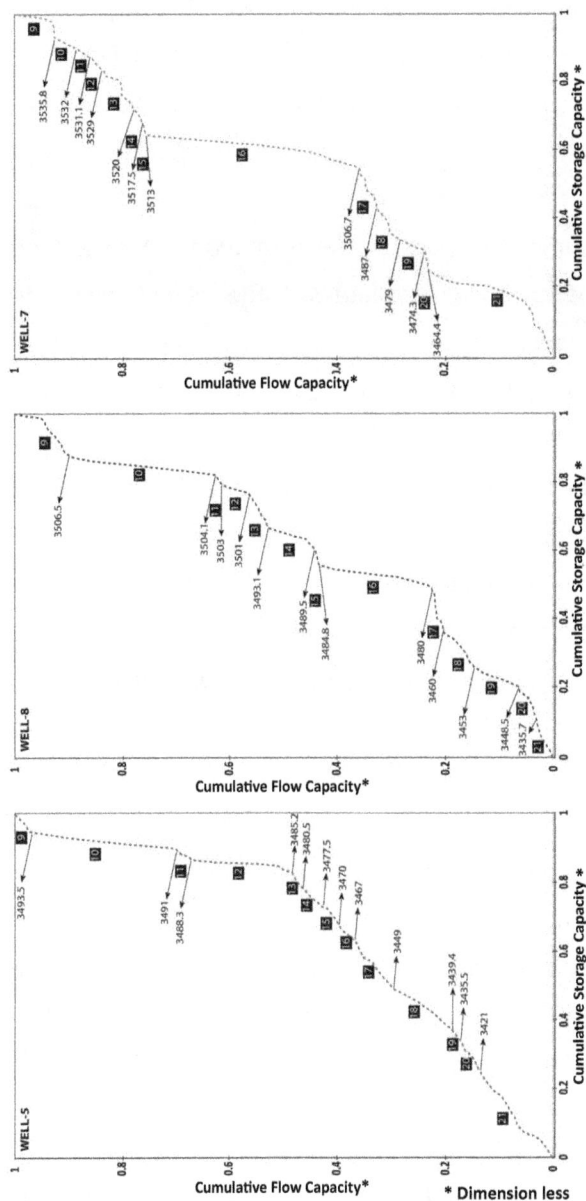

Figure 9: SML plot for carbonate parts of the Asmari formation in wells 5, 7, and 8 of the CK field.

The shaley (impermeable) and sandy (permeable) units are associated with gentle and steep slopes on the diagrams respectively (Figures 8 and 9).

Due to the significant difference in the porosity and permeability values of sandstones with carbonates, their contact is marked by a steep slope on the diagram, so it is considered as a sharp flow unit border. However, within the limy units, the variations of porosity and permeability are insignificant, so the borders of flow units are indistinct.

The carbonate-dominated part of the formation, with a low porosity and permeability variation, is marked by a gentle slope on the diagrams. In this regards, the discrimination of the flow units is rather difficult within this portion. To overcome the problem, the diagrams of the carbonate dominated portion are portrayed separately (Figure 9). Considering the slope deviation on the diagrams as the flow unit border, 13 flow units were determined in this part of the formation (numbered 9 to 21) (Figure 9).

Based on slope deviation on the diagram, 8 flow units (numbered 1 to 8) were determined in the siliciclastic-dominated parts of the formation (Figure 9). These flow units of the sandy horizons are correlatable in the 3 key wells and across the field (Figure 4).

The correlation of the flow units of the carbonate horizons across the field is not easily achievable. This is most likely related to various diagenetic features in the carbonate facies (see below). Upscaling technique is used for field-scale correlation and modeling of the flow units within the carbonate horizons of the formation [28]. For example, the combination of flow units 11, 12, and 13 resulted in the definition of a new larger scale unit which was more useful in the correlation. In this manner, reservoir 21 flow units are merged to 17.

The comparison of outcomes from the two used methods (Figure 4) shows that the contacts of flow units in the sandy parts of the formation exactly coincide with the horizons of FZI variations. Such a correlation is not observed in the carbonate parts, where the flow units show a greater internal variation in reservoir characteristics. Nevertheless, considering the thickness in the SML method makes it more precise in the flow unit analysis.

Linking Flow Units to Depositional Facies, Diagenetic Events, and Sequences

The correlations of the flow units, depositional facies, and sequences in the studied field show that the controlling parameters are different in the lower (sand–dominated) and upper (carbonate-dominated) parts of the formation (Figure 4).

In the sand-dominated part, reservoir quality is controlled by facies type and systems tracts. In other words, depositional conditions are the main controls of reservoir quality here (Figure 5). The upper shoreface shales (facies H) have the lowest reservoir quality among the siliciclastic facies, which is related to the gentle slopes on the CSC-CFC diagram (Figure 8). Therefore, the shale intervals were considered as non-reservoir zones. The barrier sandstones (facies I) and incised valley fill sandstones (facies G) have the highest reservoir quality among the siliciclastic facies, which is related to steep slopes on the CSC-CFC diagram (Figure 8). The sandy facies are the main constituents of the sequences 1, 2, and 3 of the studied formation. The sandstones of lower shoreface to offshore sub-environments (facies F) are mainly developed as the FSST of the sequence 1. They are mainly correlated with the flow unit 1 (see Figure 8). The incised valley fill sandstones/conglomerates (facies G) characterize the LST of the sequences 2 and 3. They are correlated with flow units 2 and 8. The barrier sandstones (facies I) are developed in the FSST of the sequence 2; they are correlated with flow unit 6. The sandstones of intertidal zone (facies J) characterize the TST of the sequence 2 and are correlated with flow unit 3; such associations reveal a good correlation between the flow units and depositional facies and sequences in the siliciclastic-dominated part.

The carbonate-dominated part of the formation is mainly composed of facies associations C (shoal), D (lagoon), and E (intertidal). The facies associations of A and B principally occur as intercalations of the sandy part. The flow units of the carbonate-dominated part of the formations (9 to 21) do not properly correlate with the related facies and sequences. In this part, a single flow unit may correlate with a range of facies or vice versa. The reservoir quality of the carbonate facies is mainly controlled by diagenetic features. On the basis of depositional settings, the shoal facies (ooid and bioclast grainstones) are expected to show high reservoir quality. However, some facies of this group occur in a low reservoir quality range due to the massive cementation. Conversely, some facies of the lagoon environment, which are expected to show a low reservoir quality, occur in a high reservoir quality range due to dissolution and/or dolomitization. In this regard, for the correlation of flow units with facies and sequence in the carbonate part, the diagenetic features of the facies should be taken into account.

The correlation of the flow units with siliciclastic facies and sequences (Figure 5) revealed the tiny role of post-depositional (diagenetic) processes on their reservoir quality. The insignificant role of diagenetic features on the reservoir quality of these facies is also understood from their petrographic studies. Negligible calcite and clay cements are the only porosity/permeability control diagenetic features observed in these facies.

Conversely, the results from the correlation of flow units with the carbonate facies, while combined with their petrographic, SEM, and CL studies, indicate the major role of diagenesis on their reservoir quality (Figure 10).

Association of the Flow Units with Facies Distribution, Depositional Sequences, and Diagenetic...

129

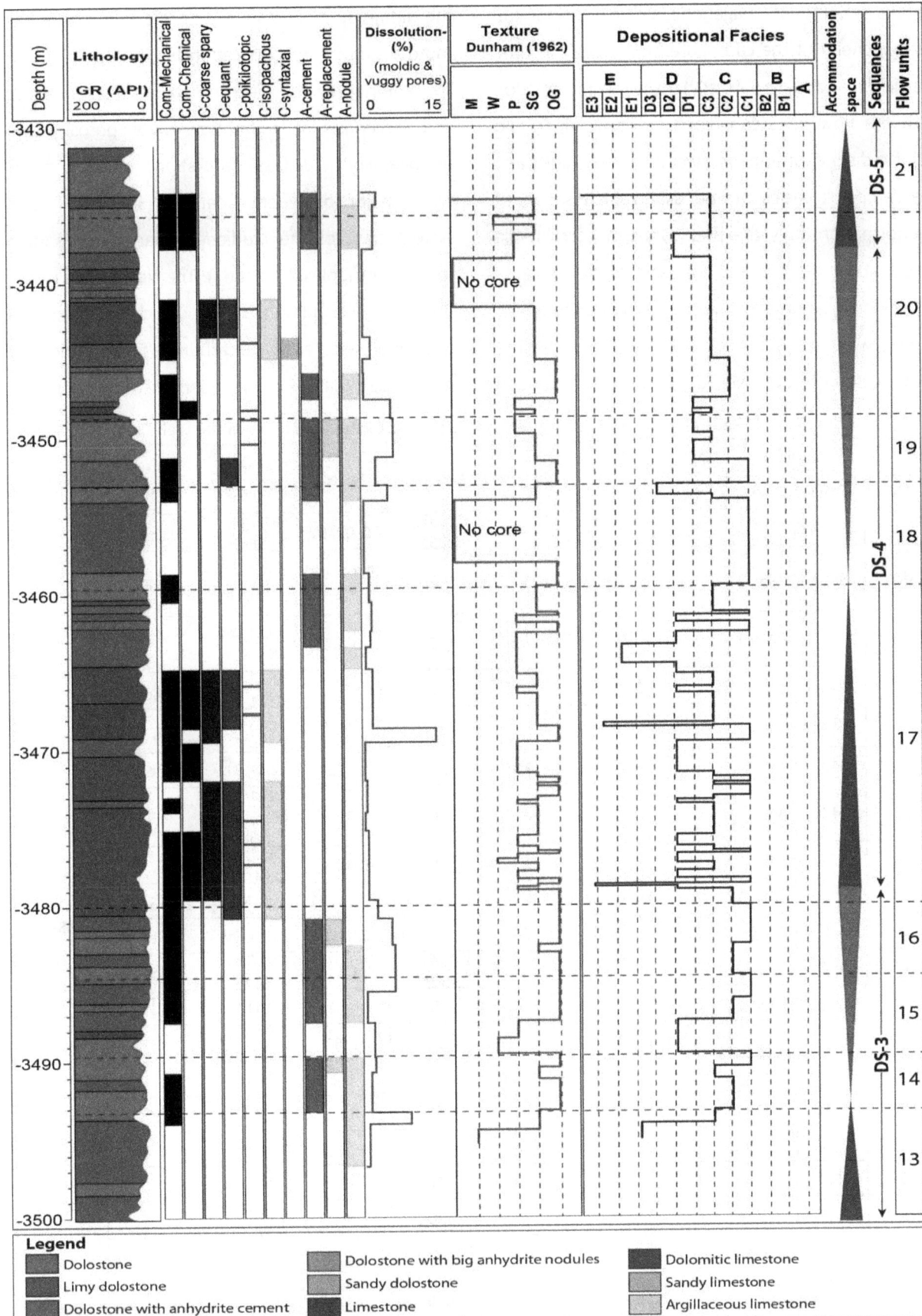

Figure 10: Correlation of flow units, depositional facies, sequences, and major diagenetic features in the carbonate-dominated part of the formation; C= Calcite; A= Anhydrite; Com= Compaction; M= Mudstone; W= Wackestone; P= Packstone; SG= Skeletal grainstone; OG=Ooid grainstone.

Various effects of diagenetic processes on the heterogeneity of the carbonate facies, both in time (vertically) and space (laterally) make their correlation with the flow units complicated (Figures 4, 9, and 10). No distinct principle between the flow units and the carbonate facies can be issued. In this part of the formation, only the boundaries of correlatable flow units across the field are used for modeling. In other words, some flow units are combined to create a new, wide scale, and correlatable flow unit. The combination of flow units 11, 12, and 13 resulted in the definition of a new larger scale unit which was more useful in the correlation (Figure 4).

To summarize the relation of facies/sequences and diagenesis with the flow units, porosity and permeability values, oil saturation, lithology, facies type, major diagenetic features, sequences, and system tract of the studied formation are shown together in Tables 2, 3, and 4. The significance of the depositional conditions of the siliciclastic facies and diagenetic features of the carbonate facies in the flow unit determination is well demonstrated here.

Table 2: Petrophysical and geological characteristics of the flow units in the well CK-5.

FU	Depth (m)		$\emptyset_{Average}$ (%)	$K_{Average}$ (mD)	S_{oil} (%)	Lithology	Facies	Diagenetic Events	Systems Tracts	Sequence
	From	To								
21	3402	3421	9.1	1.25	7.2	D (65%), L (35%)	C3, C2, D2	DLM, DIS, AN CEM	HST, TST	DS-5
20	3421	3435.5	4.8	0.486	3.75	L, D, LD, DL	C3, C1, D2, D1	DLM, AN CEM, DIS	LST, FSST	
19	3435.5	3439.4	5.07	0.62	4.13	D	C3, D2	DLM, AN CEM, DIS	FSST	DS-4
18	3439.4	3449	10.6	2.26	7.6	D	D2, C2, C1	DLM, AN CEM, DIS	HST, TST	
17	3449	3467	4.7	0.58	3.5	L (70%), D (30%)	C3, D2, C1, D3	CAL CEM, DLM	TST, LST	
16	3467	3470	10.8	2.04	9.15	D, LD	C1, C3	DLM, DIS	FSST	DS-3
15	3470	3477.5	5	0.55	3.4	D	C1	DLM, AN CEM, DIS, COM	HST	
14	3477.5	3480.5	11.3	2.26	8.8	D	C1	DLM, AN CEM, DIS,	HST, TST	
13	3480.5	3485.2	7.6	0.912	3.76	D	D1, D2, E1	DLM	TST	
12	3485.2	3488.3	9.8	10.7	3.8	L	C3, D1	CAL CEM	TST	
11	3488.3	3491	7.3	1.74	3	L (50%), D (50%)	C3	CAL CEM, DLM	TST	
10	3491	3493.5	15.08	17.4	12.2	D		DLM	TST	
9	3493.5	3502	8.5	37.03	0.072	DL, LD	D1, B1	CAL CEM, DLM, DIS	TST	
8	3502	3513	21.6	1319.3	19.09	S	G		LST	
7	3513	3518	11.4	56.95	6	D	D	DLM, DIS	FSST	DS-2
6	3518	3520	16.5	178.25	11.6	S	I		FSST	
5	3520	3526	8.7	0.5	1.43	Sh	H		FSST	
4	3526	3541.8	11.6	5.97	7.8	L/DL	B	CAL CEM, DLM, DIS	HST	
3	3541.8	3556	17.2	391.9	10.4	S and Sh	J		TST	
2	3556	3622	19.6	626	15.9	S and basal congl.	G		LST	
1	3622	Top Pabdeh	11.5	154.6	3.7	Arg. S, L	F		FSST	DS-1

D: Dolostone; DL: Dolomitic limestone; LD: Limy dolostone; L: Limestone; S: Sandstone; Sh: Shale; Arg.: Argillaceous; Congl.: Conglomerate; DLM: Dolomitization; DIS: Dissolution; AN CEM: Anhydrite cementation; CAL CEM: Calcite cementation; COM: Compaction

Table 3: Petrophysical and geological characteristics of the flow units in the well CK-8.

FU	Depth (m) From	Depth (m) To	Ø Average (%)	K Average (mD)	S$_{oil}$ (%)	Lithology	Facies	Diagenetic Events	Systems Tracts	Sequence
21	3426	3435.7	8	1.3	6.5	D (75%), L (25%)	C3	DOL, AN CEM, COM, DIS	HST, TST	DS-5
20	3435.7	3448.5	5.4	1.1	4	D, DL, LD	C3, C1, C2, D2	DOL, AN and CAL CEM, COM, DIS	LST, FSST	DS-4
19	3448.5	3453	8.2	6.6	6.9	D	C1, C2, C3, D1	DOL, AN CEM, COM	FSST, HST	DS-4
18	3453	3460	11.9	4.2	8.7	D, LD	C2, C1	DOL, AN CEM, COM	HST	
17	3460	3480	4.35	0.44	2.5	L (70%), LD (20%), D (10%)	C3, C1, D2	DOL, AN and CAL CEM, COM	TST, LST, FSST	
16	3480	3484.8	11.7	18.2	10.2	D, LD	C1, C2	DOL, AN CEM, DIS, COM	HST	DS-3
15	3484.8	3489.5	7.1	0.9	4.8	D, LD	C1, C2, D2	DOL, AN CEM, COM	HST	
14	3489.5	3493.1	11.9	9.8	9.9	D	C1, C2	DOL, AN CEM, COM	HST	
13	3493.1	3501	9	2.2	4.2	LD	E1, E2, E3, D2	DOL, AN CEM	TST	
12	3501	3503	11.7	10.9	4.8	L	No core		TST	
11	3503	3504.1	13.2	3.7	9.7	L	No core		TST	
10	3504.1	3506.5	22	55.9	19.3	D	No core		TST	
9	3506.5	3515.1	10.4	6.3	5.9	LD (60%), L (30%), D (10%)	No core		TST	
8	3515.1	3526.4	25.8	3992.5	24.2	S	G		LST	
7	3526.4	3530.45	14.6	28.5	11.95	D	D	DOL, DIS	FSST	DS-2
6	3530.45	3532.94	26.3	1721	22.1	S	I		FSST	
5	3532.94	3542	12.9 (5.5)	28.7	0.023	Sh	H		FSST	
4	3542	3553.36	14.9	4.6	10.1	L/DL	B	CAL CEM, DOL, DIS	HST	
3	3553.36	3568.3	17.8	566.9	10	S and sh	J		TST	
2	3568.3	3625.6	22.5	1178.2	18.6	S and basal congl.	G		LST	
1	3625.6	Top Pabdeh	21.5	820.8	10.4	Arg. S and L	F		FSST	DS-1

D: Dolostone; DL: Dolomitic limestone; LD: Limy dolostone; L: Limestone; S: Sandstone; Sh: Shale; Arg.: Argillaceous; Congl.: Conglomerate; DLM: Dolomitization; DIS: Dissolution; AN CEM: Anhydrite cementation; CAL CEM: Calcite cementation; COM: Compaction

Table 4: Petrophysical and geological characteristics of the flow units in the well CK-7.

FU	Depth (m)		$\emptyset_{Average}$ (%)	$K_{Average}$ (mD)	S_{oil} (%)	Lithology	Facies	Diagenetic Events	Systems Tracts	Sequence
	From	To								
21	3440	3464.4	9.6	5.7	7	D (70%), L (30%)	D2, C3	DOL, AN CEM, DIS, COM	HST, TST	DS-5
20	3464.4	3474.3	5	0.76	3.1	L	C3, C1, D2	DOL, AN/ CAL CEM, DIS, COM	LST, FSST	
19	3474.3	3479	7.9	5.7	6.1	D	C1, C3	DOL, AN CEM, DIS	FSST, HST	DS-4
18	3479	3487	11	3.5	6.6	D	C1, C2, C3	DOL, AN CEM, DIS, COM	HST	
17	3487	3506.7	5.3	0.9	3.1	L	D2, C3, C2, C1	DOL, AN/ CAL CEM, DIS, COM	TST, LST	
16	3506.7	3513	15.2	38.7	13.6	LD, L	C1, C3	DOL, AN CEM, DIS, COM	FSST, HST	DS-3
15	3513	3517.5	7.5	1.1	4.4	D, LD, L	C1	DOL, AN CEM, DIS, COM	HST	
14	3517.5	3520	13.1	3.7	11	D	No core		HST	
13	3520	3529	11.6	4	5.6	LD, D, L	D2, C1	DOL, AN CEM, COM	TST	
12	3529	3531.1	13.3	4.5	6.2	D	D2, C3	DOL, AN CEM, DIS, COM	TST	
11	3531.1	3532	5.6	0.6	0.7	L	D2, C3	DOL, AN CEM, DIS, COM	TST	
10	3532	3535.8	16.3	12	10.9	D	D2	DOL, DIS, COM	TST	
9	3535.8	3542	11.5	14.25	1.6	L, LD	D2, C2, C3	DOL, AN/ CAL CEM, COM	TST	
8	3542	3547.7	23.5	2703	20.96	S	G		LST	
7	3547.7	3553.5	14.45	9.45	10.9	D	D		FSST	DS-2
6	3553.5	3558	19.7	289.3	14.3	S	I		FSST	
5	3558	3568	10.1	5.05	0.76	Sh	H		FSST	
4	3568	3577	16.5	29.3	11.6	L/DL	B		HST	
3	3577	3592	16.6	319.2	9.5	S and sh	J		TST	
2	3592	3624.2	19.5	612.3	15.2	S and basal congl.	G		LST	

D: Dolostone; DL: Dolomitic limestone; LD: Limy dolostone; L: Limestone; S: Sandstone; Sh: Shale; Arg.: Argillaceous; Congl.: Conglomerate; DLM: Dolomitization; DIS: Dissolution; AN CEM: Anhydrite cementation; CAL CEM: Calcite cementation; COM: Compaction

CONCLUSIONS

In recent years, petroleum engineers have used different methods to determine the reservoir zones or hydraulic flow units. Winland, FZI, and SMLP methods as well as rock typing by using por-perm data are among the most common methods. Although these methods have been widely used in the oil industry, in all of them, the distribution of flow units is solely based on reservoir characteristics, and well locations are presented. On the other hand, based on previous studies and this study, pore characteristics (volume, size, type, and connectivity) in sedimentary successions is controlled by sedimentological (including facies and diagenetic) characteristics. Therefore, introducing techniques, in which reservoir properties and controlling parameters are used together, will be of great efficiency and in predicting the distribution of reservoir properties in the study area. In the current work, an integrated study was carried out on a mixed carbonate-silisiclastic succession of the Asmari formation. The results showed that the lateral and vertical distribution of some reservoir flow units (such as sandstone units 2 and 8) are strongly controlled by the geometry of depositional facies. This study showed that the correlation and modeling of flow units, solely on the basis of lithology and thickness, and regardless of facies and its geometry, will cause different facies (such as coastal and channel-filled sandstones) to be placed in a single flow unit. However, the lateral and vertical distribution and reservoir quality of these coastal and channel-filled sandstones in the Asmari formation are quite different. Also, in sequence stratigraphic framework, these sandstones were placed in different sequences (sequences 2 and 3) and systems tracts (FRST and LST). In the carbonate parts of Asmari formation, the effect of diagenetic processes on reservoir quality is much higher than the facies. Thus, the LST limestones of unit 17, as a result of calcite cementation, were changed to a thick, distinct, correlatable, and barrier unit. On the other hand, dolomitic intervals which have not been affected by anhydrite cementation have formed porous and permeable carbonate reservoir units (such as units 18 and 21).

In this integrated geological-petrophysical case study, for each flow unit, reservoir characteristics, facies, diagenetic events, and system tracts were presented. Thus, according to the relationships between these characteristics in the studied field, the prediction of reservoir properties in any given situation will be possible, which will be covered in our future studies.

REFERENCES

1. Amaefule J. O., Altunbay M., Tiab. D., Kersey D. G., and Keelan D. K., "Enhanced Reservoir Description: Using Core and Log Data to Identify Hydraulic (Flow) Units and predict permeability in uncored intervals/wells," SPE 68, Annual Technical Conference and Exhibition, Houston, Texas, **1993.**

2. Orodu O. D., Tang Z., and Fei Q., "Hydraulic (Flow) Unit Determination and Permeability Prediction: A Case Study of Block Shen-95, Liaohe oilfield, North-East China," *Journal of Applied Sciences*, **2009,** *9*, 1801-1816.

3. Noori Al-Jawad S., Saleh A. H., Al-Dobaj A., and Al-Rawi Y. T., "Reservoir Flow Unit Identification of the Mishrif Formation in North Rumaila Field," *Arabian Journal of Geosciences*, **2014,** *7*, 2711-2728.

4. Nabikhani N., Moussavi-Harami R., Mahboubi A., Kadkhodaie A., and Yosefpour M. R., "The Evaluation of Reservoir Quality of the Sarvak Formation in One of Oil Fields of the Persian Gulf," *Journal of Petroleum Science and Technology*, **2012,** *2*, 3-15.

5. Rahimpour-Bonab H., Mehrabi H., Navidtalab A., and Izadi-Mazidi E., "Flow Unit Distribution and

Reservoir Modeling in Cretaceous Carbonates of the Sarvak Formation, Abteymour oil Field, Dezful Embayment, SW Iran," *Journal of Petroleum Geology*, **2012**, *35*, 213-236.

6. Ebanks W. J. Jr., Scheihing M. H., and Atkinson C. D., "Flow Units for Reservoir Characterization," In: D. Morton-Thompson, A. M. Woods (Eds.), Development Geology Reference Manual, American Association of Petroleum Geoplogists, *Methods in Exploration Series*, **1992**, *10*, 282-284.

7. Moore C. H., "Carbonate Reservoirs: Porosity Evolution and Diagenesis in a Sequence Stratigraphic Framework. Developments in Sedimentology," Amsterdam, *Elsevier*, **2001**, *55*, 444.

8. Hosseini nia T., "Well Completion Report of Well Cheshmeh Khush-9," Geology Department, Iranian Central Oil Fields Company, Internal Report, **2006**.

9. Van Buchem F. S. P., Allan T., Lausen G. V. , Lotfpour M., et al., "Sequence Stratigraphy and Sr Isotope Stratigraphy of the Oligo-miocene Deposits in the Dezful Embayment (Asmari and Pabdeh Formations, SW Iran)-implications for reservoir characterization," 1st International Petroleum Conference, European Association of Geoscientists and Engineers (EAGE), Shiraz, Iran, **2010**.

10. Jong-Se L., "Reservoir Properties Determination Using Fuzzy Logic and Neural Networks from Well Data in Offshore Korea," *Journal of Petroleum Science and Engineering*, **2005**, *49*, 182-192.

11. Helle H. B., Bhatt A., and Ursin B., "Porosity and Permeability Prediction from Wireline Logs Using Artificial Neural Networks: a North Sea Case Study," *Geophysical Prospecting*, **2001**, *49*, 431-444.

12. Gunter G. W., Finneran J. M., Hartmann D. J., and Miller J. D., "Early Determination of Reservoir Flow Units Using an Integrated Petrophysical Method," *Society of Petroleum Engineers*, SPE Annual Technical Conference and Exhibition, San Antonio, Texas, **1997**, 373-380.

13. Honarmand J., "Sedimentological and Diagenetic Controls on Reservoir Properties of the Asmari Formation, Cheshmeh Khush Oil Field, SW Iran," Ph.D. Thesis, School of Geology, University of Tehran, Tehran, **2013**.

14. Thomas A. N., "Facies Variation in Asmari limestone," 18th Inter. Geol. Cong., **1952**, 74-82.

15. Adams T. D., "The Asmari Formation of Lurestan and Khuzestan Provinces, Exploration Directorate," *National Iranian Oil Company*, **1969**, Report No. 1151.

16. McCoard D. R., "Regional Geology of the Asmari of Ahwaz and Marun Areas," Exploration Directorate, *National Iranian Oil Company*, Internal Report, **1974**.

17. Pairaudeau J. G., "Environmental Analysis of Ahwaz Sands in Lower Asmari Marun Field," OSCO Technical note, *National Iranian Oil Company*, Internal Report, **1978**.

18. Seyrafian A. and Hamedani A., "Microfacies and Depositional Environment of the Upper Asmari Formation (Burdigalian) North-Central Zagros Basin, Iran," *Journal of Geological Palaeontology Abh*, **1998**, *210*, 129–141.

19. Seyrafian A. and Hamedani A., "Microfacies and Palaeoenvironmental Interpretation of the Lower Asmari Formation (Oligocene), North-Central Zagros Basin, Iran," *Journal of Geological Palaeontology Abh*, **2003**, *3*, 164–174.

20. Aqrawi A. A. M., Keramati M., Ehrenberg S. N., Pickard N., et al., "The Origin of Dolomite in the Asmari Formation (Oligocene-Lower Miocene), Dezful Embayment, Southwest Iran," *Journal of Petroleum Geology*, **2006**, *29*, 381-402.

21. Vaziri-Moghaddam H., Kimiagari M., and Taheri A., "Depositional Environment and Sequence Stratigraphy of the Oligocene-Miocene Asmari

Formation in SW Iran," *Facies*, **2006**, *52*, 41-51.

22. Honarmand J. and Amini A., "Diagenetic Processes and Reservoir Properties in the Ooid Grainstones of the Asmari Formation, Cheshmeh Khush Oil Field, SW Iran," *Journal of Petroleum Science and Engineering*, **2012**, *81*, 70-79.

23. Buxton M. W. N. and Pedley H. M., "A Standardized Model for Tethyan Tertiary Carbonate Ramps," *Journal of the Geological Society*, **1989**, *146*, 746-748.

24. Flugel E., "Microfacies of Carbonate Rocks: Analysis, Interpretation and Application," Berline, Springer Verlag, **2010**, 976.

25. Ehrenberg S. N., Pickard N. A. H., Laursen G. V., Monibi S., et al., "Strontium Isotope Stratigraphy of the Asmari Formation (Oligocene–Lower Miocene), SW Iran," *Journal of Petroleum Geology*, **2007**, *30*, 107-128.

26. Hunt D. and Tucker M. E., "Stranded Parasequences and the Forced Regressive Wedge Systems Tract: Deposition during Base-level Fall," *Sedimentary Geology*, **1992**, *81*, 1-9.

27. Catuneanu O., "Principles of Sequence Stratigraphy," Amsterdam, *Elsevier*, **2006**, 375.

28. Lucia F. J., "Carbonate Reservoir Characterization: An Integrated Approach," Berlin, Springer-Verlag, **2007**, 336.

11

The Effect of Fault Plane on the Horizontal In Situ Stresses Orientation: A Case Study in one of Iranian Oilfield

Mohammadreza Zare Reisabadi and Seyyed Saeed Ghorashi*

Faculty of Research and Development in Upstream Petroleum Industry, Research Institute of Petroleum Industry (RIPI), Tehran, Iran

ABSTRACT

Knowledge of the orientation of horizontal in situ stresses is important to some areas of oil and gas field development plans. Borehole breakouts observed in image logs and drilling-induced fractures are the main parameters for the determination of the stresses' directions in situ.

In this work, the orientations of borehole breakouts were investigated as a function of depth in oil wells A and B in Lali oilfield, in the Southwest of Iran. Borehole breakouts were detected from FMI logs. By the statistical analysis of the borehole breakouts' orientation in the foregoing two wells, it was found that, while a mean orientation of minimum horizontal stress in well A is NE-SW, the azimuth of breakout in well B is different with a mean azimuth of 312°±10°. The result reveals that the orientation must be different in these two wells due to some geological abnormality. Therefore, accurate and reliable geomechanical analyses are crucial steps toward minimizing the costs of drilling and completion programs and mitigating borehole instability problems.

Keywords: In Situ Stresses, Borehole Breakout, Drilling-induced Fractures, Lali Oilfield

INTRODUCTION

Borehole breakouts and drilling-induced fractures (DIF's) are important indicators of horizontal stress orientations [1]. Knowledge of the orientation of horizontal in situ stresses derived from the analysis of borehole breakouts is important for planning hydrocarbon exploration strategies; developing production strategies; engineering reservoirs, drilling, and wellbore mechanics; and studying crustal stress and rock mechanics [1].

Borehole breakouts are stress-induced enlargements of the wellbore cross-section [2].

When a borehole is drilled, the material removed from the subsurface is no longer supporting the surrounding rock. As a result, the stresses become concentrated across the surrounding rock (including the wellbore wall). Borehole breakouts occur when the stresses around the borehole exceed the compressive strength of the borehole wall [3]. The enlargement of the wellbore is caused by the development of intersecting conjugate shear planes that cause spalling of the fragments of the formation being drilled. Around a vertical borehole, stress concentration is generally greatest in the direction of the minimum horizontal stress. Hence the long axes of borehole breakouts are oriented approximately perpendicular to the maximum horizontal stress orientation [4].

*****Corresponding author**
   Seyyed Saeed Ghorashi
   Email: ghorashis@ripi.ir

Borehole Failures Analysis by Image Logs

Borehole breakouts can be determined by using image logs [4, 5]. There are currently a wide variety of imaging tools that fall into two general categories of acoustic and resistivity tools. In this study, formation micro imager (FMI) logs were used to determine borehole breakouts and fault planes. FMI is a resistivity imaging tool which provides an image of the wellbore wall based on resistivity contrasts [6,7].

Figure 1a illustrates borehole breakouts and tensile induced fractures as dark bounds and dark lines respectively [8], while Figure 1b is showing the transverse induced fracture in one of the Iranian oil wells. As it can be seen in Figure 1b, discriminating between transverse induced fractures and natural fractures or bedding is so difficult that an accurate interpretation would run into difficulty without the knowledge of a comprehensive geomechanics model. Indicating such high conductivity fractures, Barton and Zoback have presented a method for distinguishing induced fractures from natural fractures by using reservoir geomechanical models and fitting a flexible sinusoid to the pair of fractures [9].

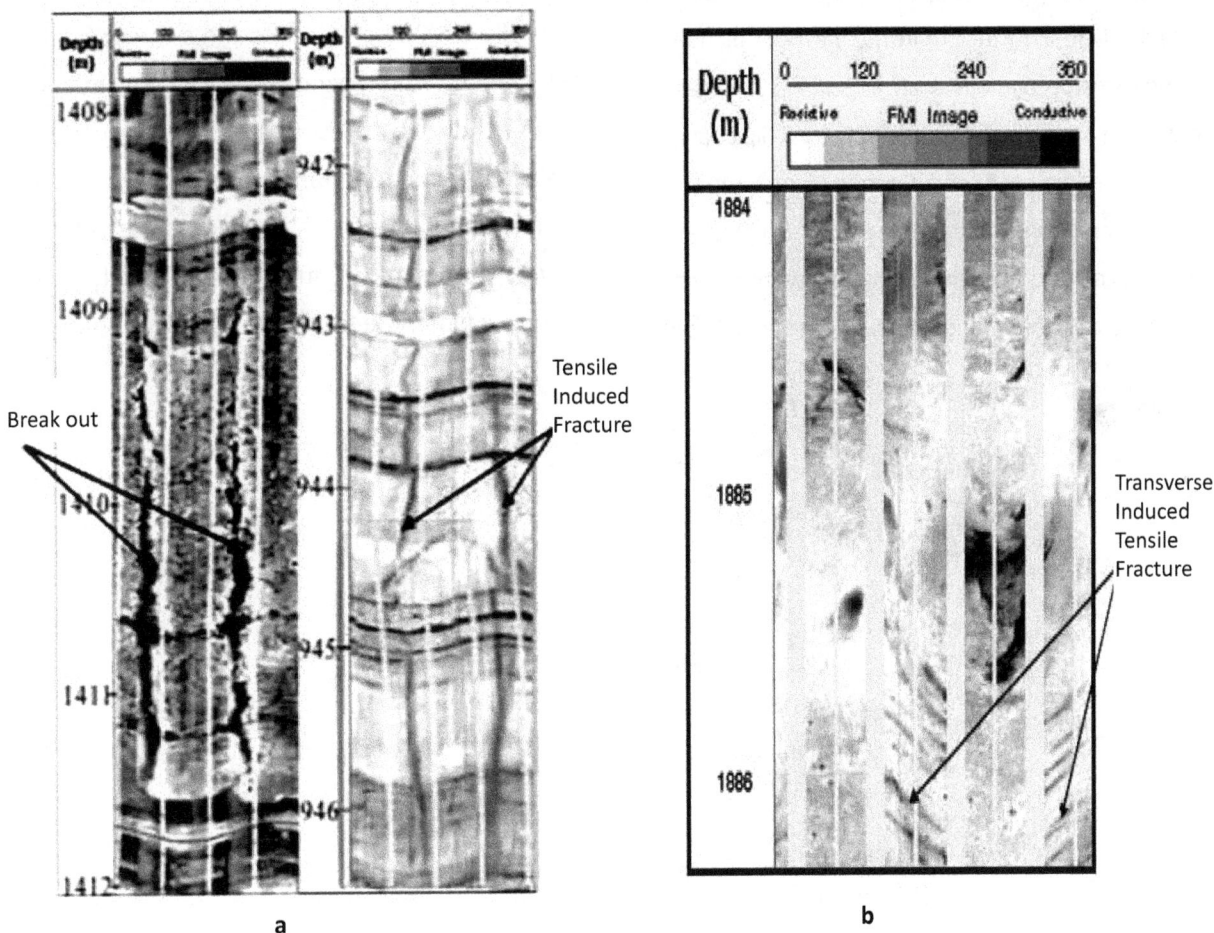

a

b

Figure 1: (a) An FMI log example of borehole breakout and tensile induced fracture [8], and (b) an FMI log example of transverse induced tensile fracture in one of Iranian oil wells.

As tensile induced fractures are controlled by the orientation of tectonics stresses and are not formed continuously around the borehole wall, they propagate only in tensile regions of borehole wall where the hoop stress exceeds the rock Itensile strength [9]. The best approach to distinguish induced fractures from natural fractures is comparing the results of image logs with the fractures observed on the core samples.

Detection and determination of faults are also one of the most important applications of image logs in the field of structural geology. Fault is a displaced planar fracture or discontinuity in a volume of rock. If the displacement across the plane of a fault is lower than the wellbore diameter, we can then observe it on the image log directly. However, if the displacement across the fault plane is further than the wellbore diameter, the indirect observations such as sudden changes in the dip and azimuth of layering (structural dip change), truncated bedding, the concentration of fractures in the area of sheared zones, and the development of borehole breakouts can be employed. Figure 2 shows the effect of normal fault on the FMI log in one of Iranian oil fields.

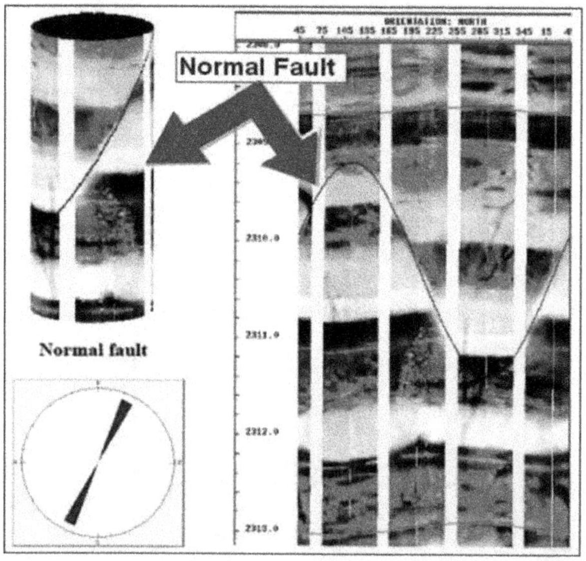

In this paper, borehole breakouts detected by FMI log in wells A and B are investigated. Although wells A and B are not relatively far from each other, the orientation of borehole breakouts or horizontal in situ stresses is significantly different. Here, we will discuss the reasons for this event.

Stress Orientation in Lali Oil Field

Figure 3 illustrates the maximum horizontal stress orientation map of Iran [10]. Each symbol indicates the type of stress measurement and each color shows the corresponding stress regime. As Figure 3 shows, the only method reported for stress measurement in Iran is the focal mechanism. In this figure, red color indicates normal faulting (NF), green color indicates strike-slip (SS), blue color indicates thrust faulting (TF), and "U" is an unknown tectonic regime [10]; most of them are thrust faulting regimes. Since a significant number of Iranian oil and gas reservoirs are located in the south and southwest of Iran, in which the state of stress is the mainly thrust faulting regime, the borehole instability problems,

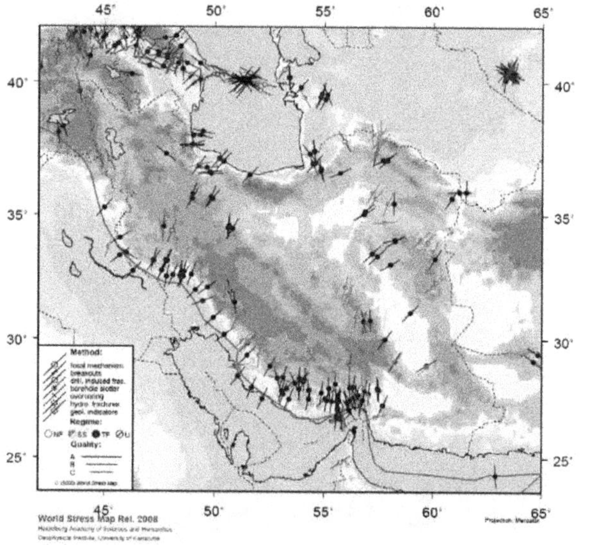

Figure 2: The effect of normal fault on the FMI log in one of Iranian oil wells.

Figure 3: Map of maximum horizontal stress orientations in Iran for all the data with different stress measurements on the regional topographic base [10].

such as, stuck pipe, lost circulation, sand production, and casing collapse frequently occur at the borehole wall while drilling and later during production. Therefore, accurate and reliable geomechanical analyses are crucial steps toward minimizing costs of drilling and completion programs and mitigating the borehole instability problems.

The anticline of the Lali Oilfield is located in the southwest of Iran. It has an asymmetric structure, in which the dip of southern flank is more than the northern one. According to underground contour (UGC) map, the maximum dip of southern flank is 35°. Also, the axis of this anticline in the West tends to North and in the East, tends to South. The Asmari reservoir of this field, for the first time, was explored in 1938 and drilled to the depth of 1768 m (383 m in Asmari formation). So far, 27 wells have been drilled in Lali oilfield. Only two wells A and B have available data. Unlike well A, which was drilled into the Asmari and the Pabdeh formations at the depth ranging from 2072 to 2439 m, well B was penetrated only through the Asmari formation between the depths of 1544 to 2066 m. The dominant lithology of the Asmari formation in the Lali Oilfield is limestone. In some areas, it is composed of dolomite limestone and lime shale with seams of anhydrate [11].

By interpreting and observing the image logs in the vertical interval of wells A and B, borehole breakouts without any tensile induced fracture were detected in both wells. Figure 4 shows the borehole breakout recorded by FMI log in well A (Figure 4a) and well B (Figure 4b).

Breakouts are seen as the vertical stripes of dark colors in FMI logs. As it can be seen from Figure 4, the orientations of borehole breakouts in well A are almost NE–SW, whereas in well B, it appears to be along NW–SE orientation.

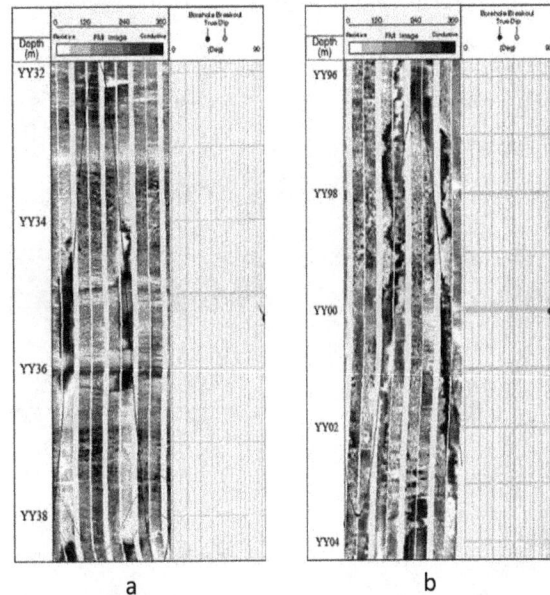

Figure 4: (a) FMI log of 6 m in well A and (b) 8 m in well B; the dark color in image logs indicates the location of low reflected amplitude (low resistivity) caused by breakout zone.

Figures 5a and 6a show the depth of occurrence and the orientations of borehole breakouts, defined by gray points, in wells A and B respectively. As illustrated in these figures, the orientations of borehole breakouts remain the same with depth in both wells. Figures 5b and 6b illustrate the frequency of breakout dip angle and orientation in wells A and B respectively. Figure 7 also shows the rose diagram of induced fracture in well B. As it can be seen from Figure 7, the induced fractures are vertical in well B (a vertical well). In addition, since a mean dip direction is derived near vertical (90°) and borehole breakouts offset by 180° at borehole wall (Figure 5 and Figure 6), the plane of principal stresses must be horizontal in both wells. It is found by rose diagram that a mean azimuth of borehole breakout in well A is 37°±15°(Figure 5b), and it is, however, 312°±10° in well B (Figure 6b).

According to Figure 3, which shows the orientation of horizontal in situ stresses in Iran, it would be expected that the orientations of borehole breakouts in both wells are similar with a mean azimuth of about NE-SW. This principal is observed in well A, but as Figures 5 and 6 show that well B is a special case, and there is no correspondence between the horizontal in situ stresses direction in this well and the world stress map of Iran.

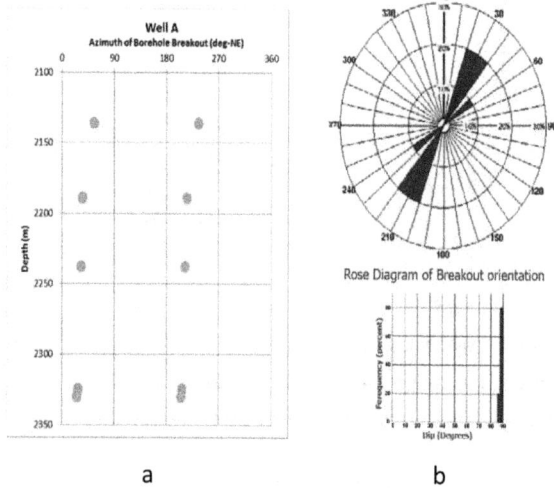

a   b

Figure 5: (a) Depth versus the borehole breakout orientation (gray points) at a depth of 250 m of well A; (b) (upper) the rose diagram of 5 data record and (lower) the frequency of dip direction of breakout.

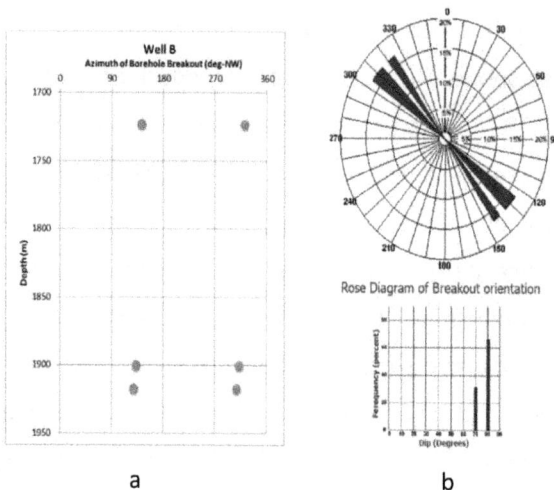

a   b

Figure 6: (a) Depth versus the borehole breakout orientation (gray points) at a depth of 250 m of well B; (b) (upper) the rose diagram of 3 data record and (lower) the frequency of dip direction of breakout.

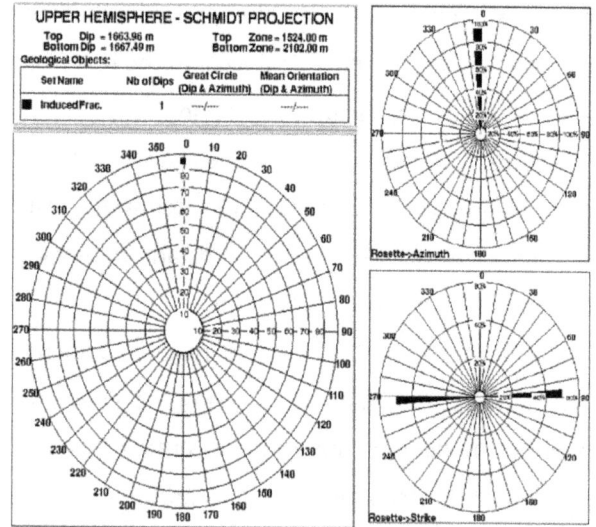

Figure 7: Rose diagram of induced fracture in well B.

Image log analyses suggest that significant changes of the borehole breakout orientation with a depth in accordance with the borehole axis or between neighboring wells are explained by the influence of geological heterogeneities; such as, faults, folds, and salt diaper; however, they are often due to slip on fault, and the variations may be seen while approaching a fault [12]. In this investigation, the result reveals that the orientations must be different in these two wells due to some geological abnormality.

The Effect of Faults on Borehole Breakout Orientations

Many faults have been detected by the interpretation of FMI logs in well B. Figure 8 shows the major fault plane in well B, since the distinct changes have been observed in layering structure in the wellbore wall from top to the depth of 1816 m. As it can be seen from Figure 8, the upper layers have a dip in the direction to S35°W, while gradually to the bottom, the dip is increasing and at depth of 1816 m, the dip direction has reversed suddenly to NE.

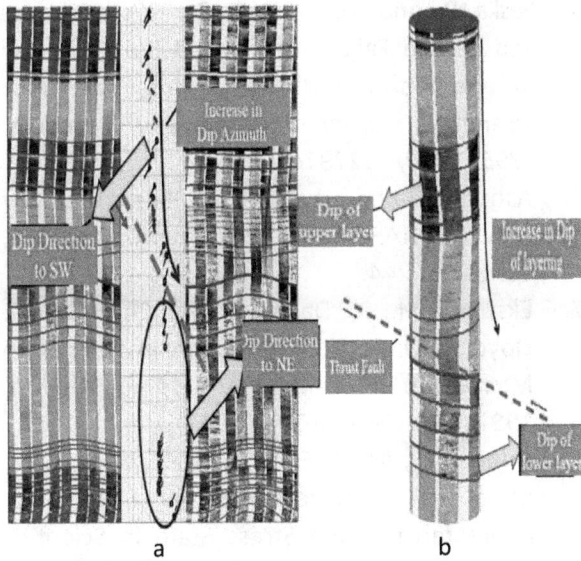

a   b

Figure 8: (a) The effect of the major fault plane at a depth of 1816 m in well B on the FMI log and (b) 3D view of the FMI log.

So, the depth of 1816 m could be considered as the location of the major fault plane. To detect a fault, indicators are required. All the factors such as conditions governing the status of drilling, partial knowledge of regional geological condition, seismic profiles, and even pressure data inside the well could help to recognize a fault in a region or on the well scale. In this case, the downward movement of the cap rock and the Asmari formation compared to the drilling forecast program was another reason for detecting this event.

Figure 9a shows the depth of occurrence and orientations of faults, defined by gray points in well B respectively. As shown in this figure, the orientation of the fault is constant with depth. Figure 9b illustrates the frequency of the fault orientation in well B. It is found by a rose diagram that a mean azimuth of the fault in well B is 300°±10°. By the determination of faults strike in well B, it is clear that the fault strike is different from the one in well A. Therefore, the change in the orientation of borehole breakouts or minimum horizontal in situ stresses in well B is justifiable.

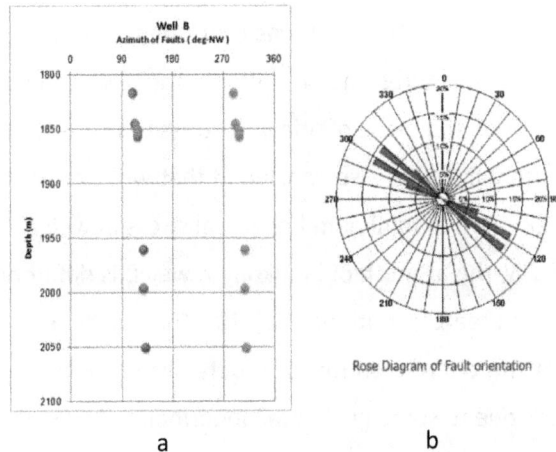

a   b

Figure 9: (a) Depth versus the fault orientation (gray points) at a depth of 300 m of well B; (b) the rose diagram of 7 data illustrates that a mean azimuth of faults is 300°(s.d.=10°).

Applications

Where in situ stresses are disturbed around large scale fractures, such as, faults the direction of principal stresses can deviate from the regional trend to a completely new direction [13]. This can have a great impact on engineering design and planning. For well B, the results indicate this point. Knowledge of principal in situ stresses orientation is crucial in some areas of petroleum industry. In this case, drilling reports have shown that lost circulation is negligible in well A, but, for well B, it is about 40 bbl/hr from the depths of 1544 to 2066 m. This event in well B is due to the lack of the knowledge of geomkechanical modeling (in situ stresses orientation, stresses magnitude, rock mechanic properties, etc.), which causes borehole instability problems during the drilling operation. By knowing the horizontal in situ stress orientation in a region, engineers can determine the optimum deviation and azimuth for drilling nearby planned wells in a way to improve borehole stability.

CONCLUSIONS

In this work, the orientation of borehole breakouts and induced fractures were investigated as a function of depth in oil wells A and B in the Lali oilfield in the southwest of Iran. It was concluded that drilling induced fractures are vertical in the Lali oilfield; therefore, the plane of principal stresses are vertical-horizontal. It was also concluded that while a mean orientation of minimum horizontal stress in well A is NE-SW, the azimuth of breakout in well B is different with a mean azimuth of 312°±10°. The results reveal that the orientation must be different in these two wells due to some geological abnormality. This study has indicated that a drilled well becomes closer to the plane of existing faults in the field, and this causes stress perturbation. This has also caused the direction of the minimum horizontal stress to be changed significantly, about 90°, in accordance with another nearby well.

ACKNOWLEDGEMENTS

Authors thank National Iranian South Oilfield Company's staffs and advocates for their help and support during this work and providing sufficient datasets used herein.

REFERENCES

1. Prensky S., "Borehole Breakouts and In Situ Rock Stress: a Review," *The Log Analyst*, **1992**, *33,* 304-312.
2. Bell J. S. and Gough D. I., "Northeast-Southwest Compressive Stress in Alberta: Evidence from Oil Wells," *Earth and Planetary Science Letters*, **1979**, *45,* 475-482.
3. Zoback M. D., Moos D., Mastin L. G., and Anderson R. N., "Well Bore Breakouts and In Situ Stress," *Journal of Geophysical Research*, **1985**, *90,* 5523-5530.
4. Plumb R. A. and Hickman S. H., "Stress Induced Borehole Elongation: a Comparison between the Four-arm Dipmeter and the Borehole Televiewer in the Auburn Geothermal Well," *Journal of Geophysical Research*, **1985**, *90,* 5513–5521.
5. Peska P. and Zoback M. D., "Compressive and Tensile Failure of Inclined Borehole and Determination of In Situ Stress and Rock Strength," *Journal of Geophysical Research,* **1995**, *100,* 12791–12811.
6. Asquith G. and Krygowski D. "Basic Well Log Analysis", *AAPG Methods in Exploration,* **2004**, Series 28, 244.
7. Ekstrom M. P., Dahan C. A., Chen M. Y., Lloyd P. M. et al., "Formation Imaging with Microelectrical Scanning Arrays," *Log Analyst,* **1987**, *28,* 294-306.
8. Hung J. H., Ma K. F., Wang C. Y., Song S. R., et al., "Subsurface Structure, Fault Zone Characteristics and Stress State in Scientific Drill Holes of Taiwan Chelungpu Fault Drilling Project," *Scientific Drilling*, No 1, **2007**, 55–58.
9. Barton C. A. and Zoback M. D., "Discrimination of Natural Fractures from Drilling Induced Wellbore Failures in Wellbore Image Data Implications for Reservoir Permeability," Paper SPE 58993, Presented at the SPE International Petroleum Conference and Exhibition in Mexico, Villahermosa, Mexico, **2000**, 1–3 February.
10. world-stress-map.org, The World Stress Map (WSM) Project, **2008**.
11. Ghorbani Ghashghai I., "Geological Study a Static Model Construction of Asmari Reservoir in Lali Oilfield," **2008**, *NISOC*, 6409.
12. Brudy M., and Kjoholt H., "Stress Orientation on the Norwegian Continental Self-derived from Borehole Failures Observed in High-resolution Borehole Imaging Logs," *Tectonophysics*, **2000**, *337,* 65–84.
13. Yaghoubi A. and Zeinali M., "Determination of Magnitude and Orientation of the In Situ Stress from Borehole Breakout and Effect of Pore Pressure on Borehole Stability: Case Study in Cheshmeh Khush Oil Field of Iran," *Journal of Petroleum Science and Engineering,* **2009**, *67,* 116-126

Recovery of Oil in Limestone with Cationic Surfactant: Investigation of the Adsorption, Wettability, and Spontaneous Imbibition

Vanessa Cristina Santanna*, Tereza Neuma Castro Dantas, Allan Martins Neves, Juliana Rocha Dantas Lima, and Camila Nóbrega Pessoa

Department of Petroleum Engineering, Federal University of Rio Grande do Norte, Natal, Brazil

ABSTRACT

The performance of petroleum recovery methods is dependent on the characteristics of the reservoir rock, oil quality, and formation water. One fundamental feature for considering petroleum recovery methods is the wettability of the rock reservoir. Also, this feature impacts the displacement of fluids in the rock reservoir. In the case of the limestone, which normally features oil wettability, enhanced recovery methods are responsible for altering surface wettability. In this study, the effect of concentration of a cationic surfactant in the oil recovery from limestone is shown. In these tests, finite-bath technique was used to investigate adsorption by varying the amount of adsorbent and contact time. By carrying out contact angle assays, the wettability of the limestone was assessed. In addition, the oil recovery was effected by spontaneous imbibition. In this study, the limestone samples with solutions of CTAB in the concentrations of 0.8596 g/L and 1.2280 g/L, using a 2% KCl solution are treated. The results showed enhanced adsorption of CTAB at the CMC. The nature of this phenomenon did not induce changes in the wettability of the rock, which was confirmed by the contact angle measurements. In the spontaneous imbibition, more oil could be recovered when using a CTAB below the CMC, possibly due to less intense adsorption by the limestone.

Keywords: Imbibition, Surfactant, Oil Recovery, Adsorption, Limestone

INTRODUCTION

Rock wettability is a decisive parameter in the production and recovery of oil for it affects the economic viability of potential projects. It can occur in diverse environments such as low-permeability and stratified limestone reservoirs, naturally fractured reservoirs, and even high-permeability sandstone reservoirs.

The great majority of carbonate rocks presents wettability that favors oil and presents low produtivity. Therefore, many studies seek to invert the wettability in this type of rock to increase productivity. The use of surfactants in advanced recovery processes in carbonate rock aims to alter the rock's wettability, promoting oil flow, and consequently, increasing recovery. The use of

*Corresponding author

Vanessa Cristina Santanna

Email: vanessasantanna@yahoo.com.br

surfactants in the process leads to an increase in surface wettability lowering water surface tension and promoting adsorption in the solid surfaces. Literature offers several reports on the study of the effects of wettability and heterogeneity of the rock on the production and recovery of oil [1,2,3,4,5,6]. However, when researching this subject, many gaps and inconsistencies can be found.

Imbibition experiments at 40 °C with 100% oil saturated cores, using CTAB surfactant concentration in the range of 0.13 to 5.0 wt.% were carried out by Standnes and Austad in 2000 [7]. It is observed that the imbibition rate and the oil recovery increase with increasing concentration of cationic surfactants above their critical micelle concentration (0.2 wt.%) [7]. However, concentrations much higher than the CMC from 1.0 to 5.0 wt.% impair the recovery. Moreover, this has been ascribed to the solubilization of recovered oil in the micelles. Based on this premise, this work uses CTAB concentrations 30 % below its CMC (0.8596 g/L) and at the actual CMC (1.2280 g/L).

Spontaneous imbibition assays in limestone plugs, both saturated with oil (100 %) and containing initial water (Swi > 0) were carried out by Babadagli and Boluk in 2005 [8]. Anionic, nonionic, and cationic surfactants in concentrations below and above their CMC's were tested by them [8]. Also, it was verified that for Swi = 0, the nonionic IGEPAL and the cationic CTAB were more efficient at low concentrations.

Moreover, the spontaneous imbibition in limestone plugs saturated with oil was investigated by Høgnesen in 2006 [9]. The ability of CTAB was again examined above its CMC. Then, it was confirmed that surfactant concentration directly affected recovery rates. Observations show that

when higher interfacial tensions were generated between oil and CTAB solutions, capillary forces were dominant in the oil displacement [9]. For the most part, gravity forces governed oil displacement in lower interfacial tensions.

The spontaneous imbibition of limestone samples saturated with oil (S_{wi}=0) was examined by Babadagli in 2006 [10]. Nonionic, anionic, and cationic surfactants were used in concentrations below and above their CMC's. The results showed that, in general, when using concentrations above the CMC, the recovery rates are lower than when using brine, with only two exceptions for such behavior: using a fatty quaternary amine salt and the anionic hexylene glycol sulfonate.

The surfactant effects on dolomite rocks were studied by Jarrahian et al in 2012 [16]. Also, three surfactants: C_{12}TAB (cationic), TritonX-100 (nonionic), and SDS (anionic) were studied by them. Based on the results, the cationic surfactant C_{12}TAB presented the highest wettability, promoting ionic interactions, and leading to increased water wettability.

Dodecyltrimethylammonium chloride (DTAC) below and above the CMC in the spontaneous imbibition of limestone samples saturated with oil was used by Pons-Jiménezo et al in 2014 [11]. Then it was observed that the recovery rate was directly proportional to the surfactant concentration, achieving the highest percentage with the highest amount of surfactant used.

A methodology was presented by Wang et al in 2015 to determine the static adsorption of the dodecyl hydroxypropyl sulfide (DSB) over limestone using HPLC and zeta potential measures [17]. Results showed that increasing NaCl concentration caused a decrease in DSB adsorption in limestone surface due to an increase in zeta potential. In contrast, the DSB adsorption started with a decrease and gradually increased with the increase in $CaCl_2$, in response to the zeta potential action in the surface. The wettability alteration in carbonate rock by combining brine water with surfactant during water flooding was studied by Karimi et al in 2016 [12]. Different brine formulations were tested through contact angle measurements and spontaneous imbibition experiments in oil-wet limestone core samples, both in the presence and absence of surfactant (dodecyltrimethylammonium bromide -DTAB). Contact angle measurements revealed that all low salinity brine solutions changed the wettability of oil-wet surface towards a more water-wet state. In the presence of cationic surfactant, utilizing both, with the wettability influencing ions, more water-wet surfaces are resulted. The spontaneous imbibition results show that the combination of the effect of wettability influencing ions (Mg^{+2} and SO_4^{-2}) along with the cationic surfactant can result in a remarkable oil production.

In the present work, the influence of the cationic surfactant cetyltrimethylammonium bromide ($C_{16}TAB$) concentration on oil recovery from limestone by wettability, adsorption, and spontaneous imbibition is examined. Traditionally, literature describes these parameters separately, but when studying them combined one can achieve interesting results.

EXPERIMENTAL PROCEDURES
Materials and Method
Fluids and Rock

The chemicals used in this investigation were the cationic surfactant cetyltrimethylammonium bromide (CTAB), purchased from VETEC, with a purity of 99 %; and potassium chloride, provided by Synth, with a purity of 99 %. Petroleum samples were collected from the Ubarana field (Brazil) and kindly supplied by Petrobras (oil = 0.8344 g/mL specific mass and 2.90 cP viscosity). Limestone samples were extracted from the Jandaíra formation, in the State of Rio Grande do Norte (Brazil), classified as calcite rock.

Corresponding to the brine solution used in the experiments, 2 % (w/v) KCl aqueous solutions were prepared. This saline solution was then used as the solvent of two surfactant systems which were tested, with CTAB concentrations of 0.8596 g/L (below the CMC) and 1.2280 g/L (at the CMC).

Adsorption Assays

Limestone samples were calcined (at 250 °C for 6 hours), crushed, and sieved to be used in the adsorption assays as 100-mesh particles. Analyses of the adsorption phenomena were made with the finite-bath technique in a Dubnoff waterbath. First, the mass of adsorbent (limestone) was varied in samples of 16 g, 24 g, 32 g, 40 g, and 48 g for each solution (2% KCl without CTAB, and solutions of CTAB in 2% KCl at concentrations of 0.8596 g/L and 1.2280 g/L), for 2 hours, under constant stirring at 30 °C. Then, these samples were filtered out and surface tension measurements were acquired with a SensaDyne tensiometer. Following, the surfactant concentration was determined in each sample by plotting an analytical curve with surface tension (η) versus concentration (ln C) data [13]. The best-fit equation was $\eta = -1.7927$ (ln C) + 41.417, with $R^2 = 0.9834$. The adsorption capacity (q, in mg/g) was calculated by equation 1,

$$q = [V(C_0 - C_e)]/m \qquad (1)$$

where: C_0 is the initial surfactant concentration (mg/mL), C_e is the concentration of surfactant in the filtrate (mg/mL); V is the solution volume (mL); and m is the mass of adsorbent (g).

The equilibrium mass was determined for each solution by varying the contact time in intervals of 10, 20, 30, 60, 90, and 120 minutes, under constant stirring, at 30 °C. For each experiment carried out at a different time interval, all samples were filtered out and the surface tension of the filtrate was assessed to determine the surfactant concentration in it. Then, it was possible to establish the equilibrium time for the adsorption. Langmuir and Freundlich isotherms were compared with the experimental data to verify which adsorption model provides the best fit. These theoretical models obey the linearized equations 2 and 3 respectively.

$$C_e/q = 1/(K_L q_m) + C_e/q_m \qquad (2)$$
$$\ln (q) = \ln (K_F) + (1/n) \ln (C_e) \qquad (3)$$

where: q_m (mg/g) is the adsorption capacity; K_L (mL/mgr) is the equilibrium constant; and K_F and n are empirical constants, indicating adsorption capacity and intensity of the adsorption energy respectively.

Contact Angle Measurements

Calcined limestone plugs (at 250 °C for 6 hours) were crushed in fine particles, which were sieved and collected in a 200-mesh plate. By using a hydraulic press, cylindrical tablets were prepared with the aim of reducing roughness, which consists of surface irregularities in a small scale on the limestone surface, and aiding contact angle measurements.

The saturation of the rock tablets was performed with crude oil with density of 0.8344 g/mL and viscosity of 2.90 cP. The tablets were allowed to interact with the oil for 48 hours in an oven kept at 50 °C. After saturation, the tablets were cleansed

with toluene and n-heptane, and their dry masses were registered before and after saturation with oil [14]. Then imbibition was carried out with CTAB solutions in 2% KCl at surfactant concentrations of 0.8596 g/L and 1.2280 g/L, and also in 2 % KCl brine (without surfactant), for 30 minutes, at room temperature (25 °C). For reference, the conditions under which the tablets were tested are shown in Table 1.

Table 1: Imbibition fluids for each tablet.

Tablet	Imbibition fluid
1	KCl 2%
2	
3	
4	
5	CTAB 0.8596 g/L
6	
7	
8	
9	CTAB 1.2280 g/L
10	
11	
12	

Contact angle measurements were acquired with the imbibed tablets using the 2 % KCl solution and crude oil in a Krüss goniometer model DSA100.

Imbibition Assays

Limestone cylindrical plugs with 4.0 cm diameter and 4.0 cm length were previously calcined at 250 °C for 6 hours with the objective of removing all humidity and removing organic matter. Their average porosity was 42%. For the imbibition assays, the plugs were also saturated with oil for 48 hours in an oven at 50 °C. Later, the plugs were cleansed with toluene and n-heptane and, finally, allowed to dry and weighed [14]. Plugs dry masses were also registered before and after saturation with oil.

The spontaneous imbibition assays were carried out with a simple glass cell, vertically disposed and attached to a graduated tube at a higher level. During the imbibition assays, oil-saturated plugs were immersed in the cell containing brine (2% KCl) to simulate conventional oil recovery, with surfactant solutions (CTAB) at the concentrations listed previously (0.8596 g/L and 1.2280 g/L in 2 % KCl) to simulate enhanced oil recovery. All assays were carried out at 25 °C. The volume of oil that was displaced by capillary and gravity forces was obtained by readings along the graduated tube (to an error of 0.1 mL), and recorded every 5 days, during 60 days. The total oil recovery rate (RR), in percentage (%), was calculated with equation 4:

$$RR = (V_{or}/V_{ooip}) * 100 \% \qquad (4)$$

where: V_{or} is the volume of residual oil and V_{ooip} is the original oil in place (in mL).

RESULTS AND DISCUSSIONS
Adsorption

Figure 1 shows the adsorption data acquired when using the surfactant below and at the CMC.

It indicates that adsorption in finite-bath experiments with varying adsorbent mass is more effective at the CMC. According to [15], when studying the adsorption of a cationic surfactant, Cetylpyridinium chloride (CPC), and an anionic surfactant, sodium dodecyl sulfate (SDS), in carbonate materials, verified that the adsorption of cationic surfactants in natural carbonates is linked to the silica composition present in the carbonate samples. There is a strong electrostatic interaction between CPC and the negative poles of silica and clay.

The equilibrium adsorption masses for both surfactant concentrations were established as 40 g of rock sample with 0.8596 g/L and 1.2280 g/L (Figure 1). The equilibrium is observed in the point where the adsorption practically stops varying with the increase in adsorbent mass. After this finding, new finite-bath experiments were devised with the optimized adsorbent mass by changing the contact time. This is depicted in Figure 2, which shows that the variation in the amount of adsorbed matter is minimal within the studied time interval, and therefore the contact time is not a much influent parameter in the investigation.

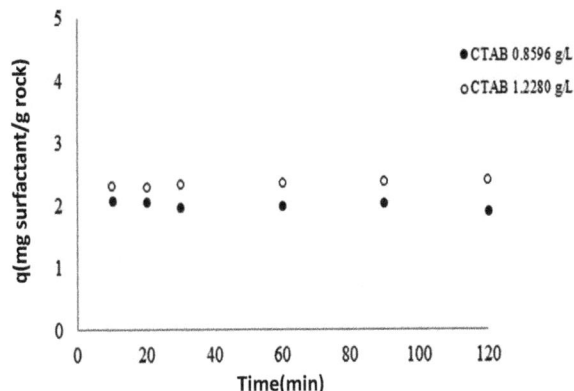

Figure 1: Adsorption data when using CTAB below the CMC (0.8596 g/L) and at the CMC (1.2280 g/L).

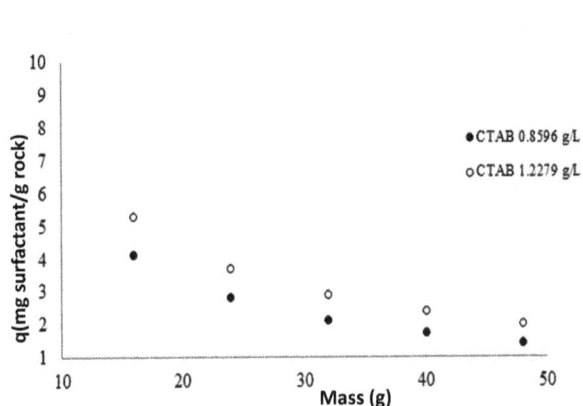

Figure 2: Adsorption of CTAB below and at the CMC as a function of contact time.

Langmuir and Freundlich isotherms were applied to the experimental adsorption data with the surfactant solutions in the limestone samples. Figures 3 and 4 show the fittings of such data with both models.

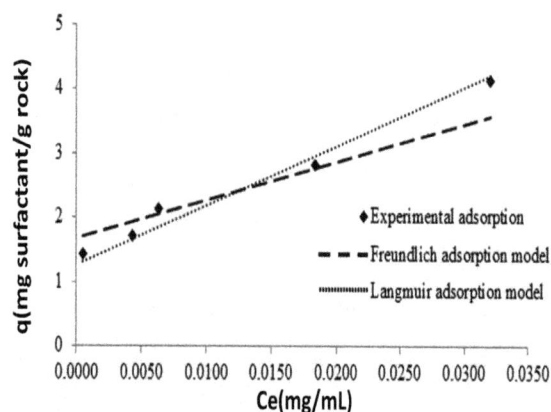

Figure 3: Langmuir and Freundlich isotherms for CTAB below the CMC (0.8596 gr/L).

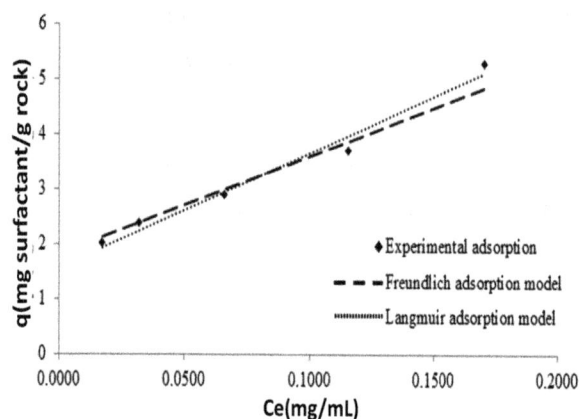

Figure 4: Langmuir and Freundlich isotherms for CTAB at the CMC (1.2280 g/L).

Table 2 shows the equations of the isotherms for the CTAB samples at concentrations below the CMC (0.8596 g/L) and at the CMC (1.2280 g/L).

Table 2: Non-linear isotherm equations for CTAB.

Concentration	Equation	Correlation coefficient
Below the CMC (0.8596 g/L)	Langmuir $q = (4.4228*188.4167*C_e)/(1+188.4167*C_e)$	0.9165
	Freundlich $q = 7.6860*(C_e^{0.2364})$	0.8439
CMC (1.2280 g/L)	Langmuir $q = (6.3694*17.4444*C_e)/(1 + 17.4444*C_e)$	0.8863
	Freundlich $q = 9.3933*(C_e^{0.3948})$	0.9380

The correlation coefficients shown in Table 2 and the plots depicted in Figures 3 and 4 confirm that the experimental data for the surfactant concentration of 0.8596 g/L are better adjusted with the Langmuir model (in this adsorption model, a continuous monolayer of adsorbate molecules is formed surrounding a homogeneous solid surface), also, for the surfactant concentration of 1.2280 g/L, the Freundlich isotherm is a better fit. One can conclude that surfactant adsorption on the reservoir possibly occurs as monolayers in lower concentrations, as suggested by the Langmuir model, with surfactant molecules behaving as monomers. When more surfactant is added, the formation of multilayers is favored, which is suggested by the Freundlich model. One can also point out that micelles were probably formed under such circumstance, since the CMC was reached.

Contact Angle

In the contact angle assays, the tablets were saturated with oil. The experiments analyzed twelve different samples. The amounts of remaining oil in each experiment are shown in Table 3, and the corresponding contact angles are shown in Table 4.

Table 3: Oil volumes in the tablets.

Tablet	Initial mass (g)	Mass after saturation (g)	Mass of remaining oil (g)	Oil volume (mL)
1	0.4094	0.4293	0.0199	0.0240
2	0.3990	0.4281	0.0291	0.0351
3	0.4150	0.4445	0.0295	0.0355
4	0.3917	0.4163	0.0246	0.0296
5	0.3890	0.4150	0.0260	0.0313
6	0.4115	0.4413	0.0298	0.0359
7	0.3865	0.4124	0.0259	0.0312
8	0.4124	0.4423	0.0299	0.0360
9	0.3895	0.4179	0.0284	0.0342
10	0.4014	0.4285	0.0271	0.0326
11	0.3982	0.4269	0.0287	0.0346
12	0.3968	0.4244	0.0276	0.0332

Table 4: Contact angles for the assays with the limestone tablets.

Tablet	Imbibed fluid	Fluid used in the contact angle assay	Contact angle
1	Brine	Brine	126.20
2	Brine	Brine	109.40
3	Brine	Oil	23.00
4	Brine	Oil	26.98
5	CTAB 0.8596 g/L	Brine	119.77
6	CTAB 0.8596 g/L	Brine	119.45
7	CTAB 0.8596 g/L	Oil	26.77
8	CTAB 0.8596 g/L	Oil	25.74
9	CTAB 1.2280 g/L	Brine	118.69
10	CTAB 1.2280 g/L	Brine	120.99
11	CTAB 1.2280 g/L	Oil	27.39
12	CTAB 1.2280 g/L	Oil	20.50

These results show that the rock wettability was not much affected by the surfactant solutions. Moreover, better results would have been obtained if the wettability of oil-wet limestone had changed to water-wet conditions because this would enhance oil recovery by imbibition. For tablets imbibed with 2% KCl, the values indicate that the limestone is oil-wet, since the angles are higher than 90º when brine is used and much lower than 90º when oil is used in the contact angle assays. When imbibition is promoted by the surfactant solutions, the contact angles remained the same, which is an indication that the wettability of the rock is not altered. The ionic character of the surfactant is possibly the most relevant parameter to justify such findings. Adsorption on the rock surface may mainly occur through the ionic moiety of the surfactant molecule; therefore, leaving the nonpolar tails away from more intense interactions with the surface. Then the rock wettability maintains its original oil-wet feature.

Imbibition

In the imbibition assays, it could be verified that the oil recovery is enhanced with CTAB solutions below the CMC, as shown by Figure 5. Oil production could only be observed from the tenth day after imbibition started, and was kept constant until the end of the study.

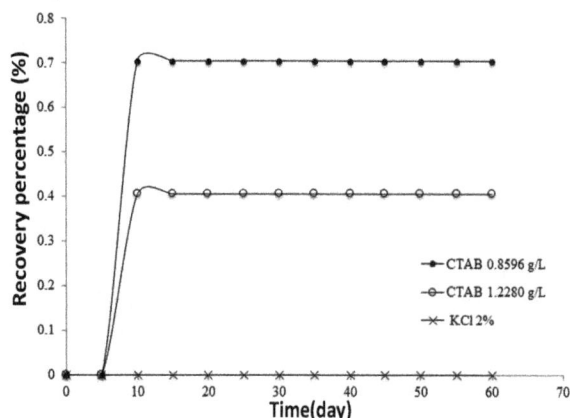

Figure 5: Spontaneous imbibition assays with the limestone in different solutions.

The oil recovery rates were 0.7 % (with CTAB below CMC), 0.4 % (with CTAB at the CMC), and 0 % (with 2 % KCl brine), because fewer surfactant molecules are adsorbed on the rock surface below the CMC, thereby increasing the concentration of free surfactant molecules. The presence of these free surfactant molecules help reduce surface tension. Ultimately, this favors oil recovery because the surfactant at the CMC (1.2280 g/L) is in the micellar form and cannot further decrease surface tension.

CONCLUSIONS

Surfactant concentration is an important parameter to be considered when its effects on adsorption which take place in oil recovery is investigated. This work typically establishes the limits for enhanced surfactant adsorption with increasing surfactant concentration in the case of CTAB-limestone

systems. Surfactant monolayers are formed below the CMC, but multiple layers are favored when more surfactant is added. However, the rock wettability barely is affected by surfactant concentrations, and, in the case of limestone, its original oil-wet character is maintained. In another approach, imbibition assays could confirm that oil recovery is actually affected by surfactant concentration, in that more oil could be extracted below the CMC, possibly due to less intense adsorption between the surfactant molecules and the rock surface.

ACKNOWLEDGMENTS

The authors would like to thank the Brazilian agencies ANP–PRH-PB 221 for the financial support.

NOMENCLATURES

CMC   : Carboxymethyl Cellulose
CTAB : Cetyl Trimethylammonium Bromide
KCl   : Potassium Chloride

REFERENCES

1. Zhou X., Morrow N. R., and Shouxiang M., "Interrelationship of Wettability, Initial Water Saturation, Aging Time, and Oil Recovery by Spontaneous Imbibition and Waterflooding," *Society of Petroleum Engineers*, **2000**, *5*, 199-207.

2. Høgnesen E. J., Standnes D. C., and Austad T., "Scaling Spontaneous Imbibition of Aqueous Surfactant Solution into Preferential Oil-wet Carbonates," *Energy Fuels*, **2004**, *18*, 1665-1675.

3. Babadagli T., "Analysis of Oil Recovery by Spontaneous Imbibition of Surfactant Solution," *Oil and Gas Science and Technology*, **2005**, *60*, 697-710.

4. Karimaie H. and Torsæter O., "Effect of Injection Rate, Initial Water Saturation and Gravity on Water Injection in Slightly Water-wet Fractured

Porous Media," *Journal of Petroleum Science and Engineering,* **2007,** *58,* 293-308.

5. Agbalaka C., Dandekar A. Y., Patil S. L., Khataniar S., and et al., "The Effect of Wettability on Oil Recovery: A Review," *Asia Pacific Oil and Gas Conference and Exhibition,* Perth, Australia, **2008,** 1–13.

6. Al-Attar H. H., "Experimental Study of Spontaneous Capillary Imbibition in Selected Carbonate Core Samples," *Journal of Petroleum Science and Engineering,* **2010,** *70,* 320-326.

7. Standnes D. C. and Austad T., "Wettability Alteration in chalk 2. Mechanism for Wettability Alteration from Oil-wet to Water-wet Using Surfactants," *Journal of Petroleum Science and Engineering,* **2000,** *28,* 123-143.

8. Babadagli T. and Boluk Y., "Oil Recovery Performances of Surfactant Solutions by Capillary Imbibition," *Journal of Colloid and Interface Science,* **2005,** *282,* 162-175.

9. Høgnesen E. J., Olsen M., and Austad T., "Capillary and Gravity Dominated Flow Regimes in Displacement of Oil from an Oil-wet Chalk Using Cationic Surfactant," *Energy Fuels,* **2006,** *20,* 1118-1122.

10. Babadagli T., "Evaluation of the Critical Parameters in Oil Recovery from Fractured Chalks by Surfactant Injection," *Journal of Petroleum Science and Engineering,* **2006,** *54,* 43-54.

11. Pons-Jiménez M., Cartas-Rosado R., Martínez-Magadán J. M., Oviedo-Roa R., and et al., "Theoretical and Experimental Insights on the True Impact of C12TAC Cationic Surfactant in Enhanced Oil Recovery for Heavy Oil Carbonate Reservoirs," *Colloids and Surfaces A: Physicochemical and Engineering Aspects,* **2014,** *455,* 76-91.

12. Karimi M., Al-Maamari R. S., Ayatollahi S., and Mehranbod N., "Wettability Alteration and Oil Recovery by Spontaneous Imbibition of Low Salinity Brine into Carbonates: Impact of Mg^{+2}, SO_4^{-2} and Cationic Surfactant," *Journal of Petroleum Science and Engineering,* **2016,** *147,* 560–569.

13. Castro Dantas T. N., Ferreira Moura E., Scatena Júnior H., Dantas Neto A. A., and et al., "Micellization and Adsorption Thermodynamics of Ovel Ionic Surfactants at Fluid Interfaces," *Colloids and Surfaces A: Physicochemical and Engineering Aspects,* **2002,** *207,* 243–252.

14. Santanna V. C., Castro Dantas T. N., Borges T. A., Bezerril A. R., and et al., "The Influence of Surfactant Solution Injection in Oil Recovery by Spontaneous Imbibition," *Petroleum Science and Technology,* **2014,** *32,* 2896–2902.

15. Ma K., Cui L., Dong Y., Wang T., and et al., "Adsorption of Cationic and Anionic Surfactants on Natural and Synthetic Carbonate Materials," *Journal of Colloid and Interface Science,* **2013,** *408,* 164–172.

16. Jarrahian K., Seiedib O., Sheykhanc M., Vafaie Sefti M., and et al., "Wettability Alteration of Carbonate Rocks by Surfactants: A Mechanistic Study," *Colloids and Surfaces A: Physicochemical and Engineering Aspects,* **2012,** *410,* 1–10.

17. Wang Y., Ge J., Zhang G., Jiang P., and et al., "Adsorption Behavior of Dodecyl Hydroxypropyl Sulfobetaine on Limestone in High Salinity Water," *Royal Society of Chemistry,* **2015,** *5,* 59738–59744.

13

Prediction of Electrofacies Based on Flow Units Using NMR Data and SVM Method

Mahdi Rastegarnia[1], Mehdi Talebpour[2], Ali Sanati[3], and Seyed Hassan Hajiabadi[3]*

[1] Department of Petrophysics, Pars Petro Zagros Engineering and Services Company, Tehran, Iran
[2] Department of Petroleum Engineering, Islamic Azad University, Science and Research Branch, Tehran, Iran
[3] Faculty of Petrochemical and Petroleum Engineering, Hakim Sabzevari University, Sabzevar, Iran

ABSTRACT

The classification of well-log responses into separate flow units for generating local permeability models is often used to predict the spatial distribution of permeability in heterogeneous reservoirs. The present research can be divided into two parts; first, the nuclear magnetic resonance (NMR) log parameters are employed for developing a relationship between relaxation time and reservoir porosity as well as introducing the concept of relaxation group. This concept is then used for the definition of electrofacies in the studied reservoir. A graph-based clustering method, known as multi resolution graph-based clustering (MRGC), was employed to classify and obtain the optimum number of electrofacies. The results show that the samples with similar NMR relaxation characteristics were classified as similar groups. In the second part of the study, the capabilities of nonlinear support vector machine as an intelligent model is employed to predict the electrofacies and permeability distribution in the entire interval of the reservoir, where the NMR log parameters are unavailable. SVM prediction results were compared with laboratory core measurements, and permeability was calculated from stoneley wave analysis to verify the performance of the model. The predicted results are in good agreement with the measured parameters, which proves that SVM is a reliable tool for the identification of electrofacies through the conventional well log data.

Keywords: Flow Zone Index, Electrofacies, Support Vector Machine, Nuclear Magnetic Resonance, Conventional Petrophysical Data

INTRODUCTION

Geological and petrophysical properties play a significant role in fluid flow in the petroleum reservoirs. Hydraulic flow units are defined as correlatable and mappable zones within a reservoir which control fluid flow [1]. Flow units can be used to correlate important reservoir petrophysical parameters such as porosity and permeability and to facilitate field scale reservoir modeling tasks. Laboratory core measurement is the oldest techniques for obtaining porosity and permeability. Expensive costs and long times are the disadvantages of laboratory measurements. Meanwhile, it presents discontinuous information from subsurface formations. On the other hand, well logs can provide a continuous measurement of petrophysical characteristics of the subsurface

*Corresponding author

Seyed Hassan Hajiabadi
Email: s.h.hajiabadi@hsu.ac.ir

formations along the wellbore, and a wide range of geological characteristics of formations can be extracted from well log data. The present research aims to benefits from the advantage of well logs instead of expensive and time consuming core measurements. The identification of flow units through well log data is a cost-effective and potent method. As opposed to petrophysical parameters, lithology characterizations accompany with various challenges where a high degree of heterogeneity in rock formations is observed. In other words, the higher the heterogeneity of the reservoir is, the harder the predictions would be. Several researchers have identified flow units based on modeling workflows using the capabilities of intelligent models such as neural network (NN) models, principal component analysis (PCA), clustering methods, fuzzy logic (FL), hierarchical techniques, classification methods, and optimization algorithms [2-9]. Among various intelligent techniques, the present research selects support vector machines (SVM) to design a predictor model.

Studying previous literature shows that support vector machines (SVM) have demonstrated good generalization performance in many real-life applications [10-18]. This technique is a supervised machine learning algorithm based on the statistical learning theory developed by Vapnik [19]. SVM generates mapping functions from a set of labeled training data. On the other hand, SVM method is based on structural risk minimization principle, which in turn is based on the statistical learning theory. Statistical learning enhancement gives better generalization abilities by minimizing the testing error. SVM method can be used for nonlinear classification where kernel functions are used to map the input space into a higher-dimensional feature space. This effectively maps the non-linearity of the relationship to a linear one. SVM learning algorithm allows the sparse representation of models by considering only

a portion of available training points which generate the support vectors [20]. Quadratic programming optimization is used to solve SVM formulation in dual space and nonlinear decision, and indicator hyper surfaces are then constructed in a functional form to classify new data and predict corresponding values [21]. SVM has been successfully used in a number of applications such as radar target detection [10], face detection [22], hand writing [11], text detection [13], speech recognition [14], financial time series prediction [15], porosity prediction [23], and lithology classification [16]; permeability reconstruction based on well log data [18], [23, 24] used SVM to estimate porosity using several well-log measurements, and it was demonstrated that SVM could be used instead of back-propagation neural network to predict permeability. For further details, the basic concepts of SVM are explained in the next section.

Nuclear magnetic resonance (NMR) log is one of the potent tools made a revolution in the measurement of reservoir properties which had been impossible to measure previously by logging instruments. Permeability and free and bulk fluid of rock, which was previously measured through time consuming laboratory measurements, can now be identified by NMR log without the necessity of expensive operation of coring.

In this research, the NMR responses are used to obtain an optimum number of electrofacies based on flow units. Data used in this study are related to Asmari reservoir in the Cheshmeh Khush oil field, Southern Iran (Figure 1). There are 3 wells, namely well A, B, and C. Well A has NMR and petrophysical evaluation results, and wells B and C have conventional data and petrophysical evaluation results. In wells B and C, permeability was calculated from stoneley wave analysis and core data.

The electrofacies modeling was first applied to well

A. Then, the defined electrofacies in well A were propagated in wells B and C by SVM method. For verifying these electrofacies, the results of SVM were compared with the results of petrophysical evaluation and core data as well as stoneley permeability in wells B and C; good agreement was achieved.

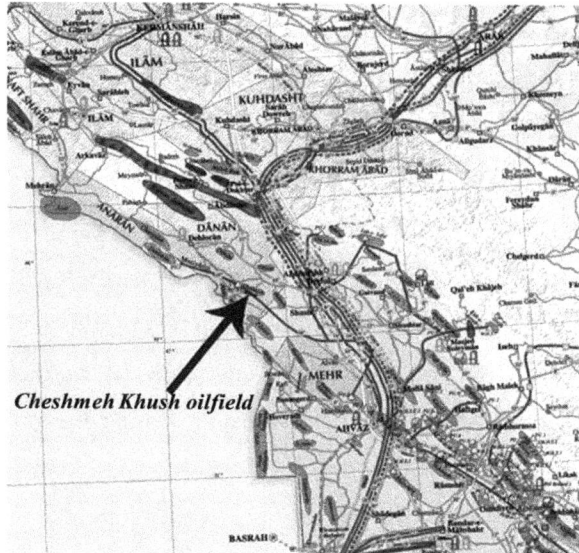

Figure 1: Location of the Cheshmeh Khush oilfield presented on the map.

EXPERIMENTAL PROCEUDRES
Support Vector Machine

SVM is a good procedure, which is able to deal with linear situations, always with the assumption that the data are separable without misclassifications by a linear hyper plane. The optimality criterion is to put the hyper plane as far as possible from the nearest samples, and keeping all the samples in their correct side. This means that maximum margin should be between the separating hyper plane and its nearest samples, and the margin hyper planes $w^T x + b = \pm 1$ must be placed into the separation margin; w is the optimum weight of the vector, and x and b are the bios of the model; T is the transpose notation. Consequently, SVM criterion is to maximize the distance between the separating

hyper plane and the nearest samples subject to the constraints. The following equation describes the performance of SVM in a linear system (in the classification, it can be so called SVC)

$$L_D = -\frac{1}{2}\sum_{i=1}^{N}\sum_{j=1}^{N}\alpha_i\alpha_j y_i y_j x_i^T x_j + \sum_{i=1}^{N}\alpha_i \qquad (1)$$

where, LD is the notation of the basic equation of SVM in the linear system; α_i and α_j are the Lagrange multipliers; y_i and y_j are the binary scalar values; x_i and x_j are the input vector samples, and N is the number of samples.

Figure 2 shows the performance of SVM as a classifier in a linear system. The basic idea is that vectors x in a finite dimension space (called input space) can be mapped to a higher (possibly infinite) dimensional Hilbert space. A linear machine can be constructed in a higher dimensional space (often called the feature space), but it stays non-linear in the input space. Most of the transformations are unknown, but the dot product of the corresponding spaces can be expressed as a function of the input vectors as $\phi(x_i, x_j) = k(x_i, x_j)$. These spaces are called reproducing kernel Hilbert spaces (RKHS), and their dot products $k(x_i, x_j)$ are called Mercer kernels. The most common kernels, which are popular and used in many problems, are shown in the Table 1 [25].

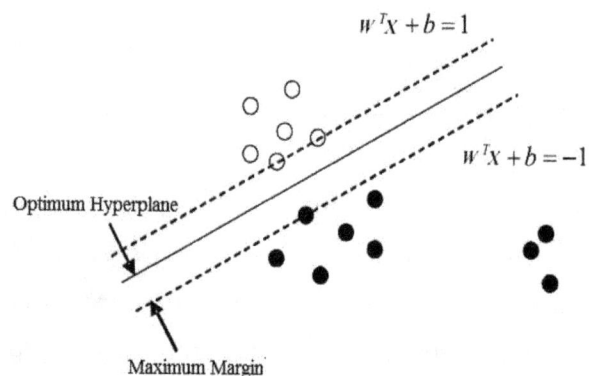

Figure 2: Performance of SVM in a linear system.

Table 1: Kernel and the type of classifier.

Kernel Function	Type of classifier
$K(x_i, x_j) = (x_i^\mathrm{T} x_j)^\rho$	Linear
$K(x_i, x_j) = (x_i^\mathrm{T} x_j + 1)^\rho$	Complete polynomial of degree ρ
$K(x_i, x_j) = e^{\frac{\|x_i - x_j\|^2}{2\sigma^2}}$	Gaussian, radial basis function (RBF)
$K(x_i, x_j) = \tanh(\gamma x_i^\mathrm{T} x_j + \mu)$	Multilayer Perceptron
$K(x_i, x_j) = \dfrac{\sin\left[(n+1/2)(x_i - x_j)\right]}{2\sin[(x_i - x_j)/2]}$	Dirichlet

Construction of a Nonlinear SVM

The solution for a linear SVM is indicated by a linear combination of the training data of $w = \sum_{i=1}^{N} y_i \alpha_i x_i$ to map the data into a Hilbert space, and then the solution is given by:

$$w = \sum_{i=1}^{N} y_i \alpha_i \varphi(x_i) \tag{2}$$

where, $\phi(x_i)$ is the mapping function. The parameter vector $\alpha_j < C$ is a combination of vectors into the Hilbert space, but many transformations $\phi(x_i)$ are unknown; however, the problem can still be solved because the machine just needs the dot products of the vector. We cannot use

$$y_i = w^\mathrm{T} \varphi(x_i) + b \tag{3}$$

since w parameters are in an infinite dimensional space; therefore, no expression exists for them. However, by substituting Equation 2 into Equation 1, the below equation is obtained.

$$y_i = \sum_{i=1}^{N} y_i \alpha_i \phi(x_i)^T \phi(x_j) + b = \sum_{i=1}^{N} y_i \alpha_i k(x_i, x_j) + b \tag{4}$$

The resulting machine can now be expressed directly in terms of the Lagrange multipliers and the kernel dot products. The kernel is used to compute this matrix $k_{ij} = k(x_i, x_j)$. When this matrix is computed, solving a non-linear SVM will be so easy. In order to compute the bias b, we can use $y_i(w^T x_i + b) - 1 = 0$, but for the nonlinear SVM, it changes to Equations 5 and 6.

$$y_i\left(\sum_{i=1}^{N} y_i \alpha_i \phi(x_i)^T \phi(x_j) + b\right) - 1 = 0 \tag{5}$$

$$y_i\left(\sum_{i=1}^{N} y_i \alpha_i k(x_i, x_j) + b\right) - 1 = 0 \tag{6}$$

for all x_j for which $\alpha_j < C$.

We just need to calculate b from the above expressions and calculate the mean value for all the samples with $\alpha_j < C$. The idea of a non-linear SVM can be seen in Figure 3.

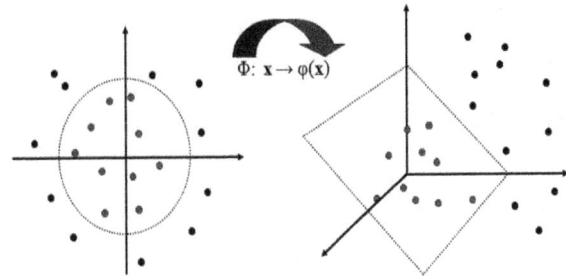

Figure 3: Original feature space can always be mapped to some higher-dimensional feature space by SVM.

Significance of NMR Logging in Reservoir Characterization

The importance and application of nuclear magnetic resonance (NMR) is not only limited to reservoir evaluation, but it is widely used in physics, chemistry, biology, and medicine. Combining the permanent magnets and pulsed radio frequencies with the concept of logging led to an important logging tool today known as NMR log in the world of petrophysicists. An applicable instrument of NMR logging was introduced by taking the benefits of a medicine magnetic resonance imaging (MRI) (6). This tool was the first NMR log that could be run into the formation rather that placing the rock sample in the instrument. Afterwards, several revisions were considered on his tool, and it was improved day by day; however, the fundamental of all NMR logs is the same, where a magnetic field magnetized formation material using timed bursts of radio-frequency energy. The transmitted energy polarizes the spin axes of unaligned formation protons in a specific direction. By removing the transmitted magnetic field the protons start to align in their original direction. The time in which protons come back to their previous alignment is known as echo time. The amplitude of spin-echo trains are measured as a function of time, which is directly related to the number of hydrogen atoms of the fluid formation. Two time-based parameters of longitudinal relaxation time (T_1) and apparent relaxation time (T_2) are obtained from this step, and they include the most crucial outputs of NMR log leading to further properties of the formation. T_1 and T_2 indicate the time in which protons relax longitudinally and transversely respectively related to the transmitted magnetic field. In fact, T_2 is the most important parameter that can be directly converted to porosity. The T_2 plot includes movable and immovable fluids of the rock, which are separated based on a cut-off value of T_2. Only fluids are visible for transmitted magnetic field on NMR tool, and this fact redounds to one of the advantages of NMR log compared to conventional logs such as sonic, neutron, and bulk-density logs. In fact, the porosity measured by NMR log is not influenced from the matrix materials. In other words, NMR porosity is lithology-independent and does not need to be calibrated with lithology in different zones and intervals of a well. Three groups of invaluable information about reservoir condition can be obtained from NMR raw data, including pore size distribution of a formation, fluids properties of pore spaces, and finally quantities of these fluids. Porosity and pore size distribution are theoretically related to permeability

in a direct relationship. Therefore, permeability and movable fluids (free fluid index) of the formation can be estimated using the aforementioned raw data. In the next section, the relation of T_2 and FZI is discussed.

Relation between NMR Log Responses and FZI

The fundamental equation which can relate the apparent relaxation rate of a single pore in the porous media is given by:

$$\frac{1}{T_2} = \frac{1}{T_{2b}} + \rho \frac{S}{V} \tag{7}$$

where, T_2 is the observed transverse relaxation time, and T_{2b} is the relaxation time of bulk fluid; ρ is relaxation surface, and S represents the surface area of the pore; V stands for the volume of the pore body. Since T_{2b} is significantly greater than the relaxation time (T_2) in the porous medium, the above equation can be reduced to:

$$\frac{1}{T} = \rho \frac{S}{V} \tag{8}$$

Georgi and Menger [26] proposed a relation between the surface area to volume ratio of the pore space, porosity, and the specific surface per grain volume ratio (S_{gv}). This relation was also used by Ohen et al. [27] and is given below:

$$\frac{S}{V} = S_{gv} \frac{1-\varphi}{\varphi} \tag{9}$$

Thus, Equation 7 can be written as:

$$\rho T_2 = \frac{\varphi}{[S_{gv}(1-\varphi)]} \tag{10}$$

or:

$$\frac{T_2}{\varphi_z} = \frac{1}{\rho S_{gv}} \tag{11}$$

Hence, Equations 10 and 11 can be related to RQI by:

$$RQI = \varphi_z.FZI = [\frac{\varphi}{1-\varphi}][\frac{1}{\tau S_{gv}\sqrt{F_s}}]$$

$$= \frac{\rho T_2}{(\tau\sqrt{F_s})} \tag{12}$$

This suggests that RQI is related to relaxation time and porosity group (ϕ_z). A logarithmic equation can be written using the above equations:

$$\log T_2 = \log(\varphi_z) + \log[\frac{1}{\rho S_{gv}}] \tag{13}$$

This equation relates relaxation time to porosity and forms the basis of relaxation group concept, which can be compared to the hydraulic unit concept. Equation 12 classifies samples which exhibit similar NMR relaxation (rock-fluid interaction) characteristics into a group. The average parameters for the group serve as calibration points for interpreting NMR logs. The factor $\frac{1}{\rho S_{gv}}$ is often recognized as relaxation product, and represents the relaxation power and textural attributes of the formation. As implied from Equation 13, the logarithmic plot of T_2 versus ϕ_z would result in a slope line, and $\frac{1}{\rho S_{gv}}$ would be constant for all data points on this slope line.

Formation rock samples or intervals with similar NMR relaxation characteristics lend themselves to the same group, with their $\log T_2$ versus $\log \phi_z$ clustering around the intercepting slope line. In this study, the samples or intervals of a formation or reservoir with similar NMR characteristics were considered in a similar relaxation group.

CASE STUDY
Methodology

Carbonate formations generally have wide pore size distributions, which ranges from microcrystalline pores to large vugs. Understanding the pore distribution and their geometries is vital for reservoir characterization. T_2 distributions are usually correlated with pore size and pore size distribution, so this fact is used for relating NMR responses to permeability. Micro, macro, and vuggy porosity are related to the NMR responses, which can be used to determine the effect of these types of porosities on permeability. NMR responses are calibrated for predicting the irreducible bound fluid volume (BVI) using a variable T_2 cutoff, which can be derived from experimental measurements. By comparing the T_2 distributions of the fully saturated cores and the cores with irreducible saturation, a T_2 cutoff separating is obtained for the pore space, which participates in fluid migration from the non-participating pore space. The original method for computing permeability from NMR well logs using a fixed T_2 cutoff of 90 milli-seconds (ms) produces unreliable results in the carbonate formation. However, this method and the presence of isolated secondary porosities such as moldic and vuggy pores in carbonate could vary from 90 ms to 700 ms due to lithology heterogeneity and complex pore reservoirs. Therefore, several measurements are used to estimate the T_2 cut-off values for well log calibration in carbonates. Studying thin section, using scanning electron microscope (SEM) and computerized tomography (CT) image analysis along with the core data provides necessary information to evaluate T_2 cut-off method. Finally, this information can be employed to characterize vuggy carbonate lithology [28].

In this research, the spectral BVI (SBVI) method has been applied to determining permeability by using NMR data in well A. In this method, each pore size observed in the 100% brine-saturated spectra is assumed to contain some bound water. In this study, a two-step approach is proposed for electrofacies prediction from conventional logs data as explained below:

a) Determining optimum number of flow units using NMR data and MRGC method in well A.

b) Predicting electrofacies based on flow units from conventional logs using a support vector machine neural network (in wells B and C).

Multi Resolution Graph-based Clustering (MRGC)

Multi resolution graph-based clustering (MRGC) is a multi-dimensional dot-pattern recognition method based on non-parametric K-nearest-neighbor and graph data representation [29]. MRGC automatically determines the optimal number of clusters; also MRGC allows the geologist to control the level of detail actually needed to define the flow units. The underlying structure of the data is analyzed, and natural data groups are formed that may have very different densities, sizes, shapes, and relative separations. In this paper, the optimum number of flow units was found by MRGC method. The flow unit analysis was performed by clustering similar permeability, porosity, and volume T_2 distributions according to $\frac{1}{\rho S_{gv}}$.

Figure 4 shows the results obtained from MRGC clustering technique for limestone. Figure 4 presents the variations of relaxation time for each extracted flow unit.

Figure 5 shows the facies obtained after MRGC clustering method applied to the NMR data of well A. As shown, there is acceptable agreement among permeability log estimated by full wave sonic (PERM-ST), oil content, water saturation, and T_2, indicating the reliability of the technique used for clustering. Since the studied reservoir is a complex formation, clustering was separately used for both carbonate and sandstone lithological units. The variation of each parameter is shown in Figures 6 and 7.

Figure 8 displays the variation of grain size (T2LM) in carbonate and sandstone intervals. Facies No. 5 and No. 4 of carbonate intervals can be merged together as they have a suitable overlapping. This can also be considered for facies No. 5 and No. 6 of sandstone interval. Figure 9 shows the variation of grain size (T2LM) versus effective porosity in carbonate and sandstone intervals, where relaxation groups are distinguished.

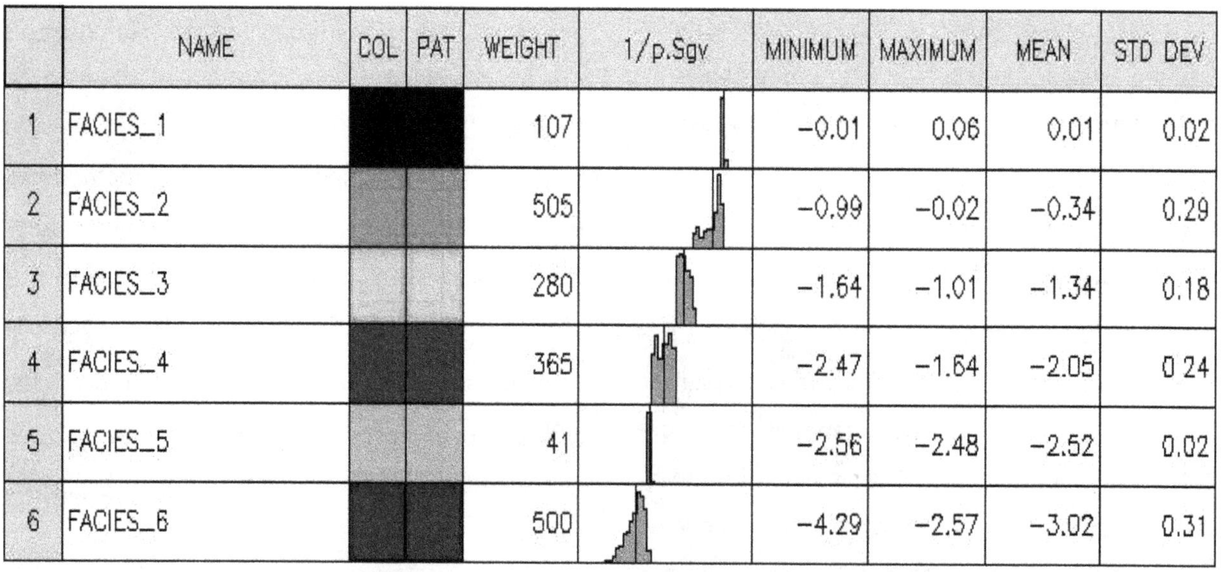

	NAME	COL	PAT	WEIGHT	1/p.Sgv	MINIMUM	MAXIMUM	MEAN	STD DEV
1	FACIES_1			107		−0.01	0.06	0.01	0.02
2	FACIES_2			505		−0.99	−0.02	−0.34	0.29
3	FACIES_3			280		−1.64	−1.01	−1.34	0.18
4	FACIES_4			365		−2.47	−1.64	−2.05	0 24
5	FACIES_5			41		−2.56	−2.48	−2.52	0.02
6	FACIES_6			500		−4.29	−2.57	−3.02	0.31

Figure 4: Result of MRGC clustering in well A.

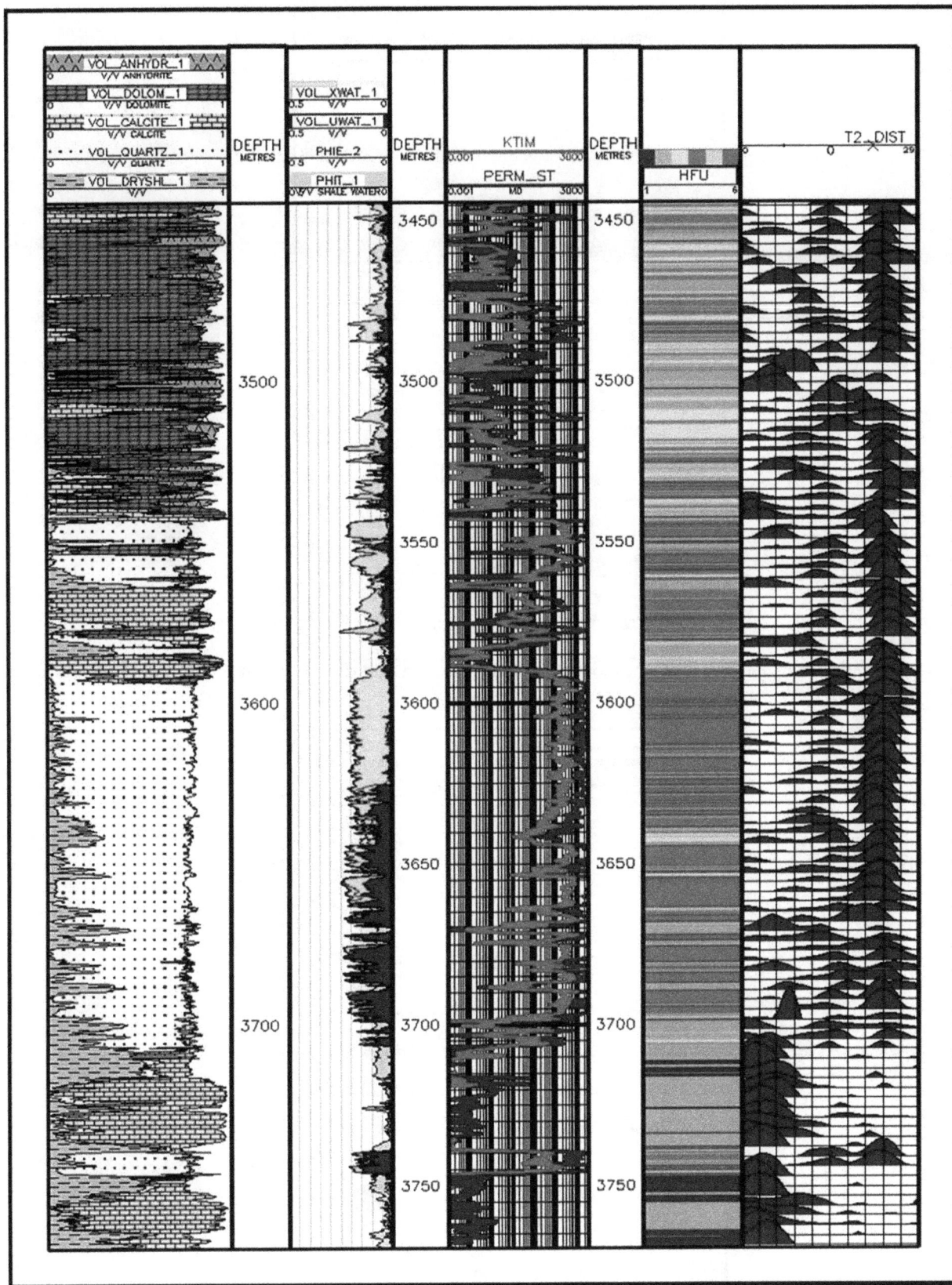

Figure 5: Facies obtained after applying MRGC clustering to well A.

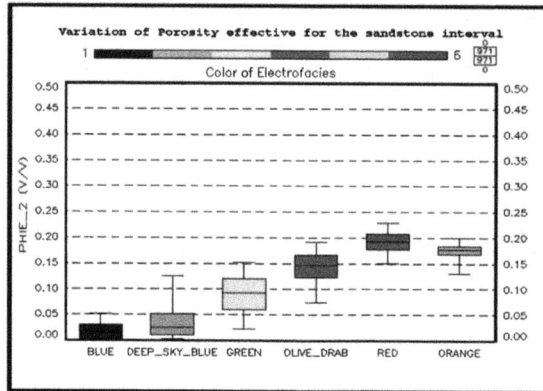

Figure 6A: Variation of effective Porosity for sandstone interval showing that facies No. 5 (Orange) and No. 6 (Red) can be merged together as they have a suitable overlapping.

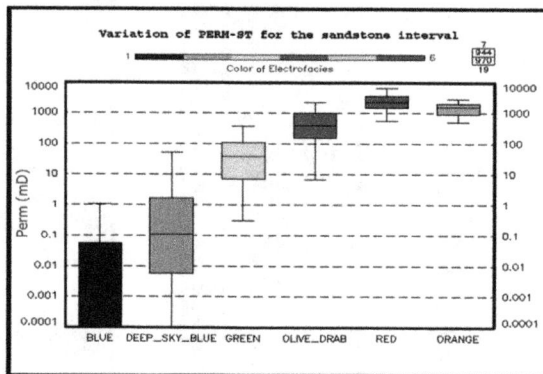

Figure 6B: Variation of permeability for sandstone interval showing that that facies No. 5 (Orange) and No. 6 (Red) have high permeability in value.

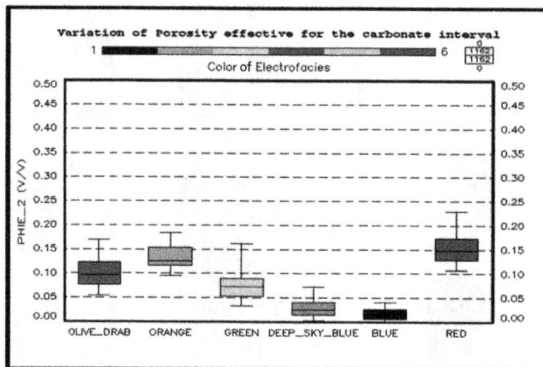

Figure 7A: Variation of effective Porosity for the carbonate interval showing that facies No. 5 (Orange) and No. 4 (Olive_Darb) can be merged together as they have a suitable overlapping.

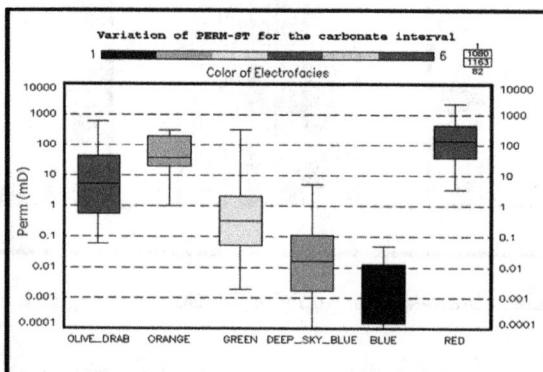

Figure 7B: Variation of permeability for the carbonate interval showing that facies No. 5 (Orange) and No. 4 (Olive_Darb) have high permeability in value.

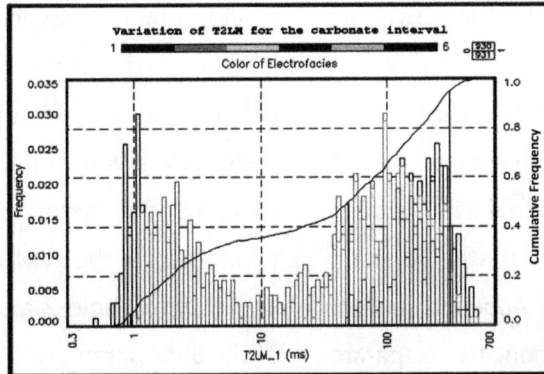

Figure 8A: Variation of grain size (T2LM) for the carbonate interval showing that facies No. 5 (Orange) and No. 4 (Olive_Darb) have a suitable overlapping.

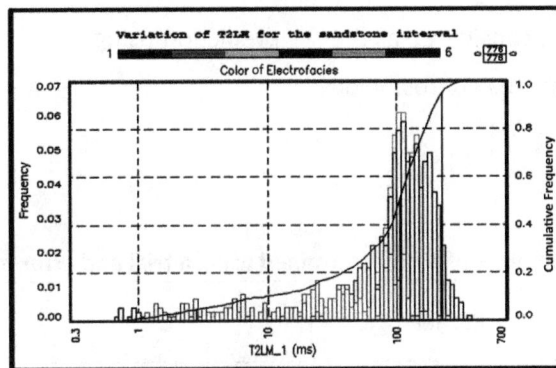

Figure 8B: Variation of grain size for the sandstone showing that facies No. 5 (Orange) and No. 6 (Red) have a suitable overlapping.

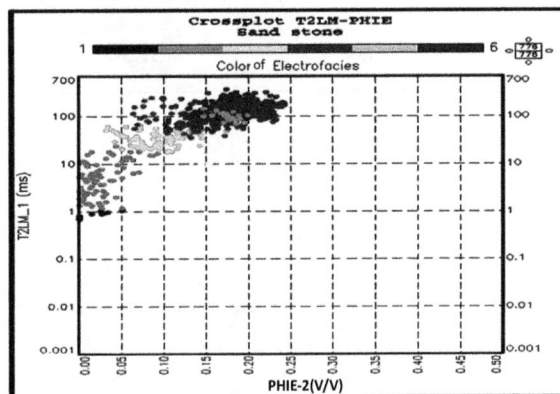

Figure 9A: Cross plot of MRGC clustering results of NMR log measurements to identify facies for sandstone interval.

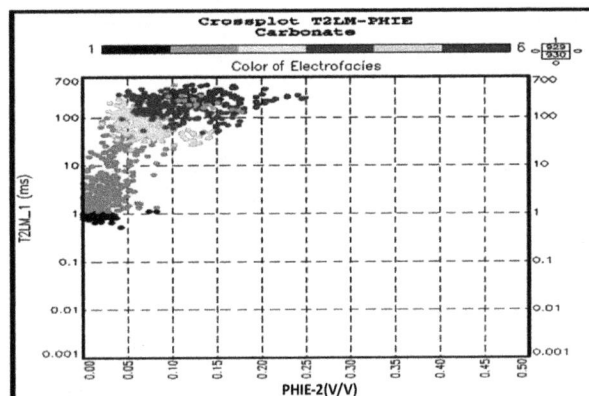

Figure 9B: Cross plot of MRGC clustering results of NMR log measurements to identify facies for carbonate interval.

Prediction of Electrofacies Using Support Vector Machine

In this step, those facies obtained during clustering by MRGC were taken into consideration to be predicted by support vector machine (SVM). Since the background of SVM has been extensively presented in the literature, we brought the background in the Appendix section. To run the SVM for this prediction, two separate models were built for the sandstone and carbonate intervals. For this purpose, porosity log (PHIE), total porosity (PHIT), and DT logs were considered as the inputs to carbonate intervals whereas porosity log

(PHIE), total Porosity log (PHIT), DT log, and RHOB log were taken as the inputs for the estimation of electrofacies in sandstone intervals. Available data were normalized to be in the range of 0 and 1. A trial and error method was used to determine the parameters of the SVM, including parameters C and kernel type. Table 2 tabulates the optimum value of the SVM parameters determined during the trial and error methods. Tables 3 and 4 respectively list the confusion matrix showing the results of prediction using the SVM method.

Table 2: Parameters of SVM determined using a trial and error method.

Type reservoir	Kernel	C	Accuracy
Sand stone	Gaussian	50	83.62%
Carbonate	Gaussian	3200	78.81%

Table 3: Confusing matrix of carbonate interval showing the prediction results obtained by SVM.

		Carbonate				
		Predicted				
		Electrofacies 1	Electrofacies 2	Electrofacies 3	Electrofacies 4	Electrofacies 5
Actual	Electrofacies 1	8	6	0	0	0
	Electrofacies 2	2	51	3	2	0
	Electrofacies 3	0	3	21	6	1
	Electrofacies 4	0	0	4	24	1
	Electrofacies 5	0	0	0	4	15

Table 4: Confusing matrix of sandstone interval showing the prediction results obtained by SVM.

		Sand stone				
		Predicted				
		Electrofacies 1	Electrofacies 2	Electrofacies 3	Electrofacies 4	Electrofacies 5
Actual	Electrofacies 1	1	1	0	0	0
	Electrofacies 2	0	11	3	0	0
	Electrofacies 3	0	1	8	3	0
	Electrofacies 4	0	0	2	16	4
	Electrofacies 5	0	0	0	5	61

In addition, the accuracy and recall values of each electrofacies predicted by SVM were calculated and are presented in Tables 5 and 6.

After achieving the suitable results during the training and testing of SVM in the prediction of electrofacies, the SVM was used to predict the electrofacies of wells B and C. Figures 10 and 11 show the results of SVM prediction in wells B and C. The results of prediction in wells B and C were calibrated using the results obtained using core and full wave sonic data as well as petrophysical evaluation. The results show acceptable agreements between the results of SVM prediction and those of the petrophysical evaluations.

Table 5: Accuracy and recall values of each electrofacies in the carbonate interval predicted by SVM.

Carbonate	Electrofacies 1	Electrofacies 2	Electrofacies 3	Electrofacies 4	Electrofacies 5
Precision	0.8	0.85	0.75	0.667	0.882
Recall	0.57	0.864	0.677	0.827	0.789

Table 6: Accuracy and recall values of each electrofacies in the sandstone interval predicted by SVM.

Sand stone	Electrofacies 1	Electrofacies 2	Electrofacies 3	Electrofacies 4	Electrofacies 5
Precision	1	0.846	0.615	0.667	0.8938
Recall	0.5	0.786	0.667	0.727	0.924

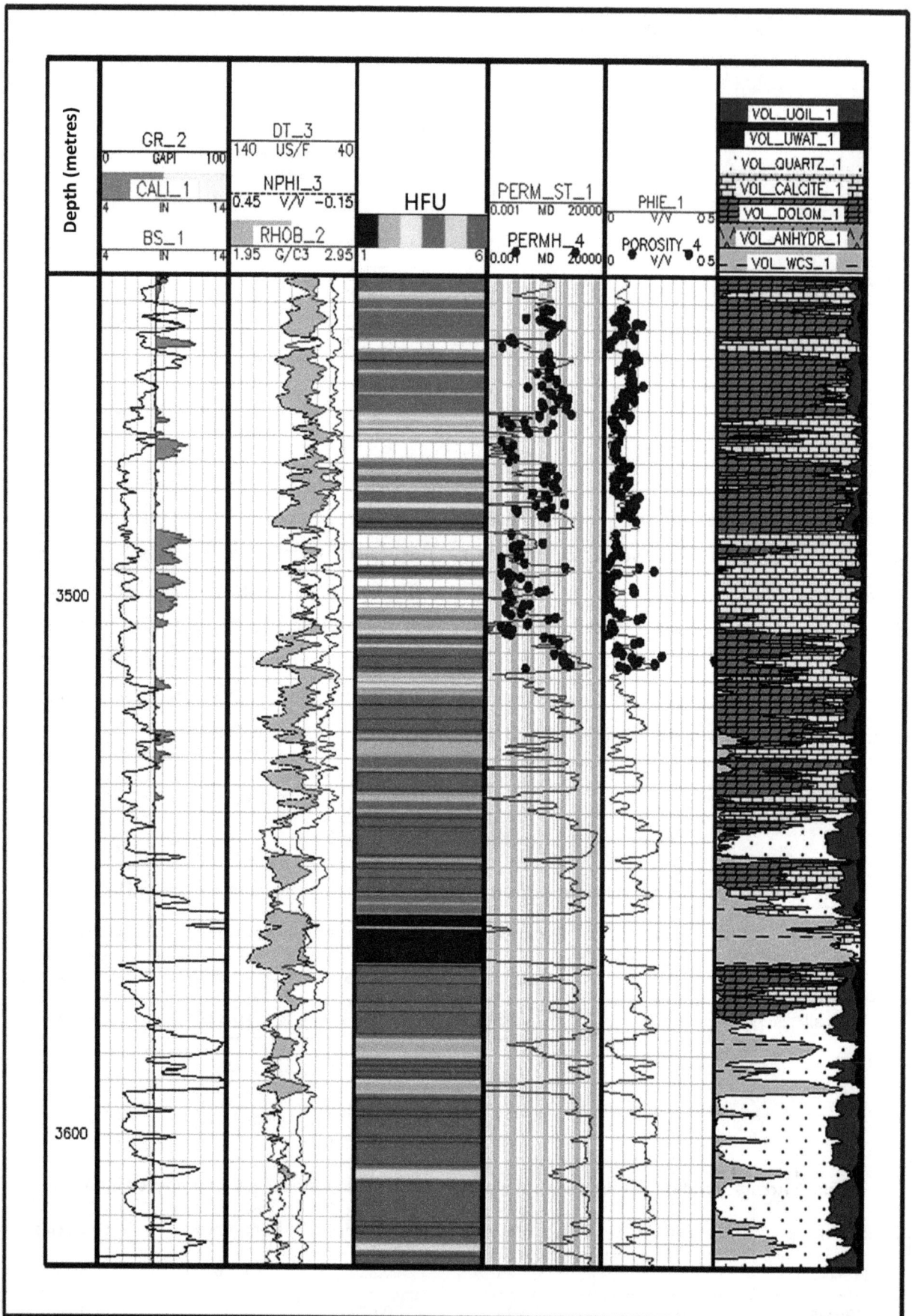

Figure 10: Electrofacies predicted by SVM in well B.

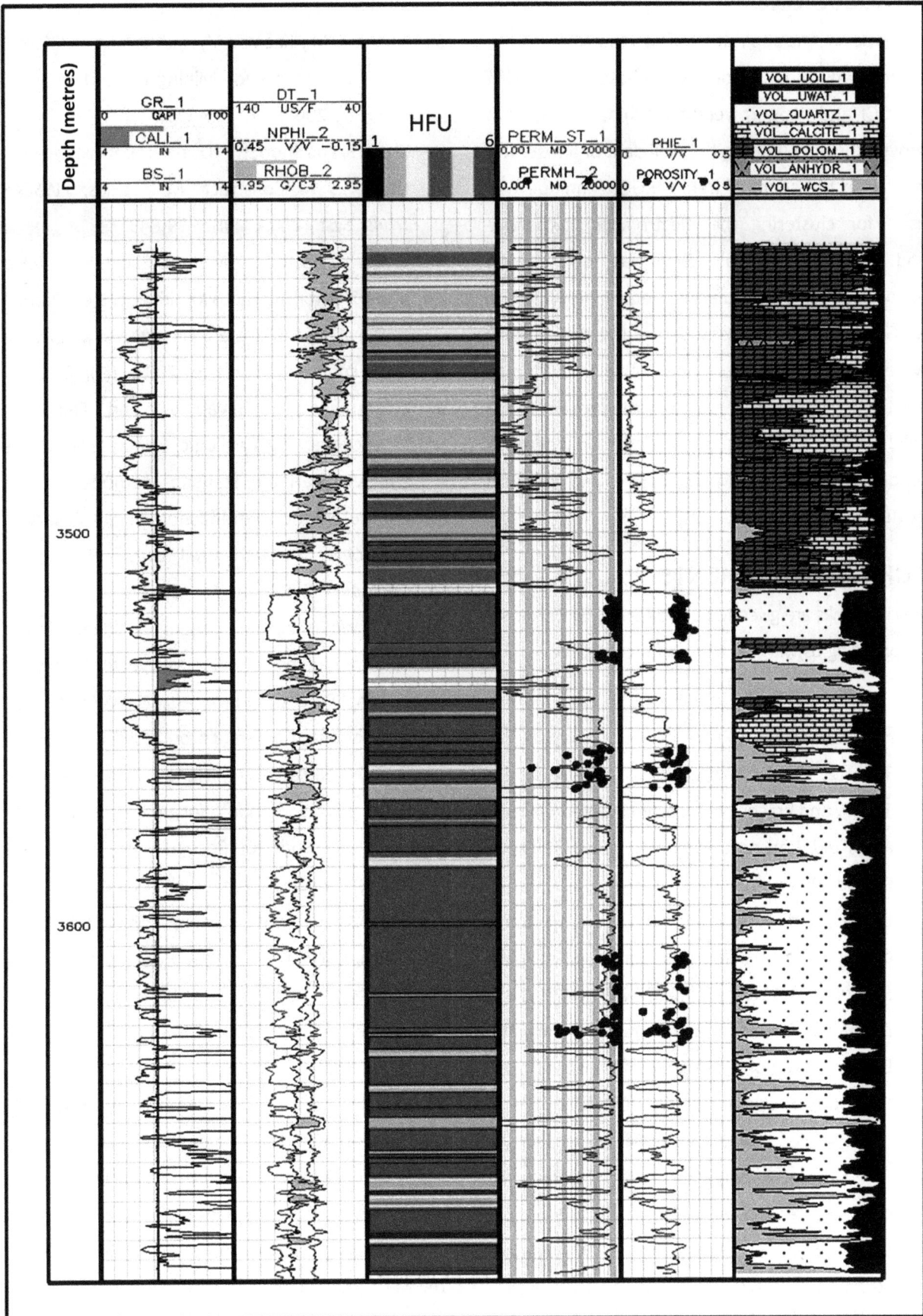

Figure 11: Electrofacies predicted by SVM in well C.

CONCLUSIONS

In this paper, the application of two robust methods, namely multi-resolution graph-based clustering (MRGC) and support vector machine (SVM), was proven through utilizing the well data of a reservoir located in the south of Iran. MRGC technique was used for clustering the electrofacies carbonate and sandstone intervals. SVM was then utilized to predict the electrofacies determined by the MRGC technique. Porosity, sonic, and density logs were considered as the input to the prediction of electrofacies, and the results have proved to be valid since there was a good match between the prediction results of SVM and those of petrophysical evaluation.

ACKNOWLEDGEMENTS

Special thanks are given to Iranian Central Oil Company for providing us with the data and cooperation.

REFERENCES

1. Tiab D. and Donaldson E. C., "Petrophysics Theory and Practice of Measuring Reservoir Rock and Fluid Transport Properties," *Elsevier Press*, Oxford, **2004**, 112-132.

2. Busch J., Fortney W., and Berry L., "Determination of Lithology from Well Logs by Statistical Analysis," *SPE Formation Evaluation*, **1987**, *2*(04), 412-418.

3. Delfiner P., Peyret O., and Serra O., "Automatic Determination of Lithology from Well Logs," *SPE Formation Evaluation*, **1987**, *2*(03), 303-310.

4. El-Sheikh T. S. and Syiam M., "An Efficient Technique for Lithology Classification," *Geoscience and Remote Sensing*, **1989**, *27*(5), 629-632.

5. Lim J. S., Kang J. M., and Kim J., "Multivariate Statistical Analysis for Automatic Electrofacies Determination from Well log Measurements," in *Asia Pacific Oil & Gas Conference & Exhibition*, **1997.**

6. Huang T. M., Kecman V., and Kopriva I., "Kernel Based Algorithms for Mining Huge Data Sets," *Springer*, **2006**, *17*, 125-173.

7. Chikhi S. and Shout H., "Using Probabilistic Neural Networks to Construct Well Facies," *WSEAS Trans. Syst.*, **2003**, *2*(4), 839-843.

8. Chikhi S., Batouche M., and Shout H., "Hybrid neural Network Methods for Lithology Identification in the Algerian Sahara," *International Journal of Computational Intelligence*, **2005**, *1*(1), 25-33.

9. Carrasquilla A., Silvab J., and Flexa R., "Associating Fuzzy Logic, Neural Networks and Multivariable Statistic Methodologies in the Automatic Identification of Oil Reservoir Lithologies through Well Logs," *Revista de Geologia*, **2008**, *21*(1), 27-34.

10. Li Z., Weida Z., and Licheng J., "Radar Target Recognition Based on Support Vector Machine," in Signal Processing Proceedings, WCCC-ICSP, 5th International Conference on. 2000. IEEE, **2000**.

11. Choisy C. and Belaid A., "Handwriting Recognition Using Local Methods for Normalization and Global Methods for Recognition in Document Analysis and Recognition," Proceedings. 6th International Conference on. 2000. IEEE, **2001**.

12. Gao J., Harris C. J., and Gunn S. R., "On a Class of Support Vector Kernels Based on Frames in Function Hilbert Spaces," *Neural Computation*, **2001**, *13*(9), 1975-1994.

13. Shin C., Kim K. I., Park M. H., and Kim H. J., "Support Vector Machine-based Text Detection in Digital Video," in *Neural Networks for Signal Processing X*, Proceedings of the 2000 IEEE Signal Processing Society Workshop, **2000**.

14. Ma C., Randolph M. A., and Drish J., "A Support Vector Machines-based Rejection Technique for Speech Recognition," in *Acoustics, Speech, and*

Signal Processing, Proceedings.(ICASSP›01). 2001 IEEE International Conference on. 2001. IEEE, **2001**.

15. Van Gestel T., Suykens J. A. K., Baestaens D. E., Lambrechts A., et al., "Financial Time Series Prediction using Least Squares Support Vector Machines within the Evidence Framework," *Neural Networks*, IEEE Transactions on, **2001**, *12*(4), 809-821.

16. Al-Anazi A. and Gates I., "On the Capability of Support Vector Machines to Classify Lithology from Well Logs," *Natural Resources Research*, **2010**, *19*(2), 125-139.

17. Al-anazi A. F. and Gates I. D., "Support-vector regression for permeability prediction in a heterogeneous reservoir: A comparative study," *SPE Reservoir Evaluation & Engineering*, **2010**, *6*(10), 21-18.

18. Al-Anazi A. and Gates I., "Support Vector Regression for Porosity Prediction in a Heterogeneous Reservoir: A Comparative Study," *Computers & Geosciences*, **2010**, *36*(12), 1494-1503.

19. Vapnik V., "The Nature of Statistical Learning Theory," Springer Science & Business Media, **2000**.

20. Wang L., "Support Vector Machines: Theory and Applications," Springer Science & Business Media, **2005**:

21. Al-Anazi A., Gates I., and Azaiez J., "Fuzzy Logic Data-driven Permeability Prediction for Heterogeneous Reservoirs," Paper SPE 121159 Presented at *the EUROPEC/EAGE Conference and Exhibition*, **2009**.

22. Lu J., Plataniotis K., and Ventesanopoulos A., Face Recognition Using Feature Optimization and V-support Vector Machine," *IEEE Neural Netw. Signal Process*, **2001**, 373–382.

23. Wong K.W.,Fung C. C., Ong Y. S., and Gedeon T. D., "Reservoir Characterization Using Support Vector Machines," in Computational Intelligence for Modelling, Control and Automation, 2005 and International Conference on Intelligent Agents, Web Technologies and Internet Commerce, International Conference on. 2005. IEEE., **2005**.

24. Al-Anazi A. and Gates I., "Support Vector Regression to Predict Porosity and Permeability: Effect of Sample Size," *Computers & Geosciences*, **2012**, *39*, 64-76.

25. Suykens J. A., "Data Visualization and Dimensionality Reduction Using Kernel Maps with a Reference Point," *Neural Networks*, *IEEE Transactions*, **2008**, *19*(9), 1501-1517.

26. Georgi D. and Menger S., "Reservoir Quality, Porosity and Permeability Relationships," *in Proc. 14th Mintrop Seminar*, Münster. **1994**.

27. Ohen H. A. and Ajufo A., "A Hydraulic (Flow) UMR Based Model for the Determination of Petrophysical Properties from nmr Relaxation Measurements," SPE Annual Technical Conference and Exhibition, Dallas, Texas, **1995**.

28. Parra J., Hackert C. L., Collier H. A., and Bennett M., "NMR and Acoustic Signatures in Vuggy Carbonate Aquifers," in SPWLA 42nd Annual Logging Symposium, Society of Petrophysicists and Well-Log Analysts, **2001**.

29. Ye S. J. and Rabiller P., "A New Tool for Electro-Facies Analysis: Multi-resolution Graph-based Clustering," in SPWLA 41st annual logging symposium, Society of Petrophysicists and Well-Log Analysts, **2000**.

14

Enhanced Gas Recovery with Carbon Dioxide Sequestration in a Water-drive Gas Condensate Reservoir: a Case Study in a Real Gas Field

Hossein Zangeneh[1]* and Mohammad Amin Safarzadeh[2]

[1] Ahwaz Research Center, Research Institute of Petroleum Industry (RIPI), Ahvaz, Iran
[2] Tehran Energy Consultants Company (TEC), Tehran, Iran

ABSTRACT

Gas reservoirs usually have high recovery due to high mobility and low residual gas saturation, although some of them producing under water-drive mechanism have low recovery efficiency. Encroachment of water into these reservoirs traps considerable amount of gas and increases the maximum residual gas saturation, which results in the reduction of gas and condensate production. Generally, the recoveries of water-drive gas reservoirs vary between 35-75%, whereas depletion-drive reservoirs exhibit recoveries near 85%. In this work, a method was proposed for reducing water encroachment, sweeping reservoir gas content effectively, and enhancing the hydrocarbon recovery consequently. To this end, a condensate gas reservoir model, located in the south of Iran, was chosen to study the process. The injection was performed above the bottom-up aquifer from two horizontal wells, and the base gas was produced by four vertical wells. Three cases of inactive aquifer (Case I), active aquifer (Case II), and active aquifer with CO_2 injection (Case III) were studied subsequently. The proposed gas-gas displacement method increases the recovery of reservoir especially the recovery of heavier components composing the main part of the condensate. Moreover, the injection of a huge volume of CO_2 without significant CO_2 production can be interesting from an environmental point of view and can be considered as a CO_2 sequestration process.

Keywords: Water-drive, Gas Condensate Reservoir, Carbon Dioxide, Enhanced Gas Recovery, Sequestration

INTRODUCTION

The recovery efficiency of gas reservoirs which are not in association with aquifers (depleting gas reservoirs) are usually as high as 70–85% of the original gas in-place (OGIP), without considerable water production [1-2]. However, recoveries of water-drive gas reservoirs range between 35-75% due to both relatively low sweep efficiency and the entrapped gas in the water-invaded zone [2-3]. Initially, physical properties such as the residual gas saturation (S_{gr}) behind the water

front govern ultimate recovery of these reservoirs [4]. By increasing production and pressure drop, water moves to pores and throats filled with gas, and the water displaces the gas incompletely. Capillary pressure and relative permeability effects halt the flow of gas. Thus, water just passes through the rock volume [5]. Some methods have been proposed and used for increasing the recovery of water-drive gas reservoirs, i.e. blow-down technique (producing gas at a high rate to exceed the rate of water invasion into

*Corresponding author
Hossein Zangeneh
Email: zangeneh@ripi.ir

the gas zone) and up-dip production (producing water from down-dip and gas from up-dip) [6-7]. Also, the problem of high water-cut of these reservoirs are solved by reperforating the production wells in an uninvaded zone. However, these methods have some disadvantages like water conning and sand production for blow-down method and high operational cost of up-dip technique [6]. In many cases, the reperforating causes high cost of workover. Moreover, this solution does not prevent the gas entrapment in the water-invaded zone. Matthews et al. reported that the rapid de-pressurization of gas condensate reservoirs can re-mobilize up to 10% of the original gas in place (OGIP) entrapped due to water influx [8].

The last descriptions of water-drive reservoirs demonstrate the low recovery and necessity of employing enhanced gas recovery methods in these reservoirs. One of the effective methods proposed to enhance the ultimate recovery of these reservoirs is gas-gas displacement method. Enhancing gas recovery by gas-gas displacement can be achieved economically in several situations. For mature volumetric gas reservoirs suffering from low productivity due to low reservoir pressure, the injection of waste gas increases the ultimate gas recovery by maintaining gas production rates and preventing premature well abandonment. For water-driven gas reservoirs, pressure maintenance by gas injection will serve to retard the influx of aquifer and to partially mitigate water coning caused by excessive pressure drawdown [9]. The crucial aspect of designing the gas-gas displacement technique is selecting a proper displacing gas for EGR process. Carbon dioxide is an appropriate candidate for injection into the gas reservoirs. CO_2 is denser than natural gas (2 to 6 times depending on reservoir conditions) and segregates in the production zone

by the gravity. It has lower mobility than natural gas due to its higher viscosity, which creates a high displacing efficiency in EGR process [10]. The other reason for choosing CO_2 for EGR process is sequestering this gas in a geological structure. Gas reservoirs are appropriate candidate for CO_2 storage due to high capacity and integrity. Therefore, the process of carbon dioxide injection can be employed as a multi-objective solution.

The data collection of various types shows that global surface temperature have increased by 0.7 Kelvin since the 19th centuries, which is the result of an increase in greenhouse gas concentration, mainly CO_2 [11]. The measurement of the atmospheric CO_2 concentrations for the last 250 years illustrates its increase from 270 to more than 370 ppm. Experts report that carbon dioxide emissions account for about two third of the potential global warming [12]. One of the effective solutions for decreasing the emission of CO_2 is direct capturing and storing it in deep geological formations, which is known as carbon capture and storage (CCS) [13]. The injection of carbon dioxide into natural gas reservoirs is a promising technique for reducing anthropogenic greenhouse gas emission and increasing the ultimate recovery of natural gas [14-15].

Several studies have been reported on the numerical simulation and investigation of the CO_2 storage process in oil and gas reservoirs with the aim of improving recovery especially to discuss the displacement process [2, 9, 14-22].

In this work, the process of carbon dioxide (CO_2) injection into a water-drive gas condensate reservoir was studied. The aim of this process is sweeping reservoir base gas and the reduction of water encroachment into reservoir. The reservoir is located in the south of Iran and contains condensate with an active aquifer. This process can deplete the base gas before water entrapment. Moreover, the injection of CO_2 stores a large amount of CO_2, which is interested from an environmental point of view these days.

Reservoir Description

The reservoir studied in this work is located in the south of Iran and contains gas condensate. It has IGIP of 1.80 trillion SCF (50.98 billion SCM). The reservoir is represented using an 84×78×19 compositional simulated model and has an area of 4047.6 acre (16.37 km²) and an average thickness of 224.2 m. The average porosity of the reservoir rock is 10.63% with an average net to gross (NTG) ratio of 0.823. Table 1 represents the general data of the reservoir. It is composed of single porosity carbonate rock and its relative permeability curves were modeled using Corey method [23]. The reservoir was sealed from the top by an evaporate formation, but the boundaries of the reservoir were permeable. This anticline has the lithology of mainly limestone and 4 vertical zones with heterogeneous rock properties (porosity, permeability etc.). From top of the reservoir, the second zone has weaker reservoir properties (permeability) than the others.

The reservoir has 4 vertical production wells, and 2 horizontal injection wells were proposed for EGR scenario. The production wells were perforated in the whole reservoir thickness at a distance of water/oil contact. The horizontal part of the injection wells were perforated with a length of about 3.8 km, and the average horizontal distance between the injection and vertical wells is about 1.7 km. Figure 1 depicts the 3D view of the reservoir and its wells. The reservoir fluid was characterized using three-parameter Peng-Robinson equation of state (EOS). The composition of the reservoir fluid is represented in Table 2.

3D View of Reservoir, Its Production and Proposed Injection Wells

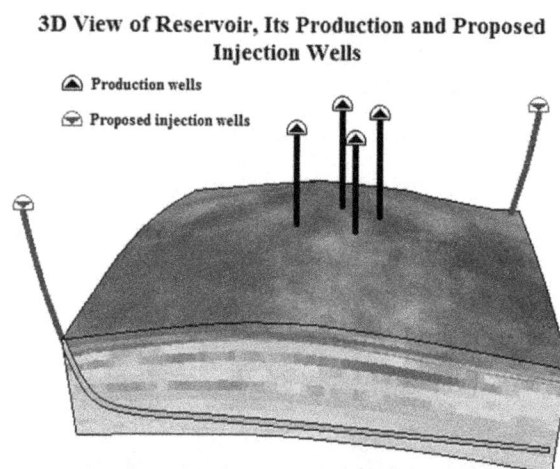

Figure 1: 3D view of reservoir model and their wells.

Table 1: The reservoir general data.

Parameter	Unit	Value
Average porosity	%	10.63
Average horizontal permeability	md	4.98
Average vertical permeability	md	0.5726
Average NTG	-	0.823
Initial pressure	bar	270.8
Initial temperature	°C	73
Reservoir area	acre	4047.6
Average thickness	m	224.2
Depth of water/oil contact (WOC)	m	2750
Reference depth	m	2750
Reservoir top depth	m	2486

Table 2: The composition of reservoir fluid.

Component	Mole Fraction (%)
Nitrogen (N_2)	9.69
Methane (CH_4)	85.14
Ethane (C_2H_6)	1.66
Propane (C_3H_8)	0.94
Heavy components	2.57

This reservoir has a bottom-up aquifer which is modeled using Carter-Tracy aquifer model. Table 3 shows the aquifer properties.

Table 3: The aquifer properties.

Property	Unit	Value
Porosity	%	30
Permeability	md	400
Water compressibility	bar^{-1}	0.000155
External radius	m	400
Thickness	m	100
Angle of influence	°	360
Top limit	m	-2750

Production and Injection Scenarios

In this work, three cases have been studied in order to investigate the effect of aquifer activity on cumulative gas and condensate production. In the first case (Case I), the effect of aquifer on reservoir was omitted by deactivating it. It is crucial to note that this case is an imaginary scenario and does not happen in reality. This case is similar to up-dip production in an infinite water production condition to cancel the aquifer effects. The production time is 9 years with a constant bottomhole pressure of 80 bar. The water-cut limitation of 20% controls the water production of the reservoir. In the second case (Case II), the production process by depletion scenario was studied. This case is the base scenario of the production in this reservoir. The comparison of the Case I and Case II illustrates the effect of aquifer activity on gas and condensate production. Finally, in the last case (Case III), the CO_2 injection process was simulated. In this case, the injection scenario was begun at the first year of production at a rate of 4 million standard cubic meters per day (MMSCMD) per each injection wells. The injection process was continued until the pressure reached the 90% of the reservoir initial pressure for lowering the risk of caprock fracturing. The objective of this case was investigating the effect of CO_2 injection on the reduction of aquifer water encroachment and on the increase of hydrocarbon production. The production conditions of the two last cases are the same as the first one.

RESULTS AND DISCUSSION

In this work, the process of CO_2 injection into a water-drive gas condensate reservoir was studied. The aim of this process is sweeping reservoir base gas and the reduction of water encroachment into reservoir. This process can deplete the base gas in the rock pores before water entrap them. Furthermore, the injection of CO_2 stores a large amount of CO_2 which is interesting from an environmental point of view.

Comparison of Gas and Condensate Production

Comparison of hydrocarbon recoveries shows the effectiveness of the enhanced hydrocarbon recovery technique. Figure 2 compares the cumulative gas production of Cases I, II, and III. This figure shows that aquifer activity (Case II) decreases the cumulative gas production in comparison with inactive aquifer (Case I) by 21%. It also illustrates that injecting CO_2 has a significant effect on cumulative gas production and increases the gas recovery as an inactive aquifer case.

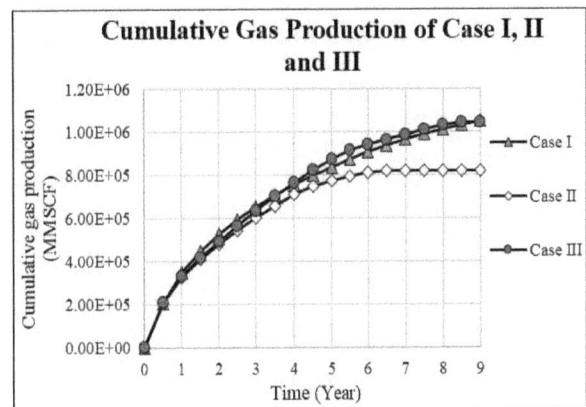

Figure 2: Comparison of cumulative gas production of Cases I, II, and III.

The comparison of cumulative condensate production of three cases is shown in Figure 3. This figure demonstrates that encroachment of water in the reservoir increases cumulative condensate production (cumulative condensate production of Case I is about 0.87 times of Case II). Considering this fact, it can be concluded that aquifer activation has positive effects on condensate production, but this point should be considered that the amount of the reduction in gas production is economically higher than the amount of the rise in condensate production. Therefore, the water movement into the reservoir zone causes profit loss, especially in this case. Moreover, this figure illustrates that the injection of CO_2 in Case III increases the condensate production by 58 and 82% in comparison with Cases II and I.

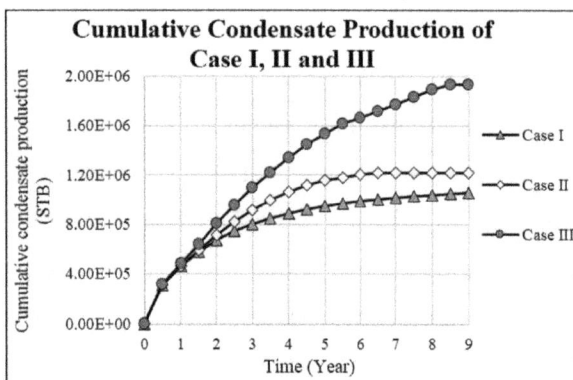

Figure 3: Comparison of cumulative condensate production of Cases I, II, and III

Actually, the condensate production increase in Cases II and III were happened due to pressure maintenance of these cases. Figure 4 depicts the reservoir pressure curve of three cases in which the pressure maintenance process of the Case III is demonstrated.

In gas condensate reservoirs, by declining the reservoir pressure to the first dew point, gas condenses in porous media. This process decreases the condensate content of producing gas, which is projected in condensate-gas ratio (CGR). Hence, maintaining the pressure in the water-drive gas condensate reservoirs has another advantage of the inhibition of condensing gas in pay zone. Comparing Figure 2 and Figure 3 shows that the CGR of Case I decreases gradually with a pressure decline from 1.5 to 1.0 STB/MMSCF although the CGR's of Case II and III are constant and equal to about 1.5 STB/MMSCF.

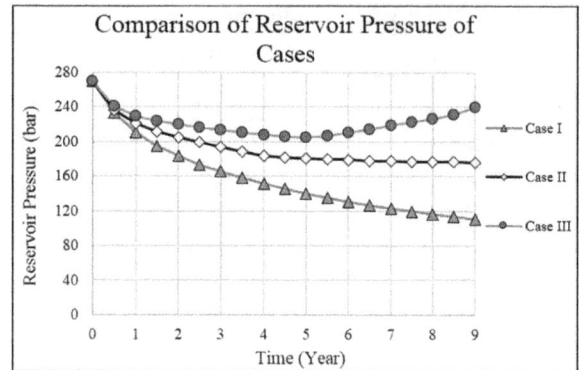

Figure 4: Comparison of reservoir pressure of Cases I, II, and III.

CO_2 Production and Injection of Case III

The amounts of injecting and reproducing CO_2 are crucial in the case of CCS. Huge reproduction of injected CO_2 may fail the process due to low net CO_2 sequestration. Figure 5 depicts the cumulative CO_2 injection of Case III in million tons. It also shows the CO_2 production in terms of the weight percent of CO_2 production to injection. As it is clear in this figure, the amount of CO_2 production is zero until the second years. The percent of CO_2 production to injection is about 0.0003 in the worst condition, which is negligible. This fact demonstrates that a huge amount of CO_2, i.e. 47.3 million tons in 9 years, can be sequestered using this process.

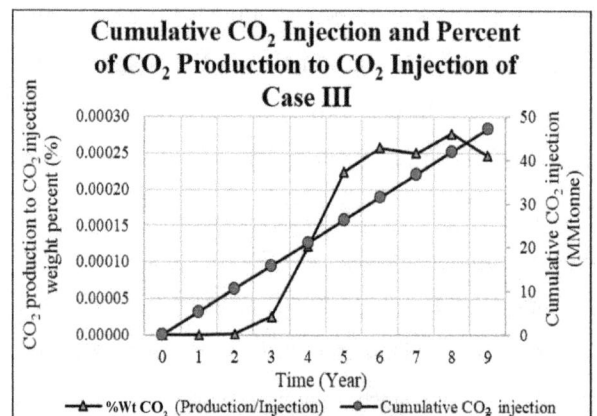

Figure 5: Cumulative CO_2 injection and the percent of CO_2 production to CO_2 injection of Case III.

Figures 2-5 illustrate that the injection of CO_2 in a gas condensate reservoir under water-drive production enhances the amount of gas and condensate recovery and profit consequently. This process stores a large amount of CO_2 with negligible CO_2 production as an interesting process for environmental experts. The injection of carbon dioxide sweeps the reservoir base gas before water invades the porous media. In fact, the injection of CO_2 enhances gas recovery and maintains the reservoir pressure, so it prevents water movement because of low pressure drawdown in the reservoir.

Comparison of Cumulative Water Influx of Cases II and III

Figure 6 compares the cumulative water influx of aquifer for Cases II and III, which approves our last conclusions. As can be seen in this figure, the volume of water influx of aquifer to reservoir in Case II is 1.5 times of Case III. This result shows the effect of CO_2 plume on water movement in reservoir.

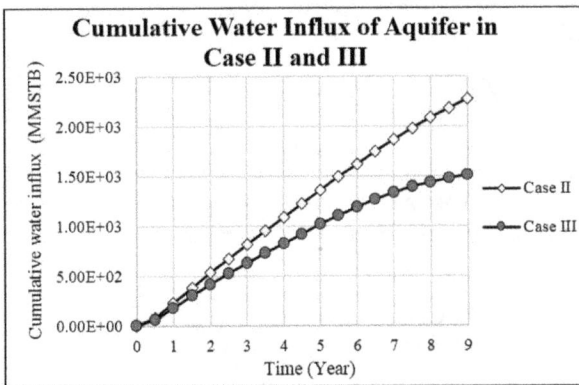

Figure 6: Comparison of cumulative water influx of aquifer for Cases II and III.

Figure 7 shows the profile of methane and carbon dioxide mole fraction of Case III and water saturation of Cases II and III versus depth. This figure demonstrates that carbon dioxide plays two roles in this process. First, the mole fraction of CO_2 and CH_4 illustrates that carbon dioxide sweeps the base gas to the production wells effectively. Second, comparing the profile of the water saturation of Case I and II shows that a CO_2 plume was composed above the aquifer, which controls the water encroachment into the reservoir. Figure 8 depicts the 3D view of CO_2 propagation in the reservoir. In this figure, the CO_2 mole fraction can be seen in the left hand side for 3 years (at end of 3rd, 6th, and 9th years). For a better understanding of the

CO_2 movement, the CO_2 plume was depicted in the left hand side, which shows the propagation of CO_2 front to the production wells and above aquifer.

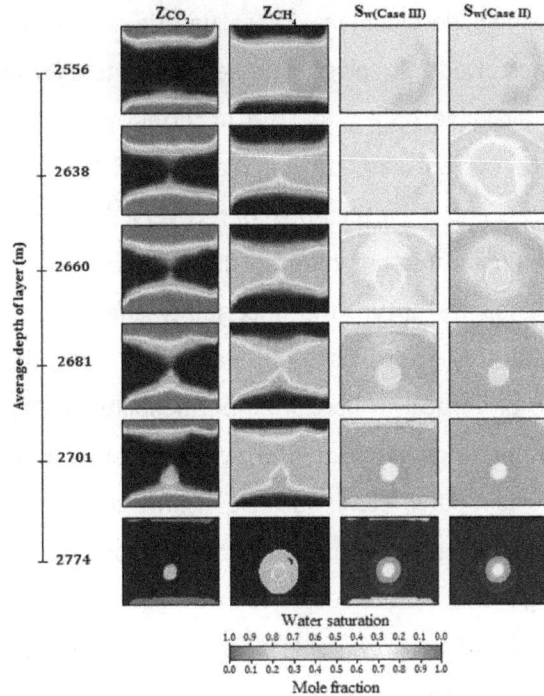

Figure 7: The profile of CO_2 and CH_4 mole fraction of Case III and water saturation of Cases II and III.

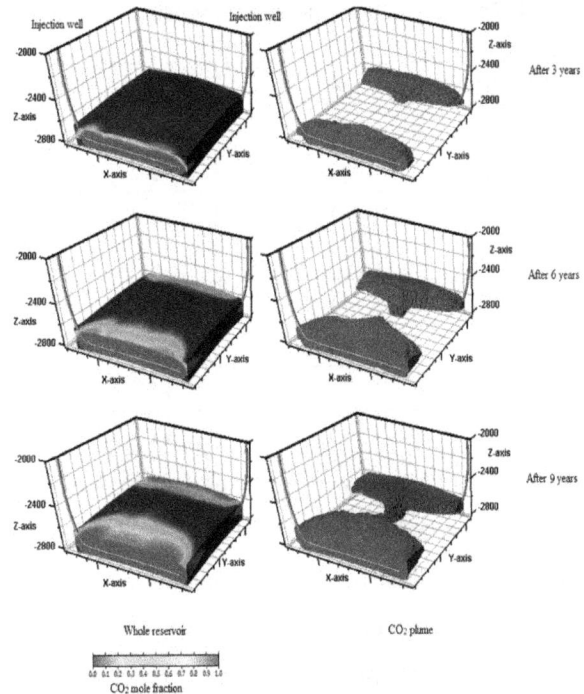

Figure 8: The 3D view of CO_2 propagation in the reservoir.

Investigating the Effect of Uncertain Reservoir Parameters on CO_2 injection Process

Uncertainties, especially in geological parameters, are always present and can be significant if associated with field development. These uncertainties could be perceived by geologists and engineers to exist in the available data [24]. The uncertainty analysis in the current case can decrease the risk of the process and show the range of recovery changes in Case III. For this aim, different scenarios have been designed to capture the uncertainty effects. Four parameters, i.e. porosity (Φ), permeability (K), the ratio of vertical to horizontal permeability (K_h/K_v), and net to gross ratio (NTG), were considered as uncertain parameters. Because of time consuming simulation runs, the partial factorial experimental design (2^{k-1}) was conducted to investigate the effect of uncertainty [25]. Table 4 shows the designed simulation cases run to capture uncertainty. The expression (-1) and (+1) shows the amount of parameters in their minimum and maximum levels.

Table 4: The partial factorial experimental design (2^{k-1}) runs.

Run number	Φ	K	K_h/K_v	NTG
1	-1	-1	-1	-1
2	+1	-1	-1	+1
3	-1	+1	-1	+1
4	+1	+1	-1	-1
5	-1	-1	+1	+1
6	+1	-1	+1	-1
7	-1	+1	+1	-1
8	+1	+1	+1	+1

Figures 9 and 10 show the comparison of gas and condensate recovery for the chosen runs in Table 4. These figures illustrate that the injection scenario is effective with different levels of uncertainty. It is notable that the aim of these figures are not discussing about the effect of different parameters

on the process, but the purpose of this section is showing that uncertainties of the reservoir parameters do not affect the success of the process. The process (Case III) in the worst case of uncertainties is effective in comparison to the other cases (Case I and II). Figure 9 illustrates the 4 and 46% recovery increase in comparison to Case II in the worst and base cases. This is about 31 and 78% for condensate recovery as can be seen in Figure 10.

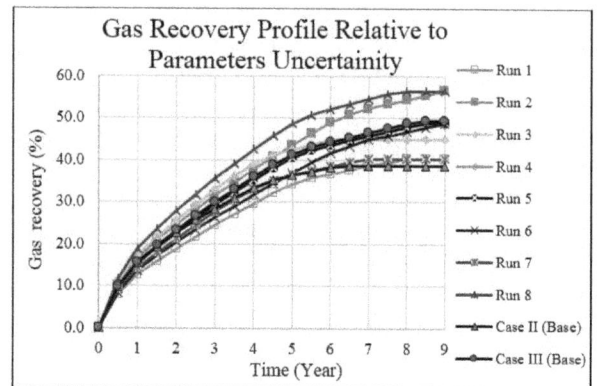

Figure 9: The gas recovery profile relative to the uncertainty of different reservoir parameters.

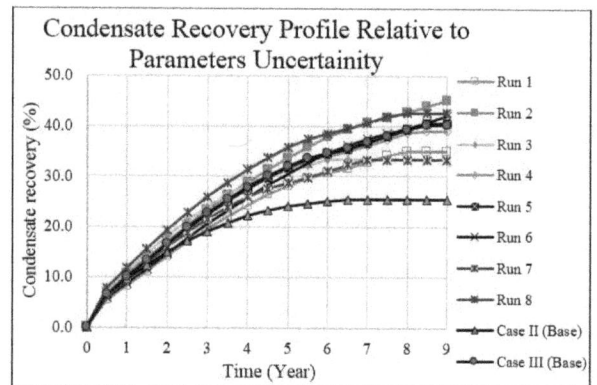

Figure 10: The condensate recovery profile relative to the uncertainty of different reservoir parameters.

CONCLUSIONS

In this work, the process of the injection of CO_2 to enhance gas recovery and control the water invasion of a water-drive gas condensate reservoir, located in the south of Iran, was studied. Considering this work, the following results can be concluded:

- The injection of CO_2 in a water-drive gas condensate reservoir increases the gas and condensate recovery by 27 and 58% respectively with negligible CO_2 production. This is because CO_2 injection prevents the encroachment of water by filling the pore volumes and effective sweeping the reservoir base gas. The uncertainty analysis of the reservoir parameters shows that in all the levels of uncertainty, the process can be beneficial.

- The injection of CO_2 in this reservoir has two significant advantages. First, it increases the recovery of gas and condensate, which raises the profit of the process. Second, it can sequester a huge amount of CO_2 (about 47.3 million tons), which is interesting from an environmental point of view.

- The pressure maintenance of the reservoir by injecting the CO_2 controls the water encroachment into the reservoir. Moreover, it prevents declining the pressure under reservoir fluid dew point and sustains the CGR constant, which results in a high recovery of condensate.

NOMENCLATURE

md   Millidarcy
M   Thousand
MM   Million
SCF   Standard Cubic Feet
SCM   Standard Cubic Meter
STB   Stock Tank Barrel

REFERENCES

1. Lee W. J. and Wattenbarger R. A., "Gas Reservoir Engineering," SPE Textbook Series, **1996**.
2. Turta A. T., Sim S. S. K., Singhal A. K., and Hawkins B. F., "Basic Investigations on Enhanced Gas Recovery by Gas-Gas Displacement," *Journal of Canadian Petroleum Technology*, **2008**, *47*(10), 39-44.
3. Mckay B. A., "Laboratory Studies of Gas Displacement from Sandstone Reservoirs Having a Strong Water Drive," *APEA Journal*, **1974**, 189-194.
4. Bassiouniz Z., "Enhanced Recovery from Water-Drive Gas Reservoirs," *Rudarsko-Geolosko-Naftni Zbornik*, **1990**, *2*, 151-159.
5. Holtz M., "Residual Gas Saturation to an Influx: A Calculation Method for 3-D Computer Reservoir Model Construction," SPE 75502, **2002**.
6. Batycky J., Irwin D., and Fish R., "Trapped Gas Saturation in Leducage Reservoirs," *Journal of Canadian Petroleum Technology*, **1998**, *37*(2), 32-39.
7. Chesney T. P., Lewis R. C., and Trice M. L., "Secondary Gas Recovery from a Moderately Strong Water-Drive Reservoir: A Case History," *JPT*, **1982**, *34*(9), 2149-2157.
8. Ali F., "Importance of Water Influx and Waterflooding in Gas Condensate Reservoir. Department of Petroleum Engineering and Applied Geophysics," Norwegian University of Science and Technology (NTNU), **2014**, 9.
9. Sim S. S. K., Brunelle P., Turta A. T., and Singhal A. K., "Enhanced Gas Recovery and CO_2 Sequestration by Injection of Exhaust Gases from Combustion of Bitumen," SPE 113468, **2008**.
10. Al-Hashami A., Ren S. R., and Tohidi B., "CO_2 Injection for Enhanced Gas Recovery and Geo-Storage: Reservoir Simulation and Economics," SPE 94129, **2005**.
11. Blyton C. A. J., "Kinetic of CO_2 Dissolution in Brine: Experimental Measurement and Application to Geologic Storage," The University of Texas at Austin, **2012**, 18-22.
12. Cakici M.D., "Co-optimization of Oil Recovery and Carbon Dioxide Storage," M. Sc. Thesis, Stanford University, **2003**, 1-2.
13. Faiz M. M., Saghafi A., Barclay S. A., Sherwood L. S. N. R., and Whitford D. J., Evaluating Geological Sequestration of CO_2 in Bituminous Coals: The Southern Sydney Basin, Australia as a Natural Analogue," *International Journal of Green-house Gas Control*, **2007**, *1*, 223-235.
14. Jikich S. A., Smith D. H., Sams W. N., and Bromhal G. S., "Enhanced Gas Recovery (EGR) with Carbon Dioxide Sequestration: A Simulation Study of Effects of Injection Strategy and Operational Parameters," SPE 84813., **2003**.
15. Oldenburg C. M., "Carbon Sequestration in Natural Gas Reservoirs: Enhanced Gas Recovery and Natural Gas Storage Proceeding," Tough Symposium, California, **2003**.
16. Forest T., "CO_2 Enhanced Oil Recovery in Strong

Water-Drive Reservoirs. Department of Petroleum Engineering and Applied Geophysics," Norwegian University of Science and Technology (NTNU), **2012.**

17. Sim S. S. K., Turta A. T., Singhal A. K., Hawkins B. F., and Council A. R., "Enhanced Gas Recovery: Factors Affecting Gas-Gas Displacement Efficiency," *Journal of Canadian Petroleum Technology*, **2009**, *48*(8), 49-55.

18. Kühn M., Förster A., Grobmann J., Meyer R., et al., "CLEAN: Preparing for a CO_2-based Enhanced Gas Recovery in a Depleted Gas Field in Germany," *Energy Procedia*, **2011**, *4*, 5520-5526.

19. Secklehner S., Arzmüller G., and Clemens T., "Tight Ultra-deep Gas Field Production Optimization – Development Optimization and CO_2 Enhanced Gas Recovery Potential of the Schoenkirchen Uebertief Gas Field," Austria. SPE 130154, **2010.**

20. Wei X. R., Wang G. X., Massarotto P., Golding S. D., et al., "Numerical Simulation of Multicomponent Gas Diffusion and Flow in Coals for CO_2 Enhanced Coalbed Methane Recovery," *Chemical Engineering Science*, **2007**, *62*, 4193–4203.

21. Zangeneh H., Jamshidi S. and Soltanieh M., "Coupled Optimization of Enhanced Gas Recovery and Carbon Dioxide Sequestration in Natural Gas Reservoirs: Case Study in a Real Gas Field in the South of Iran," *International Journal of Greenhouse Gas Control*, **2013**, *17*, 515-522.

22. Zangeneh H., Safarzadeh M. A., and Asgari A. A., "Carbon Dioxide Injection to Control Water Encroachment in Water-Drive Gas Condensate Reservoirs. TP4-23," International Gas Union (IGU) Conference, Denmark, **2014.**

23. Ahmed T., "Handbook of Reservoir Engineering," Gulf Professional Publishing, Texas, **2000**, 292-293.

24. Mesdour R., Ramsey L., and Aly A., "Optimizing Development Well Placements within Geological Uncertainty Utilizing Sector Model," SPE 12847, **2010.**

25. Montgomery D. C., "Design and Analysis of Experiments," Fifth Edition, John Wiley & Sons Inc, **2001**, 303-347.

15

Experimental Determination of the Temperature Suppression in Formation of Gas Hydrate in Water Based Drilling Mud

Ali Bakhtyari, Yasin Fayazi, Feridun Esmaeilzadeh*, and Jamshid Fathikaljahi

Department of Chemical and Petroleum Engineering, School of Engineering, Shiraz University, Shiraz, Iran

ABSTRACT

The aqueous mixtures of light gas molecules under low-temperature and high-pressure conditions are candidates to form gas hydrate clathrates. The formation of gas hydrate may lead to various problems and extra charges in natural gas production and processing. The presence of hydrate crystals forming a stable solid phase can potentially block the wells, pipes, and process facilities. To avoid such problems, hydrate formation must be studied in different conditions. In this regard, this study aims to design an experimental procedure to determine gas hydrate formation in water based drilling mud. In addition to this, the effect of different inhibitors (i.e. NaCl, methanol, and ethylene glycol) on the hydrate temperature suppression is studied. The designed apparatus is capable of determining hydrate formation conditions in both static and dynamic conditions. Based on the obtained results, among the combinations of NaCl with methanol or ethylene glycol at different concentrations, a mixture of 10 wt.% NaCl + 10 wt.% methanol shows the best inhibition effects (i.e. higher temperature suppression and longer induction time). Furthermore, the experimentally obtained temperature suppression data were fitted and compared against two different thermodynamic models. Temperature suppression variation with inhibitor concentration is described in an acceptable manner by both models.

Keywords: Gas Hydrate, Drilling Mud, Inhibitor, Sodium Chloride, Ethylene Alcohol, Methyl Alcohol, Temperature Suppression

INTRODUCTION

An important factor in drilling oil and gas wells is drilling fluid. The adverse selection of drilling fluids may lead to remarkable extra charges and irreparable defects such as derrick destruction. Drilling fluids are complex mixtures of different components, including polymers, bentonite, surfactants, weighting agents, and treatment additives beside the base fluid, which could be water or oil [1-3]. The application of water-based drilling fluids is more common due to environmental regulations [1]. Glycerol- and polyalcohol-based drilling fluids, which are classified in the water-based category, are often used in the well drilling operations [2]. Water-based muds consisting a salt-based water or fresh water are fluids used to

*Corresponding author

Feridun Esmaeilzadeh
Email: esmaeil@shirazu.ac.ir

control formation, pressure, lubrication, cooling, and removing rock fragments from the drilling well [3]. The explorations and developments of the drilling operations to deep waters, where particular conditions such as lower temperature, higher pressure, and water content are prepared, increase the possibility of gas hydrate formation. The issues of gas hydrate formation in drilling operations are serious. Hence, every attempt to guarantee to remove hydrate formation or to significantly decrease the risk of hydrate formation is appreciated.

Although gas hydrates may be considered as hydrocarbon resources and a medium for storing and transporting natural gas, they could lead to problems with the flow, wellbore collapse, uncontrollable gas release, and blowouts [4-8]. Hydrates are inclusion compounds, where small hydrophobic gas molecules such as methane and carbon dioxide appear to be inside a host molecule such as water at a low temperature and a high pressure. Consequently, gas molecules are trapped into the vacant spaces of water molecules. In other words, the simultaneous crystallization of water molecules and surrounding small gas molecules is the beginning of gas hydrate formation. The cage like crystals is thermodynamically stable in a certain range of temperature and pressure through van der Waals type interaction between the water molecules and the trapped gas molecules. Hydrocarbons such as C1 to C7 and light gases such as nitrogen, carbon dioxide, and hydrogen sulfide are possible candidates to form gas hydrates in water [4,9]. More information on the structures of hydrates could be found elsewhere [4,5]. Hydrates are generally considered a potential

hazard in drilling operations. Blocking subsea risers, obstructing drilling fluid flow and choking and killing lines, and blowing out preventer stacks may be caused by solid hydrates formed in drilling operations in deep water [9,10]. McConnell et al. reviewed the progress in evaluating gas hydrate hazards in drilling operations [11].

To prevent gas hydrate formation in drilling muds, some chemical additives called inhibitors are used. Inhibitors are classified into three types: kinetic, thermodynamic, and anti-agglomerator inhibitors. Kinetic and anti-agglomerator inhibitors can decrease hydrate formation rate and delay the formation time of hydrates, while thermodynamic inhibitors can disrupt hydrogen bonding in water phase leading to a remarkable reduction in the hydrate formation temperature. The potential capability of many salts to inhibit hydrate formation has been well studied in the literature [12-15] while more studies on the effect of other inhibitors such as alcohols and glycols should be conducted. In fact, to find a suitable formulation for drilling muds, different inhibitors and their interactions must be widely studied. In spite of the potential hazards of hydrate formation in the offshore drilling operation, a few data have been published about gas hydrate formation with and without the presence of thermodynamic inhibitors and their mixtures in drilling muds; also, it is very hard to find literature and technical references that describe real case studies of field work [12-15]. The objective of the present study is to describe a high-pressure apparatus working in both static and dynamic loops to generate gas hydrates. The effect of different thermodynamic inhibitors and their mixtures at different concentrations on gas hydration conditions in a natural water-based drilling mud is experimentally studied in static and dynamic conditions. Furthermore, two well-known models with thermodynamic backgrounds are used to correlate experimental hydrate suppression data. As far as the authors know, there is no published work in the open literature investigating the effect of different thermodynamic inhibitors on the hydrate suppression in a natural water-based drilling mud in both static and dynamic conditions.

EXPERIMENTAL PROCEDURES
Materials

The water-based drilling mud (containing bentonite, water, surfactants, and weighting agents), which is used as the aqueous phase, was donated by National Iranian Oil Company (N.I.O.C.). Carbon dioxide, methane, and nitrogen were used as the gaseous hydrate formers. Gases used in this work, their purities, and their suppliers are shown in Table 1.

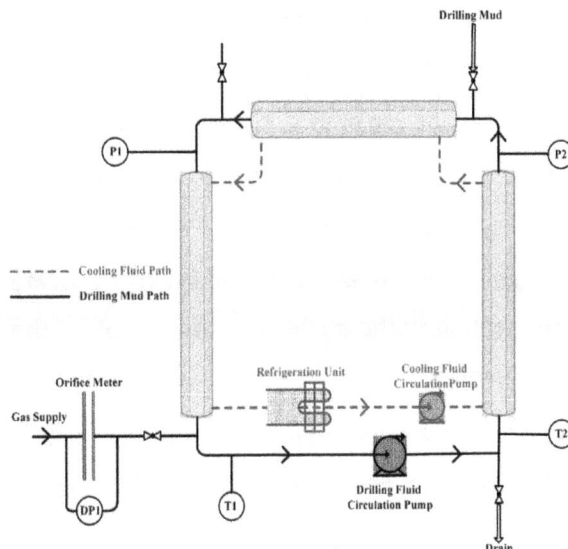

Figure 1: A schematic diagram of the pilot-scale apparatus for gas hydrate generation loop; T1 and T2: Temperature sensors; P1 and P2: Bourdon pressure gauge; and DP1: Pressure drop gauge.

The main part of the apparatus was a main closed-loop pipeline. The loop was equipped with a screw pump for supplying high-pressure condition in the system. The temperature control procedure of the loop was achieved by using a bath circulator with glycol-water fluid. Stainless steel (type-316) with an inside diameter of 2.43 cm and a length of 8 m is used to construct the pipeline. Two temperature sensors with uncertainties of ± 0.1 K were used to monitor the loop (100-ohm resistance thermometers made of platinum, i.e. T1 and T2). Three pressure indicators with uncertainties of ± 0.7 bar installed around the loop were used to control the pressure during the experiments (two Bourdon gauges, i.e. P1 and P2, in addition to a Rosemount pressure drop transducer, i.e. DP1). Moreover, a Rosemount pressure drop transducer and an orifice plate were equipped to control the rate of gas injection in the dynamic experiments. The impurities of the gas samples were eliminated using a line filter. Hence, the contents of the loop

Table 1: Details of the used gases in this study.

Gases	Purity	Supplier
N_2	99.90 %	Aboghadare Industrial Co.
CO_2	99.95 %	Aboghadare Industrial Co.
CH_4	99.95 %	Erish Gas Gostar Chemical Co.

Apparatus and Procedure

The experimental apparatus used in this study to generate hydrate in water-based drilling mud in the presence or absence of the thermodynamic inhibitors is sketched in Figure 1 [3,7].

were prevented from undesired impurities. A stainless steel bath equipped with a circulator (using ethylene glycol as the coolant) was used to decrease the temperature of the entire loop at a rate of 1.85 K/hr. In addition, the temperature of the fluid inside the loop was measured by locate thermocouples. The initial temperature of all the experiments was 300 K. In each experiment, about 4 liters of the prepared raw drilling mud was loaded to the apparatus from the valve located above the loop. It was then pressurized by using a gaseous mixture containing 26% CH_4, 27% N_2, and 47% CO_2 (on the molar basis). Once the pressure of the cell was reached the desired level (typically 150 bar), the temperature of the cell was started to decrease at a rate of 1.85 K/hr. Inherently, a decrease in the pressure of the cell was observed during the cooling process. The temperature and pressure of the system were monitored over the time. A more sharp decrease in the pressure history of the system was observed at the time of reaching thermodynamic conditions of hydrate formation. Such observation was an indication of gas enclosure in a hydrate structure. Moreover, a mild increase in the temperature of the system was observed due to the exothermic nature of hydrate formation. In the dynamic experiments, a sudden increase in the rate of gas consumption also indicates the hydrate formation. It should be mentioned that the mud with a density of 9230 kg/m³ and a viscosity of 24.2 cP, when moved at a velocity of 0.5 m/s in the flow loop, led to a Reynolds number of 4630. The gas hydrate generator described above can be utilized at pressures up to 150 bar and in the temperatures range of 273 to 300 K.

Model Description

Over the years, different researchers have tried to develop empirical and theoretical expressions for the temperature suppression of hydrate formation in the presence of various inhibitors. A trend similar to the suppression of the freezing point of water with increasing inhibitor concentration is expected for the temperature suppression of hydrate formation [16].

Yousif and Young's Model

Yousif and Young [17] suggested a simple correlation with a thermodynamic background to evaluate the temperature suppression (ΔT) of hydrate formation in drilling fluids:

$$\Delta T = a_1 x_{in}^3 + a_2 x_{in}^2 + a_3 x_{in} \qquad (1)$$

where, a_1, a_2, and a_3 are constants obtained by using the regression of the available data of hydrate formation temperature in the presence of different inhibitors. In this equation, x_{in} stands for the molar fraction of salts+inhibitor in the aqueous phase. To calculate x_{in} in a mixture containing salts, glycerol, and alcohols, the molecular weights of salts (MW) need to be corrected to obtain the apparent molecular weights (Ma_{im}). This is due to the association of the dissolved salts and the interactions with the glycerol and alcohols [14,15]:

$$Ma_{im} = \frac{MW_i}{\alpha_m (n_i - 1) + 1} \qquad (2)$$

$$\frac{1}{\alpha_m} = \sum_{i=1}^{NS} \frac{1}{\alpha_i} + 1 \qquad (3)$$

$$\alpha_{NaCl} = -3.635028 \times 10^{-5} W^3 + 1.813359 \times 10^{-3} W^2 - 0.044152W + 0.974734 \qquad (4)$$

In the above equations, MW_i is the actual molecular weight of the salt i; n_i is number of ions resulting from the salt i (i.e. n_i=1 for NaCl); α_i and α_m stand for the degree of ionization of salt i and the mixture respectively. W is the weight percent (wt. %) of the salt [16,17]. Mohammadi and Tohidi [16] and stergaard et al. [18,19] investigated the accuracy of this model and presented some hints in the application of this model.

Mohammadi and Tohidi's Model

Applying rigorous thermodynamic equations, Mohammadi and Tohidi obtained a correlation for the estimation of hydrate temperature suppression in a mixture of water and salts+organic inhibitors [16]. This model was tested against independent experimental data, and acceptable consistency was observed. They developed their model by considering the non-ideality of the system caused by both organic inhibitors and salts.

$$\Delta T = -\alpha \left[\ln \left(1 - x_{solute}\right) + \ln \gamma_w \right] \tag{5}$$

where, γ_w is the activity of aqueous phase composed of non-electrolyte and electrostatic contributions:

$$\Delta T = -\alpha \left[\ln \left(1 - x_{solute}\right) + \ln \gamma_w^{Non-Electrolyte} + \ln \gamma_w^{Electrolyte} \right] \tag{6}$$

Two-suffix Margules equation is used for the non-electrolyte contribution caused by the organic inhibitors. Electrolyte contribution caused by the presence of salts could be expressed by Debye-Hückel equation:

$$\ln \gamma_w^{Non-Electrolyte} = bx_{solute}^2 \tag{7}$$

$$\ln \gamma_w^{Electrolyte} = ch_{ws} f\left(B\sqrt{I}\right) = c\left(c_1'W + c_2'W^2 + c_3'W^3\right)f\left(B\sqrt{I}\right) \tag{8}$$

The function $f\left(B\sqrt{I}\right)$ could be eliminated if $B\sqrt{I} < 1$. Hence:

$$\ln \gamma_w^{Electrolyte} = ch_{ws} = c_1W + c_2W^2 + c_3W^3 \tag{9}$$

Substituting Equations 7 and 9 into Equation 6 results in the following equation:

$$\Delta T = -a\left[\ln\left(1 - x_{solute}\right) + bx_{solute}^2 + \left(c_1W + c_2W^2 + c_3W^3\right)\right] \tag{10}$$

where, W represents the weight percent (wt.%) of the salt; a, b, c_1, c_2, and c_3 are model constants obtaining by fitting experimental data.

The parameters of Equations 1 and 10 are determined by minimizing an appropriate objective function. Average relative deviation was selected as an appropriate objective function to obtain the parameters of Equations 1 and 10 by using a non-linear regression method such as simplex search method [20]:

$$ARD = \frac{100}{NDP} \sum_{i=1}^{NDP} \left| \frac{\Delta T_i^{model} - \Delta T_i^{exp.}}{\Delta T_i^{exp.}} \right| \tag{11}$$

RESULTS AND DISCUSSION

The experimentally obtained hydrate temperatures in the presence of different inhibitor compositions are discussed in this section. Moreover, the predicted temperature suppression of hydrate formation of the models is compared with the experimental ones.

Hydrate Formation Temperature

Hydrate formation conditions for a mixture containing 27% nitrogen, 26% methane, and 47% carbon dioxide (on a molar basis) with and without thermodynamic inhibitors were experimentally determined in static and dynamic conditions. The initial conditions, i.e. the total pressure of 150 bar and the temperature of 300 K, were set for all the experiments.

Pure salt (NaCl) and different combinations of salt and methanol (MeOH) or ethylene glycol (EG) were used as the thermodynamic inhibitors. The desired weight percent of the thermodynamic inhibitors in the solution was achieved by the injection of samples as an aqueous solution. The hydrate formation conditions are recognized in two ways: an instantaneous peak in the temperature of system due to the exothermic nature of gas hydrate formation and an increase in pressure drop in the system due to trapping gas molecules in the formed ice crystals. In addition, increasing the gas consumption rate in the dynamic experiments is a clue to gas hydrate formation.

The obtained results for the experiments in static conditions are shown in Figures 2-6. Changes in temperature and pressure versus time for the raw drilling mud (RDM) sample at different concentrations of salt (NaCl) are shown in Figures 2 and 3 respectively. The observed temperature peaks in Figure 2 are the evidence for the gas hydrate formation in the system.

The peaks observed in the temperature are coinciding with the steeper changes in the pressure of system, as shown in Figure 3. As can be seen in these figures, increasing NaCl concentration (as a thermodynamic inhibitor) not only reduces the temperature of gas hydrate formation, but also increases the induction time. A stronger bond comparing to van der Waals forces forms through interactions between the ions of inhibitor and the dipoles of present water molecules. As a result, clusters form around the molecules of the solutes [21]. Moreover, the solubility of guest molecules drops in water due to clustering. Both clustering and reducing solubility of guest molecules make the hydrate formation harder as a lower temperature is required to form hydrate crystals [21]. Figure 4 shows pressure change versus temperature at different concentrations of NaCl. A steeper change in pressure versus temperature is observed at the hydrate formation temperature.

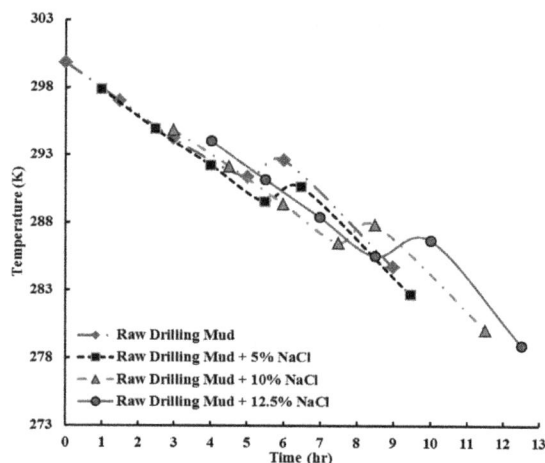

Figure 2: Temperature change versus time for the raw drilling mud sample at different concentrations (wt.%) of salt in static conditions.

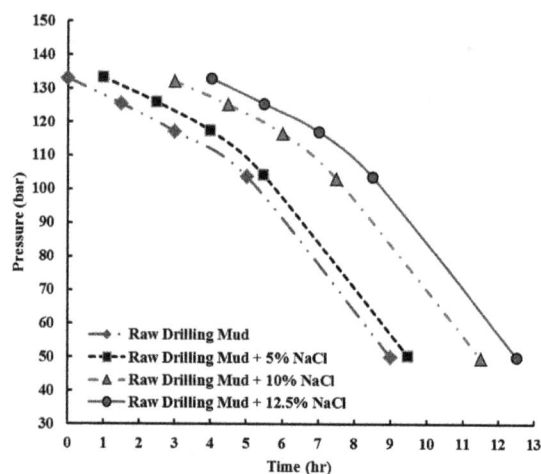

Figure 3: Pressure change versus time for the raw drilling mud sample at different concentrations (wt.%) of salt in static conditions.

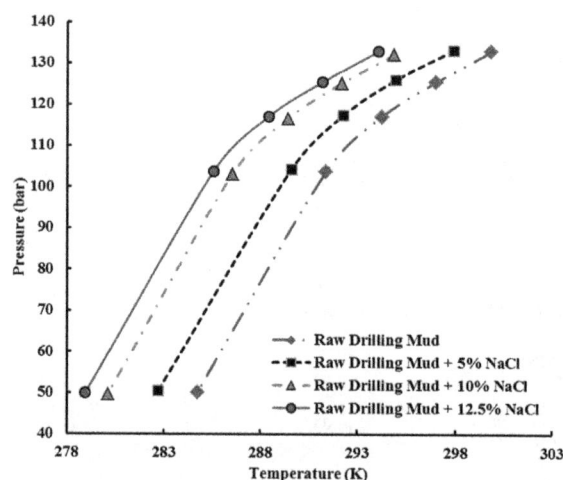

Figure 4: Pressure change versus temperature for the raw drilling mud sample at different concentrations (wt.%) of salt in static conditions.

The history curves of temperature, pressure, and time at different concentrations of NaCl and ethylene glycol (EG) or methanol (MeOH) in static conditions are shown in Figures 5 and 6. Mixtures of 5% NaCl + 5% EG and mixtures of 5% NaCl + 5% MeOH reduce the temperature of gas hydrate formation but increase the induction time. A further reduction in the induction time is observed in the mixture of 5% NaCl + 5% MeOH. Clearly, increasing inhibitor concentrations leads to a lower temperature and a longer induction time in hydrate formation. A considerable reduction in the temperature and an increase in the induction time of hydrate formation are observed for a mixture of 10% NaCl + 10% MeOH. Adding MeOH leads to a further increase in the induction time in comparison with EG. In fact, NaCl and alcohols could disrupt formed hydrogen bonding in ice crystals. As a result, hydrate formation and trapping gas molecules are postponed. Hence, increasing the inhibitor concentration causes the temperature of gas hydrate formation to reduce but increases the induction time.

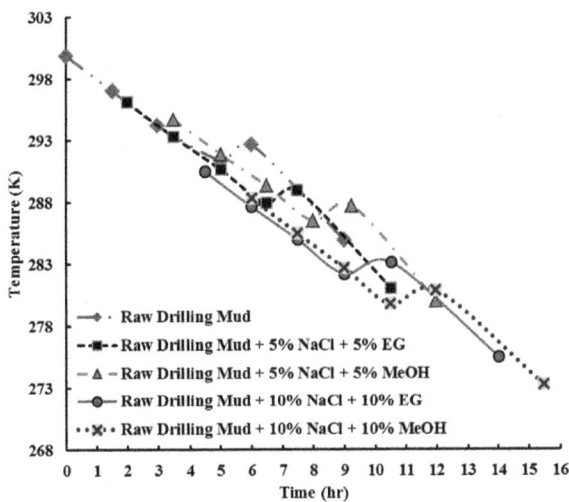

Figure 6: Pressure change versus temperature for the raw drilling mud sample at different concentrations (wt.%) of salt and EG or MeOH in static conditions.

A quantitative comparison of hydrate formation conditions at different concentrations of thermodynamic inhibitors are presented in Tables 2 and 3.

As can be seen, the addition of 5, 10, and 12.5 wt.% of pure NaCl to the RDM sample (Mix. 2, Mix. 3, and Mix. 4) leads to an increase of 0.5, 2.5, and 3.5 hrs in the induction time respectively. Additionally, a decrease of 2.1, 4.7, and 5.8 K in the temperature of hydrate formation is observed with the addition of 5, 10, and 12.5 wt.% of pure NaCl respectively. As previously mentioned, thermodynamic inhibitors reduce the temperature of gas hydrate formation by disrupting hydrogen bonding in water phase, while kinetic inhibitors decrease the hydrate formation rate and consequently increase the induction time. Accordingly, NaCl has both kinetic and thermodynamic inhibition effects. Mixtures of NaCl + MeOH show better influences on the hydrate formation condition in comparison with the mixtures of NaCl + EG. A mixture of 5 wt.% NaCl + 5 wt.% MeOH (Mix. 6) decreases the temperature of hydrate formation by 4.8 K but increases the induction time by 3 hours; however, a mixture of 5

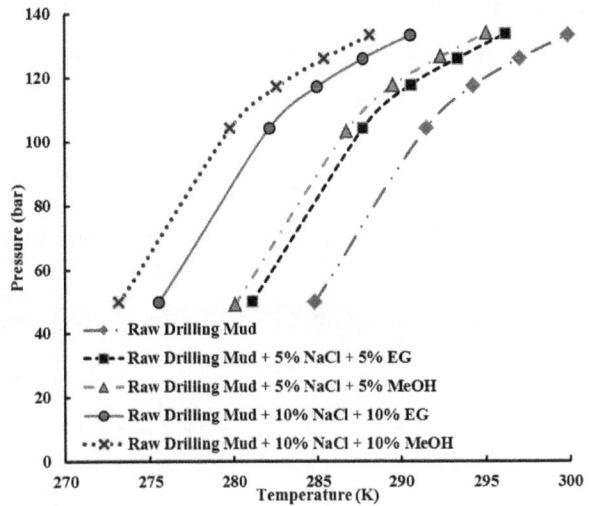

Figure 5: Temperature change versus time for the raw drilling mud sample at different concentrations (wt.%) of salt and EG or MeOH in static conditions.

wt.% NaCl + 5 wt.% EG (Mix. 5) leads to a 3.7 K and 1 hour change respectively in the temperature and induction time of gas hydrate formation. A similar trend is observed for Mix. 7 and Mix. 8 having different concentrations of EG and MeOH.

Table 2: Gas hydrate formation conditions at different inhibitor concentrations in static conditions.

Mixture	Composition (wt%)	Hydrate Formation Temperature (K)[a]	Hydrate Formation Pressure (bar)[b]	Induction Time (hr.)
Mix. 1	RDM (No inhibitor added)	292.6	106.7	5.0
Mix. 2	RDM+5 NaCl	290.5	106.7	5.5
Mix. 3	RDM+10 NaCl	287.9	106.7	7.5
Mix. 4	RDM+12.5 NaCl	286.8	106.7	8.5
Mix. 5	RDM+5 NaCl+5 EG	288.9	106.7	6.5
Mix. 6	RDM+ NaCl+5 MeOH	287.8	106.7	8.0
Mix. 7	RDM+10 NaCl+10 EG	283.2	106.7	9.0
Mix. 8	RDM+10 NaCl+10 MeOH	280.9	106.7	10.5

a: Measured with ± 0.1 K uncertainty

b: Measured with ± 0.7 bar uncertainty

Table 3: Comparing temperature suppression (ΔT) and induction time change (Δt) for gas hydrate formation at different concentrations of inhibitors in static conditions.

Mixture	NaCl (wt.%)	EG (wt.%)	MeOH (wt.%)	ΔT (K)[a]	Δt (hr.)[b]
Mix. 2	5	0	0	2.1	0.5
Mix. 3	10	0	0	4.7	2.5
Mix. 4	12.5	0	0	5.8	3.5
Mix. 5	5	5	0	3.7	1.0
Mix. 6	5	0	5	4.8	3.0
Mix. 7	10	10	0	9.4	4.0
Mix. 8	10	0	10	11.7	5.5

a: Suppression of hydrate formation temperature

b: Change in induction time compared with the raw drilling mud (no inhibitor added)

A comparison between the different mixtures reveals that the best effect in gas hydrate formation inhibition is observed for a mixture of 10 wt.% NaCl + 10 wt.% MeOH (Mix. 8) showing a 11.7 K reduction in the temperature of hydrate formation and a 5.5-hour increase in the induction time. Totally, it is observed that induction times in the presence of NaCl + MeOH would be longer and hydrate formation temperatures would considerably be lower in comparison with the mixtures in the presence of the other inhibitors. The experimentally obtained results for the dynamic system are shown in Figures 7-10. As previously mentioned, an immediate increase in the temperature of system is a criterion for gas hydrate formation. Moreover, a sudden increase in the gas consumption rate due to trapping gas molecules in the formed crystals is evidence for gas hydrate formation in the system. Consequently, gas consumption should be scanned in addition to temperature history curves.

Gas consumption and temperature profiles versus time for the RDM sample and different mixtures of NaCl are shown in Figures 7 and 8 respectively. Hydrate formation in the system could be recognized by peaks in temperature profiles as shown in Figure 7. The observed temperature peaks are coinciding with the higher gas consumption rate as shown in Figure 8. As can be seen in these figures, an increase in NaCl concentration leads to gas hydrate formation at lower temperatures and longer induction times. Accordingly, NaCl can have a remarkable kinetic inhibition effect on gas hydrate formation.

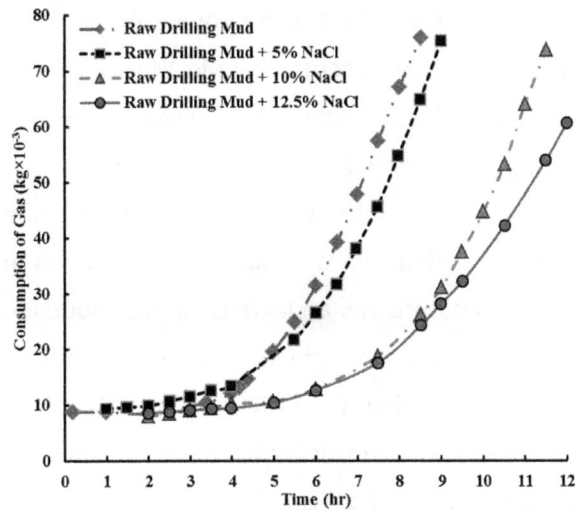

Figure 7: Gas consumption variation against time for the raw drilling mud at different concentrations (wt.%) of salt in dynamic conditions.

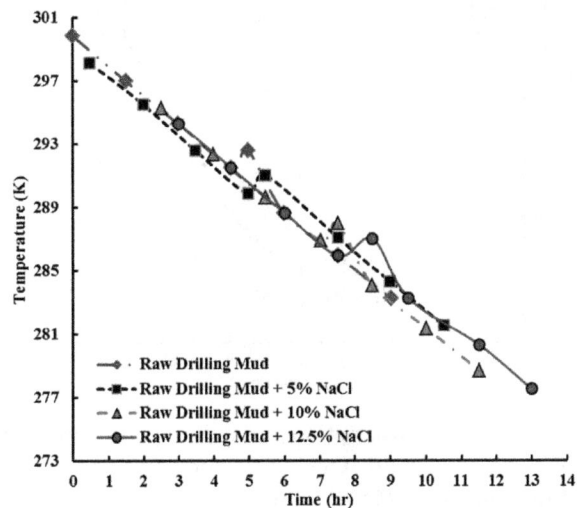

Figure 8: Temperature variation against time for the raw drilling mud at different concentrations (wt.%) of salt in dynamic conditions.

The previous curves of temperature, gas consumption, and time for the different mixtures of NaCl and EG or MeOH in dynamic conditions are shown in Figures 9 and 10. By utilizing Mix. 5 (i.e. a mixture of 5 wt.% NaCl + 5 wt.% EG) and Mix. 6 (i.e. a mixture of 5 wt.% NaCl + 5 wt.% MeOH), reduction in temperature of hydrate formation and increase in the induction time were observed in tandem. . Mixtures containing MeOH show better inhibition effects in comparison with the mixtures having EG. Increasing inhibitor concentrations lead to form gas hydrate at lower temperatures and longer induction times. A significant reduction in the temperature and an increase in the induction time of gas hydrate formation are observed for a mixture containing 10% NaCl + 10% MeOH. Adding MeOH causes more delay in the induction time compared to adding EG.

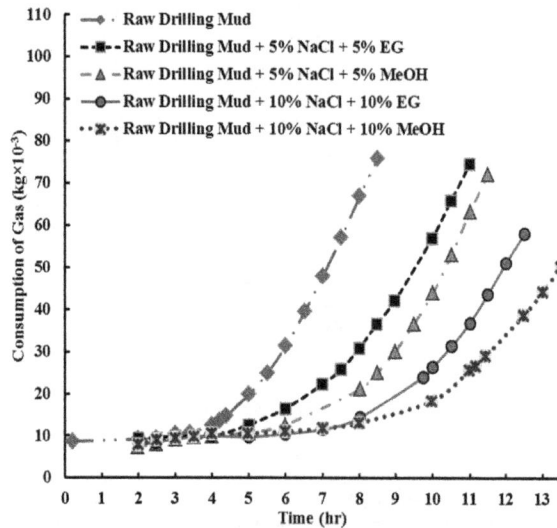

Figure 9: Gas consumption variation against time for the raw drilling mud at different concentrations (wt.%) of salt and EG or MeOH in dynamic conditions.

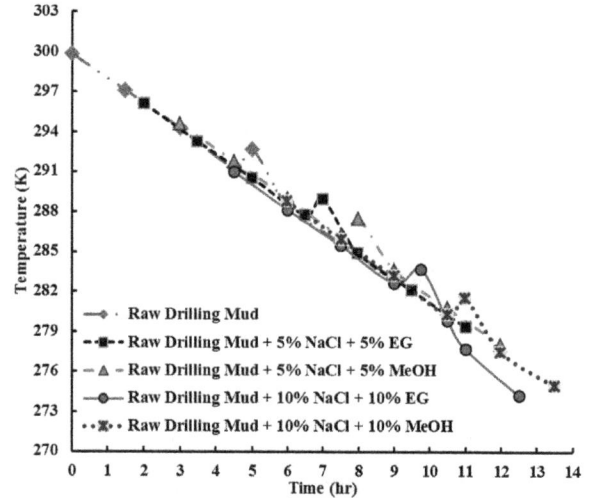

Figure 10: Temperature variation against time for the raw drilling mud at different concentrations (wt.%) of salt and EG or MeOH in dynamic conditions.

Hydrate formation conditions at the different concentrations of thermodynamic inhibitors are quantitatively compared in Tables 4 and 5.

The injection of 5, 10, and 12.5 wt.% of pure NaCl to the RDM sample in dynamic conditions (Mix. 2, Mix. 3, and Mix. 4) leads to a decrease of 1.7, 4.4, and 5.5 K in gas hydrate formation temperature respectively. Nevertheless, an increase of 0.5, 2.5, and 3 hours in the induction time is observed with the addition of 5, 10, and 12.5 wt.% of pure NaCl respectively. Consequently, both thermodynamic and kinetic inhibition effects are enhanced together by increasing NaCl concentration.

Table 4: Summary of gas hydrate formation conditions for different inhibitors in dynamic conditions.

Mixture	Composition (wt%)	Hydrate Formation Temperature (K)[a]	Induction Time (hr.)
Mix. 1	RDM (No inhibitor added)	292.6	5.0
Mix. 2	RDM+5 NaCl	290.9	5.5
Mix. 3	RDM+10 NaCl	288.2	7.5
Mix. 4	RDM+12.5 NaCl	287.1	8.0
Mix. 5	RDM+5 NaCl+5 EG	289.2	6.5
Mix. 6	RDM+5 NaCl+5 MeOH	287.9	7.8
Mix. 7	RDM+10 NaCl+10 EG	283.7	9.0
Mix. 8	RDM+10 NaCl+10 MeOH	281.5	10.5

a: Measured with ± 0.1 K uncertainty

Table 5: Comparing temperature suppression (ΔT) and induction time change (Δt) for gas hydrate formation at different concentrations of inhibitors in dynamic conditions.

Mixture	NaCl (wt.%)	EG (wt.%)	MeOH (wt.%)	ΔT (K)[a]	Δt (hr.)[b]
Mix. 2	5	0	0	1.7	0.5
Mix. 3	10	0	0	4.4	2.5
Mix. 4	12.5	0	0	5.5	3.0
Mix. 5	5	5	0	3.4	1.5
Mix. 6	5	0	5	4.7	2.8
Mix. 7	10	10	0	8.9	4.0
Mix. 8	10	0	10	11.1	5.5

a: Suppression of hydrate formation temperature

b: Change in induction time compared with the raw drilling mud (no inhibitor added)

Same as what has been observed in the static conditions, the addition of MeOH shows better results in the hydrate formation conditions in comparison with the mixtures of EG. As can be seen in Tables 4 and 5 and Figure 11, a mixture of 5 wt.% NaCl + 5 wt.% MeOH (i.e. Mix. 6) leads to 4.7 K reduction in the temperature of gas hydrate formation as well as 2.8 hours increase in the induction time. Moreover, utilizing 5 wt.% NaCl + 5 wt.% EG (i.e. utilizing Mix. 5) causes the gas hydrate temperature and induction time to change 3.4 K and 1.5 hours, respectively. Mix. 7 and Mix. 8 with different concentrations of EG and MeOH show similar trends.

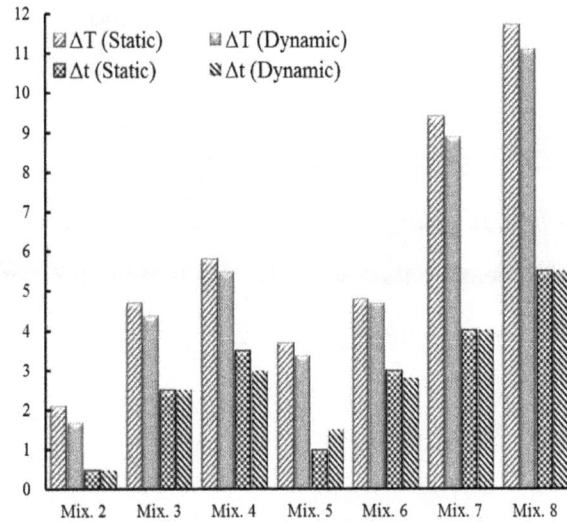

Figure 11: Comparison of changes in temperature (ΔT [K]) and induction time (Δt [hr.]) for gas hydrate formation in dynamic and static conditions.

By comparing different formulations, we could find out that the mixture of 10 wt.% NaCl + 10 wt.% MeOH (Mix. 8) with a 11.1 K reduction in gas hydrate formation temperature and a 5.5-hour increase in the induction time shows the best results in the process of inhibiting gas hydrate formation in drilling mud. The same results were obtained in static conditions.

A quantitative comparison of the obtained results in both static and dynamic conditions is illustrated in Figure 11. The trends observed in the effect of inhibitor concentration and the coincidence of the results in both static and dynamic experiments reveal the accuracy and reliability of the experiments. As can be found out from this figure, in almost all the cases, the time delay in the hydrate formation is lower in dynamic conditions. Such an observation could be explained regarding the turbulence and disturbance effects in dynamic conditions. On the other hand, turbulent conditions in the system lead to more effective contact between the gas and water molecules. Consequently, the transfer of gas molecules from gas phase to the water phase enhances. A similar observation was reported by Lee et al. in the formation of CH_4 hydrate [22]. They observed that by increasing fluid velocity, the temperature of hydrate formation increased.

Hydrate Temperature Suppression

The best-fit parameters of Equations 1 (i.e. Yousif and Young's model) are tabulated in Table 6. As can be seen, this model has a good performance in describing temperature suppressions of both static and dynamic conditions. However, higher accuracy was observed in describing temperature suppressions in dynamic conditions. The experimental results and the results of this model in both static and dynamic conditions are compared in Figure 12. Clearly, the variation of temperature suppression with the molar fraction of inhibitor is well described by this model. Consequently, this model has a capability to describe hydrate temperature suppressions of the drilling mud with a mixture of NaCl and organic inhibitors. As depicted in this figure, lower temperature suppressions are observed in dynamic conditions. In fact, hydrate crystals are formed at higher temperatures in dynamic conditions. Such an observation could be due to turbulence in the system produced by flowing fluid.

Table 6: Obtained parameters of Equation 1

a_1	a_2	a_3	%ARD	R^2
Static Conditions				
-6518.8	1162.4	57.734	5.17	0.994
Dynamic Conditions				
-8300.6	1454.1	41.496	3.44	0.997

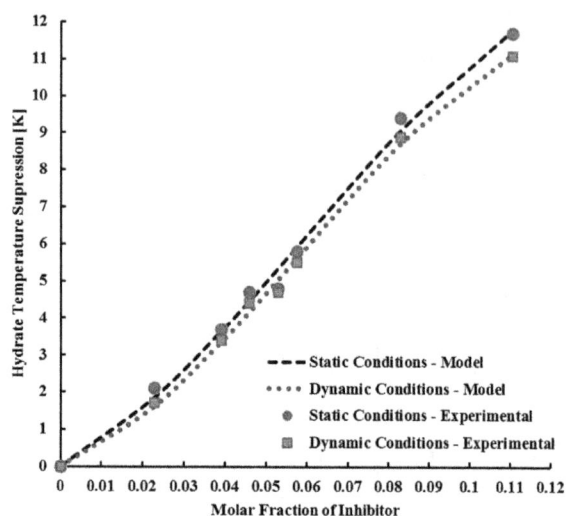

Figure 12: Comparing variation of temperature suppression with the molar fraction of inhibitor in both static and dynamic conditions.

The results obtained from the correlation of the experimental data with Equation 10 (i.e. Mohammadi and Tohidi's model) are tabulated in Table 7. According to the obtained errors, the experimental data are well correlated with this equation, and this model represents the static and dynamic experimental data pretty well. However, it shows a better performance in describing temperature suppressions obtained in dynamic conditions. The qualitative comparison of this model performance is presented in Figures 13 and 14. As can be seen, this model is practical in representing hydrate temperature suppressions of the drilling mud in the presence of NaCl and organic inhibitors.

Table 7: Obtained parameters of Equation 10

Inhibitors	a	$b \times 10^3$	$c_1 \times 10^5$	$c_2 \times 10^4$	$c_3 \times 10^5$	%ARD	R^2
Static Conditions							
NaCl+EG	41.9773	-15.5476	-81.1675	-20.7718	10.0633	5.01	0.996
NaCl+MeOH	63.6624	-22.9385	-44.3219	-13.8872	6.7291	4.72	0.998
Dynamic Conditions							
NaCl+EG	41.3252	120.3738	183.1586	-23.6088	10.8979	4.25	0.997
NaCl+MeOH	62.6185	0.9615	-16.4983	-12.7624	5.8184	2.02	0.999

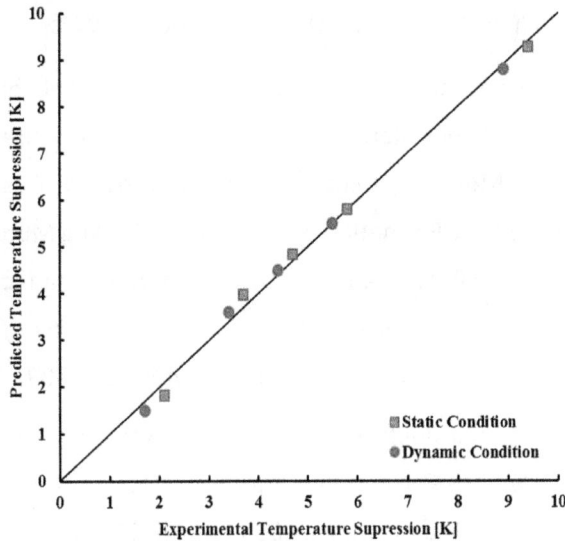

Figure 13: Comparing predictions of Mohammad and Tohidi's model with the experimental data (NaCl+EG data).

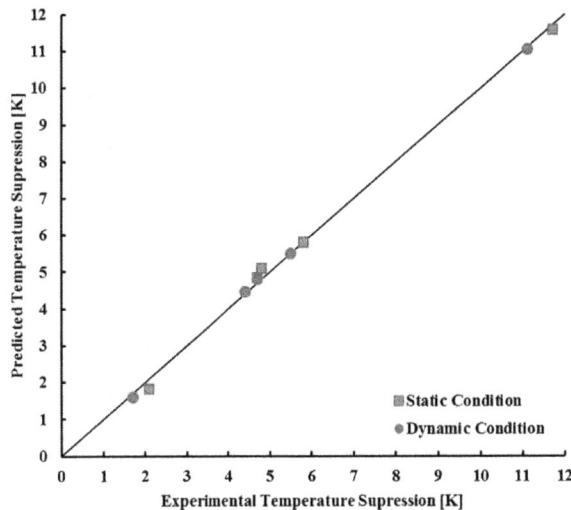

Figure 14: Comparing predictions of Mohammad and Tohidi's model with the experimental data (NaCl+MeOH data).

CONCLUSIONS

The gas hydrate formation for a gaseous mixture containing 26% CH_4, 27% N_2, and 47% CO_2 in a drilling mud sample in the presence of different concentrations of thermodynamic inhibitors such as pure NaCl and mixtures of NaCl with EG or MeOH were experimentally studied. In this regard, an apparatus working in both static and dynamic conditions was introduced. By extracting temperature, pressure, and gas consumption rate from the apparatus, the conditions of gas hydrate formation were determined. A peak in the temperature history curve and a sharp pressure change were considered as evidence for gas hydrate formation in the static experiments, while an increase in gas consumption and a peak in the temperature history curve were considered as clues to gas hydrate formation in dynamic experiments. Followings are the advancements of the present study:

• In static conditions, the hydrates were formed in a drilling mud sample without any thermodynamic inhibitors at a temperature of 292.6 K after 5 hours.

• Increasing pure NaCl concentration up to 12.5 wt.% leads to a drop of 5.8 K and 5.5 K in the hydrate formation temperature in static and dynamic conditions respectively.

At this concentration, a rise of 3.5 and 3 hrs in the induction time was respectively observed in static and dynamic conditions.

• In static conditions, the addition of 10 wt.% NaCl + 10 wt.% EG to the drilling mud causes the hydrate formation temperature to decrease by 3.2% (i.e. 9.4 K) and the induction time to increase by 80% (i.e. 4 hrs).

• In static conditions, the addition of 10 wt.% NaCl + 10 wt.% MeOH to the system suppresses hydrate formation temperature by 4% (i.e. 11.7 K) but raises the induction time by 110% (i.e. 5.5 hrs).

• In dynamic conditions, the effect of inhibitor addition was similar to the one in static conditions with slight suppression of hydrate formation temperature. The observed decrease was due to turbulence in the system.

• Two different correlations with theoretical backgrounds were used to correlate suppression of hydrate formation temperature. Based on the obtained results, both models are capable of representing the temperature suppression of hydrate formation in a drilling mud sample in the presence of NaCl and an organic inhibitor.

Authors believe that more investigations and analyses are required to determine the structure of the formed hydrate clathrates and the fraction of trapped molecules in the crystals of hydrates. Structure I is formed in the presence of small molecules such as CO_2 and CH_4, while the formation of structure II is dominant in the presence of larger molecules such as N_2.

ACKNOWLEDGEMENTS

The authors are grateful to the Shiraz University for supporting this research.

REFERENCES

1. Liyia Ch., Shenga W., and Changwena Y., "Effect of Gas Hydrate Drilling Fluids Using Low Solid Phase Mud System in Plateau Permafrost," *Procedia Engineering*, **2014**, *73*, 318-325.

2. Xiaolan L., Zuohui L., Yong Z., and Rongqiang L., "Gas Hydrate Inhibition of Drilling Fluid Additives," *7th International Conference on Gas Hydrates* (ICGH), United Kingdom, **2011**.

3. Esmaeilzadeh F., Fayazi Y., and Fathikaljahi J., "Experimental Investigation of a Mixture of Methane, Carbon Dioxide & Nitrogen Gas Hydrate Formation in Water-Based Drilling Mud in the Presence or Absence of Thermodynamic Inhibitors," *Proceedings of World Academy of Science, Engineering and Technology*, **2009**.

4. Esmaeilzadeh F., "Simulation Examines Ice, Hydrate Formation in Iran Separator Centers," *Oil & Gas Journal,* **2006**, *104*, 46-52.

5. Esmaeilzadeh F., Zeighami M. E., and Fathi J., "1-D Modeling of Hydrate Decomposition in Porous Media," *Proceedings of World Academy of Science, Engineering and Technology,* **2008**.

6. Sarshar M., Esmaeilzadeh F., and Fathikaljahi J., "Predicting the Induction Time of Hydrate Formation on a Water Droplet," *Oil & Gas Science and Technology*, **2008**, *63*, 657-667.

7. Talaghat M. R., Esmaeilzadeh F., and Fathikaljahi J., "Experimental and Theoretical Investigation of Double Gas Hydrate Formation in the Presence or Absence of Kinetic Inhibitors in a Flow Mini-Loop Apparatus," *Chemical Engineering & Technology,* **2009**, *32*, 805-819.

8. Klauda J. B. and Sandler S. I., "Global Distribution of Methane Hydrate in Ocean

Sediment," *Energy & Fuels*, **2005**, *19*, 459-470.

9. Sun J., Ning F., Wu N., Li S., and et al., "The Effect of Drilling Mud Properties on Shallow Lateral Resistivity Logging of Gas Hydrate Bearing Sediments," *Journal of Petroleum Science and Engineering*, **2015**, *127*, 259-269.

10. Zhao X., Qiu Z., Zhou G., and Huang W., "Synergism of Thermodynamic Hydrate Inhibitors on the Performance of Poly (Pinyl Pyrrolidone) in Deepwater Drilling Fluid," *Journal of Natural Gas Science and Engineering*, **2015**, *23*, 47–54.

11. McConnell D. R., Zhang Z., and Boswell R., "Review of Progress in Evaluating Gas Hydrate Drilling Hazards," *Marine and Petroleum Geology*, **2012**, *34*, 209–223.

12. Lai D. T. and Dzialowski A. K., "Investigation of Natural Gas Hydrates in Various Drilling Fluids," SPE 18637, *SPE/IADC Drilling Conference*, New Orleans, **1989**.

13. Pakulski M., Qu Q., and Pearcy R., "Gulf of Mexico deepwater Well Completion with Hydrate Inhibitors," SPE Paper No. 92971, **2005**.

14. Quigley M. S. and Hubbard J. C., "Gas Hydrates Formation in Drilling Fluids," Drilling Managers Technical Meeting at Mobil Research & Development Corp., **1990**.

15. Kotkoskie T. S., Al-Ubaldi B., Wildeman T. R., "Sloan, E. D. Jr., Inhibition of Gas Hydrates in Water-Based Drilling Muds," SPE Drilling Engineering, Society of Petroleum Engineers, SPE-20437-PA, **1992**.

16. Mohammadi A. H. and Tohidi B., "A Novel Predictive Technique for Estimating the Hydrate Inhibition Effects of Single and Mixed Thermodynamic Inhibitors," *The Canadian Journal of Chemical Engineering*, **2005**, *83*, 951-961.

17. Yousif M. H. and Young D. B., "A Simple Correlation to Predict the Hydrate Point Suppression in Drilling Fluids," SPE/IADC 25705, *SPE/IADC Drilling Conference*, Amsterdam, The Netherlands, **1993**.

18. stergaard K. K., "Gas Hydrate Stability in the Petroleum Industry and its Application in Gas-Liquid Separation," PhD Thesis, Heriot-Watt University, Edinburgh, U.K., **2000**.

19. stergaard K. K., Tohidi B., Danesh A., and Todd A. C., "Gas Hydrates and Offshore Drilling: Predicting the Hydrate Free Zone," *Annals of New York Academy of Sciences*, **2000**.

20. Lagarias J. C. and Reeds J. A., Wright M. H., and Wright P. E., "Convergence Properties of the Nelder-Mead Simplex Method in Low Dimensions," *SIAM Journal of Optimization*, **1998**, *9*, 112-147.

21. Sloan E. D. Jr., "Clathrate Hydrates of Natural Gases," Marcel Dekker Inc.: New York 10016, **1990**.

22. Lee J. H., Baek Y. S., and Sung W. M., "Effect of Flow Velocity and Inhibitor on Formation of Methane Hydrates in High Pressure Pipeline," *Journal of Industrial and Engineering Chemistry*, **2002**, *8*, 493-498.

16

A Study on the Adsorption and Catalytic Oxidation of Asphaltene onto Nanoparticles

Fatemeh Amin and Ali Reza Solaimany Nazar*

Chemical Engineering Department, University of Isfahan, Isfahan, Iran

ABSTRACT

The use of nanoparticles, including metal oxide surfaces, as asphaltene adsorbents is a potential method of removing and/or upgrading asphaltenes. The adsorption of two asphaltene types, extracted from two types of Iranian crude oil, onto nanoparticles (TiO_2, SiO_2, and Al_2O_3) are assessed and the thermal behavior of the adsorbed asphaltenes is examined under an oxidizing atmosphere through thermogravimetric and differential scanning calorimetry (TG/DSC) analyses. The extracted asphaltenes are characterized through the X-ray diffraction technique, and adsorption isotherms are measured through UV-Vis spectrophotometry of the asphaltene-toluene model solutions. The isotherm data of all the nanoparticles are adequately fitted by the Langmuir model, indicating that asphaltenes form monolayer coverage on solids surface sites. The adsorption capacities of asphaltenes onto the metal oxides follow the order of $Al_2O_3 > TiO_2 > SiO_2$. The results indicate that asphaltene with high aromaticity has more adsorption affinity, indicating the effect of the chemical structural of the asphaltenes. The results of asphaltene oxidation tests reveal that the presence of nanoparticles leads to a decrease in oxidation temperature (~100 °C) and activation energy. The effects of nanoparticles on asphaltene oxidation are catalytic.

Keywords: Asphaltene, Nanoparticles, Adsorption Isotherm, Catalytic Oxidation

INTRODUCTION

Crude oil is a very complex mixture containing up to thousands of components. The heaviest fractions of the crude oil are the asphaltenes. Most researchers agree that asphaltenes are a type of polydisperse mixture of molecules containing polynuclear aromatics and aliphatics with small amounts of dispersed heteroatoms like oxygen, sulfur, vanadium, and nitrogen. Asphaltenes are a fraction of the petroleum liquids insoluble in normal alkanes like n-heptane, while soluble in some aromatic solvents like toluene or benzene [1,2]. Asphaltenes are not chemically identifiable compounds, and their composition and structure depend on their source, the method of extraction, and the type of solvents used in extraction [3]. Asphaltene can be adsorbed onto surfaces as colloidal aggregates. Asphaltenes adsorption onto mineral surfaces and reservoir rocks create deposits which affect the oil production rate in a negative sense. The same phenomenon on the upgrading catalyst

*Corresponding author

Ali Reza Solaimany Nazar

Email: asolaimany@eng.ui.ac.ir

surface deactivates and contaminates the catalysts [4]. Researchers have applied various surfaces as a means of asphaltene removal. These surfaces include mineral, metallic, and metal oxides surfaces [5-7]. In recent practices, applying nanoparticles of metal oxide, due to their big specific surface area, fast adsorption kinetic, and appropriate dispersion of small particles, is of major concern [2,8-10]. Different nanoparticle types have different asphaltene capacities. The monolayer adsorption of asphaltene onto nanoparticles surface is reported from the toluene model solutions. Researchers have proposed various hypotheses regarding asphaltenes and surfaces bondage. The introduced hypotheses indicate that the asphaltene aromaticity and nitrogen content in its structure is related to asphaltene adsorption [11,12]. In another work [13], contrary to the findings in literature on asphaltene adsorption from model solutions onto nanoparticles, a model of sequential oxidation of adsorbed asphaltenes and multilayer adsorption of asphaltenes from heavy oil types onto prepared in situ and commercial NiO nanoparticles was proposed. It was concluded that NiO nanoparticles did not have catalytic activity towards the oxidation of adsorbed species and limited their function to better expose the adsorbed asphaltenes to the surrounding environment [14]. Researchers continue to assess the thermal oxidation of asphaltene in the presence of nanoparticles. Nassar et al. used different surfaces of Al_2O_3 [15,16], Fe_3O_4, and NiO [12,17] and reported that the nanoparticles can be applied as catalysts to asphaltene oxidation process. The presence of nanoparticles decreases at the oxidation temperature, and the activation energy of the oxidation reactions drops [16,18]. In another study [2,17], it was claimed that the use of nanoparticles for heavy oil upgrading could lead to a reduction in cost. Furthermore, the use of nanoparticles has the potential of being an environment friendly process in heavy oil recovery. The effects of surface acidity and basicity of three categories of metal oxide nanoparticles with acidic (WO_3 and NiO), amphoteric (Fe_2O_3 and ZrO_2), and basic (MgO and $CaCO_3$) surfaces on the thermodynamics of asphaltene adsorption were studied. The results indicate that the asphaltene adsorption capacity of the nanoparticles decreases in the order of NiO > Fe_2O_3 > WO_3 > MgO > $CaCO_3$ > ZrO_2, and the isotherms of the asphaltene adsorption onto the six metal oxides/salts fit the Langmuir model [19].

Nanoparticles application for heavy oil upgrading is a topic of major concern. Thermal cracking performance of an in situ preparation of alumina nanoparticles in heavy oil was assessed and a general shift towards higher API gravity was observed. It was reported that the performance was in uncertainty probably range due to agglomeration at 350 °C, which limits nanoparticle activity [20].

In this work, the adsorption and catalytic oxidation of two types of Iranian crude oil asphaltenes onto nanoparticles are studied and a model is proposed for the asphaltene kinetic decomposition. Herein, the issue in the context of exploring the potential effect of nanoparticles in heavy oil upgrading through the removal or decomposition of asphaltene is addressed; moreover, the adsorption affinity and the catalytic oxidation of asphaltene over nanoparticles are correlated to the asphaltene structural characteristics.

EXPERIMENTAL PROCEDURES
Materials

Herein, three commercially available nanoparticles of SiO_2, Al_2O_3, and TiO_2 are provided from TECNAN (Spain). The reported particle size and specific surfaces area are tabulated in Table 1. The asphaltenes are extracted from two different crude oil types both from south Iran. The SARA analyses of the crude oil types are tabulated in Table 2. The solvents used in the precipitation and extraction of asphaltene are n-heptane and toluene (analytical grade: Merck, Germany).

Asphaltene Extraction and Model Solution

The residues are extracted from crude oil using atmospheric distillation instrument according to ASTM D86-01. The asphaltenes are extracted from the residues according to ASTM D6560-00. The asphaltenes are dissolved in toluene for the preparation of an 8000 ppm stock solution. The concentration of asphaltene solutions used in the adsorption experiments are within a 100 to 7000 ppm range.

Asphaltene Characterization

Asphaltene characterization is carried out by X-ray diffraction (XRD) analysis. The XRD measurement is made through a Bruker automated diffractometer, applying the CuKα (=1.5406 Å) wavelength. A nickel detector is used to collect the diffraction signals. The diffraction angle (2θ) is scanned from 5° to 45° at a 0.005 (2θ/sec) scan rate with a 2 (sec/step) step size.

Adsorption Experiment

The amount of 0.1 g of nanoparticles is added to the 10 mL model solution of asphaltenes. The samples are shaken in a shaker incubator (NB-205V, N-Biotek instrument Korea, Inc.) at 300 rpm and at 25 °C for about 150 min in order to achieve adsorption equilibrium. The nanoparticles containing asphaltene are centrifuged (Supra 22 K, Hanil instrument Korea, Inc.) for 15 min at 5000 rpm in order to remove the suspended nanoparticles. The separated mixture of asphaltene and toluene is placed in an oven at 50 °C for 20 min in order to have any remained toluene evaporated, and the dried sample is subjected to thermal analysis. The asphaltene concentration in the supernatant is determined by an UV-visible spectrophotometer

(V570, Jasco instrument USA, Inc.). A calibration curve of UV-visible absorbance at ~ 800 nm is obtained using standard model solutions with known asphaltene concentrations. The adsorption of asphaltenes onto the nanoparticles is evaluated by measuring the concentrations of the asphaltene in the solution before and after mixing them with nanoparticles. The amount of the adsorbed asphaltenes represented by q_t (mg/m^2) is calculated through the following equation:

$$q_t = \frac{C_0 - C_t}{A \times m} V \qquad (1)$$

where, V is the sample volume (L), and m represents the mass of the nanoparticles (g); A is the specific surface area (m^2/g), and C_o stands for the initial concentration of asphaltenes in the solution (mg/L); C_t is the concentration of asphaltenes in the solution at a given time, t, (mg/L).

Thermogravimetric and Differential Scanning Calorimetry (TG/DSC) Analysis of Asphaltene

An amount of 5 mg of asphaltene and the same amount of adsorbed asphaltene onto nanoparticles are heated under air atmosphere (TGA/DSC, L81/1750, Linseis instrument, Inc.). The sample mass is kept low in order to avoid diffusion limitations. The air flow rate is set at 100 cm^3/min. The equipment cell is heated up to 1000 °C at a rate of 10 °C/min in order to get a profile of mass loss and heat changes.

RESULTS AND DISCUSSION
Asphaltene Characterization

The XRD patterns for both samples are shown in Figure 1. The volumes of the characterized parameters are obtained from the XRD measurements of the extracted asphaltene samples (see Table 3). All the terms are determined through the available equations in the related literature [1,21]. The aromaticity (f_a) is the content ratio of aromatic hydrocarbons to all hydrocarbons (aromatic and saturated hydrocarbon). The aromaticity is determined by calculating the areas of the peaks for the γ and the graphene bands [22]. The aromaticity of asphaltene type 2 is greater than that of type 1; therefore, the asphaltene type 2 is more polar than type 1. It is expected that the interaction type π-π among aromatic rings yield a higher asphaltene precipitation in the asphaltene with more aromaticity. The factor M indicates the tendency of aromatics layers aggregation, which has an essential contribution to the asphaltene aggregate formation in a solvent. The d_m, d_y, and L_c represent the distance among sequential aromatic layers, aliphatic layers, and the average diameter of the aromatic aggregates respectively.

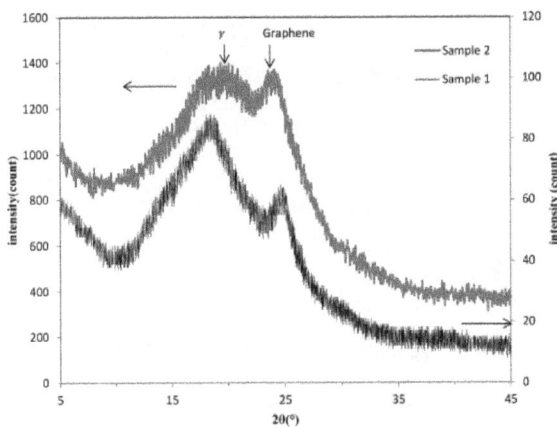

Figure 1: X-ray diffraction patterns of two Iranian crude asphaltenes.

Adsorption Isotherms

The amount of the adsorption of asphaltenes onto nanoparticles is measured at 25 °C by varying the initial concentrations of asphaltene. The adsorption isotherms of asphaltenes onto nanoparticles are shown in Figures 2 and 3. The Langmuir and Freundlich models are adopted to describe the adsorption isotherms through Equations 2 and 3 respectively [23]. The Langmuir isotherm refers to homogeneous adsorption, and the Freundlich isotherm is an empirical expression, which models the multilayer formation of a heterogeneous surface [24].

$$Q_e = Q_m \frac{K_L C_e}{1 + K_L C_e} \qquad (2)$$

$$Q_e = K_F C_e^{1/n} \qquad (3)$$

The linearized form of Langmuir and Freundlich isotherms are presented through Equations 4 and 5 respectively.

$$\frac{C_e}{Q_e} = \frac{1}{Q_m K_L} + \frac{C_e}{Q_m} \qquad (4)$$

$$Log(Q_e) = Log(K_F) + \frac{1}{n} Log(C_e) \qquad (5)$$

where, Q_e is the amount of asphaltenes adsorbed onto the nanoparticles (mg/m^2), and K_L is the Langmuir equilibrium adsorption constant related to the affinity of binding sites (L/mg); Q_m stands for the maximum adsorbed amount of asphaltenes per mass of nanoparticles for complete monolayer coverage (mg/m^2), and K_F is the Freundlich constant related to the adsorption capacity ((mg/m^2) $(L/mg)^{1/n}$); n is the adsorption intensity factor (dimensionless). The model regressed parameters of the results are listed in Table 4. The values of correlation coefficients (R^2) are used to select the best adsorption isotherm model. The R^2 value is valid as a test of the goodness of fit for one model to multiple data sets. It confirms whether the proposed linear model fits one data set better than another. According to the values of R^2, the Langmuir adsorption isotherm is an appropriate model for predicting the adsorption of asphaltenes onto nanoparticles. The Langmuir-type adsorption isotherms indicate the formation of a monolayer of asphaltenes at the toluene-nanoparticles interface. Ranking the values of Q_m of the selected metal oxides follows in the order of $Al_2O_3 > TiO_2 > SiO_2$, and K_L follows the $SiO_2 > Al_2O_3 > TiO_2$ order. Similar observations are reported by other researchers for the adsorption of asphaltenes onto different minerals, metals, and metal oxides nanoparticles [5-7, 18]. Dudasova et al. (2008)

reported that these differences can be attributed to the two physical quantities that reflect two different (but linked) phenomena, namely K_L, which contains information on interactions between the adsorbent and adsorbant (asphaltenes and nanoparticles surface) and Q_m, which describes the interactions and the conformation of asphaltenes at the interface [25].

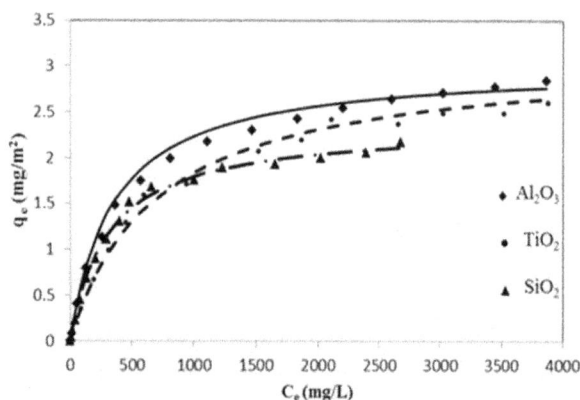

Figure 2: Adsorption isotherm of the first type of asphaltene on different nanoparticles; nanoparticle dose= 10 g/L; agitation speed= 300 rpm; T= 25 °C; the symbols are experimental data, and the solid lines show the Langmuir model.

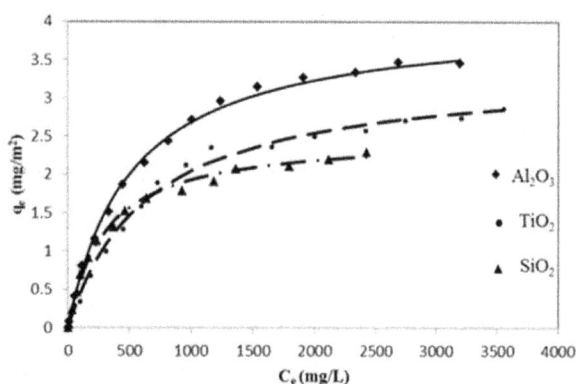

Figure 3: Adsorption isotherm of the second type of asphaltene on different nanoparticles; nanoparticle dose= 10 g/L; agitation speed= 300 rpm; T= 25 °C. The symbols are experimental data, and the solid lines show the Langmuir model.

The adsorption amount of type 2 asphaltene is more than that of type 1, which could be due to higher asphaltene aromaticity in type 2. An increase in asphaltene aromaticity increases the asphaltene polarity; consequently, the polar interactions of asphaltene-asphaltene and asphaltene-nanoparticles

cause asphaltene adsorption. Lopez-Linares et al. (2009) concluded that there exists a correlation between asphaltene adsorption and aromaticity [11]. The asphaltene adsorption behavior from asphaltene model solution to a solid interface is related to bulk aggregate characteristics. The asphaltene molecules tendency in aggregate formation, as revealed by M parameter (Table 3), is higher for the asphaltene type 2 than type 1, which suggests more adsorption capacities of the nanoparticles studied here.

Asphaltene Oxidation

The effect of nanoparticle on asphaltene oxidation is assessed by TG and DSC analyses simultaneously. In order to assess the effect of nanoparticles in the asphaltene oxidation process, the asphaltene type with a higher K_L value (type 2) is selected.

Oxidation Temperature Evaluation

The mass variation of the samples as a function of temperature is shown in Figures 4 and 5. The main change in the mass of pure asphaltenes occurs within a temperature range of 400 to 800 °C, while the oxidation of asphaltenes with nanoparticles occurs within a temperature range of 300 to 600 °C. This finding indicates a decrease in the temperature of oxidation in the presence of nanoparticles.

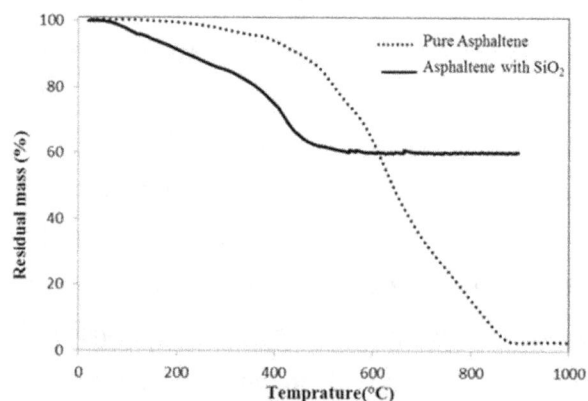

Figure 4: TG curves for the first type of asphaltene with or without SiO$_2$: plot of residual mass as a function of temperature; heating rate= 10 °C/min; air flow= 100 cm³/min.

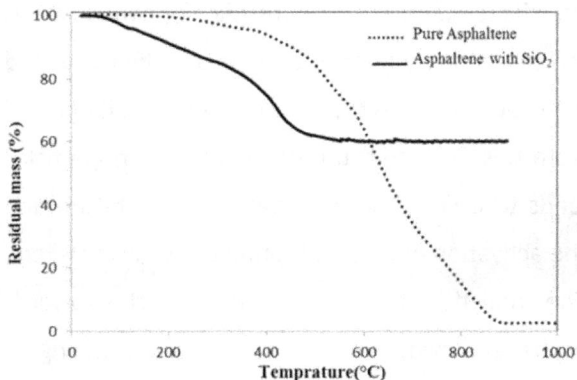

Figure 5: TG curves for the second type of asphaltene with or without nanoparticles; plot of residual mass as a function of temperature; heating rate= 10 °C/min; air flow= 100 cm³/min.

The results of DSC indicate that the thermal properties of asphaltene-adsorbent are affected by incorporating the nanoparticles in the matrix. As shown in Figures 6 and 7, the exothermic reactions for pure asphaltene occur at temperatures close to 600 °C, whereas the adsorbed asphaltene curves of SiO_2, Al_2O_3, and TiO_2 nanoparticles indicate exothermic peaks at 400, 450, and 500 °C respectively, which indicates a decrease in the oxidation temperature of such compounds with respect to these nanoparticles (Figure 7). By comparing the lower specific surface area of Al_2O_3 and TiO_2 with the higher specific surface area of SiO_2, it becomes clear that the essential sites of the adsorption and catalytic transformation of these nanoparticles are less than those of SiO_2; hence, a lesser catalytic oxidation occurs through these nanoparticles.

The TG curves of asphaltenes with nanoparticles show changes in the samples mass close to zero, and their oxidation is completed in temperatures above 550 °C. A change in the enthalpy of the samples above 550 °C is due to the adsorbent, not the asphaltene.

According to the TG and DSC results, regarding the oxidation of the pure asphaltene and the asphaltene adsorbed onto nanoparticles, it becomes evident that the oxidation reactions mechanism in the

presence of nanoparticles leads to major changes in the DSC and TG curves. These changes indicate the nanoparticles catalytic activity. The pure asphaltene oxidation reactions occur homogeneously, while in the presence of nanoparticles it can occur in a heterogeneous manner. These changes in the reaction mechanism lead to the observation of a change in the asphaltene DSC/TG curves with or without nanoparticles.

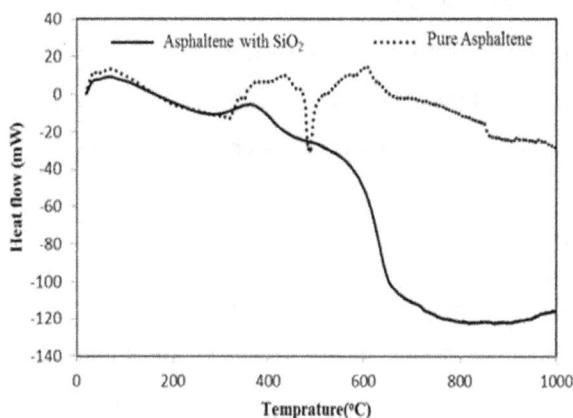

Figure 6: Figure 6: DSC curves for first type of asphaltene with or without SiO_2. Plot of enthalpy change as a function of temperature. Heating rate, 10 °C/min; air flow, 100 cm³/min.

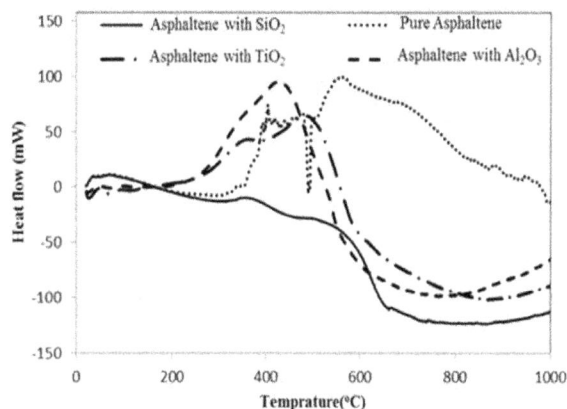

Figure 7: DSC curves for second type of asphaltene with or without nanoparticles. Plot of enthalpy change as a function of temperature. Heating rate, 10 °C/min; air flow, 100 cm³/min

Estimation of Activation Energy

Activation energy can be calculated through processing the thermogravimetric analytic data. All the methods of activation energy estimation are based on the fraction of conversion (α) and the conversion rate ($d\alpha/dt$) [26]. The fraction of conversion is defined by:

$$\alpha = \frac{m_0 - m_t}{m_t - m_\infty} \qquad (6)$$

where, m_0 is the initial sample mass, and m_t is the sample mass at any given time, t; m_∞ is the final sample mass.

The most commonly used equation in describing the conversion rate in the non-isothermal decomposition kinetics is expressed as the product of independent temperature and reaction mechanism functions:

$$\frac{d\alpha}{dt} = k(T) f(\alpha) \qquad (7)$$

Arrhenius equation is considered as a temperature function $(K(T))$ and $f(\alpha)=(1-\alpha)^n$ describing the mechanism of reaction(s), where, n is the order of reaction. According to the Coats-Redfern method [27], by replacing the above definitions in Equation 7 and then by performing an integration, Equations 8 and 9 are yielded for cases $n=1$ and $n>1$ respectively.

$$In \frac{-\ln(1-\alpha)}{T^2} = \ln\left(\frac{AR}{\beta E_a}\right) - \frac{E_a}{RT} \quad n=1 \qquad (8)$$

$$In \frac{1-(1-\alpha)^n}{(1-\alpha)T^2} = \ln\left(\frac{AR}{\beta E_a}\right) - \frac{E_a}{RT} \quad n>1 \qquad (9)$$

where, A is the Arrhenius constant, and T represents the temperature (K); β is the heating rate (dT/dt), and R stands for the ideal gas constant (8.314 J/mol.K); E_a is the activation energy.

A plot of the left-hand-side of Equation 8 (assuming $n=1$) versus $1/T$ represents a straight line with a slope of E_a/R. The fraction of conversion (α) for the second type of asphaltene with and without nanoparticles as a function of temperature is shown in Figure 8. At any

specific temperature, the presence of nanoparticles increases the amount of conversion. The calculated activation energy is tabulated in Table 5. It is inferred from this table that the presence of nanoparticles leads to a decrease in oxidation temperature and the activation energy, indicating the catalytic effect. The catalytic activity of the nanoparticles towards asphaltene oxidation is confirmed by estimating an activation energy lower than the one in the absence of the nanoparticles. The active sites of nanoparticles catalyze the oxidation reaction, thereby lowering the oxidation temperature of the adsorbed asphaltene significantly.

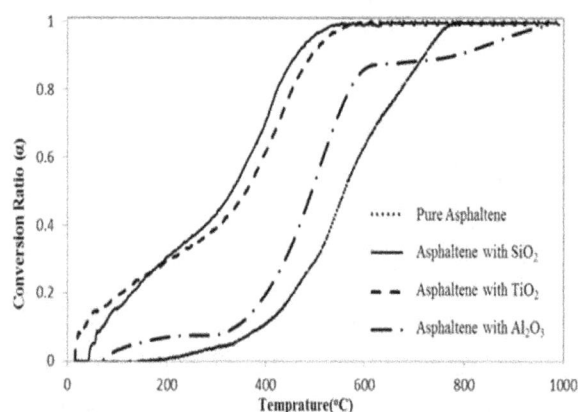

Figure 8: Fraction of conversion (α) for the second type of asphaltene with or without nanoparticles as a function of temperature.

To better understand whether the presence of nanoparticles in the oxidation process could convert the heavy molecules of asphaltene into light weight molecules, more analytical experiments like X-ray photoelectron spectra (XPS) are necessary for further studies.

CONCLUSIONS

The application of TiO_2, Al_2O_3, and SiO_2 nanoparticles in the adsorption and catalytic oxidation processes of asphaltene in asphaltene-toluene model solution is assessed. The adsorption study confirms that the adsorption isotherms of the asphaltene types extracted from two Iranian crude oil types follow the Langmuir models, so the adsorbed asphaltene on nanoparticles surfaces are monolayer. The Al_2O_3 has the highest adsorption capacities in both of the samples. The aromaticity of the asphaltene type 2 is more than that of asphaltene type 1, indicating a higher polar interaction between asphaltene and nanoparticles with a more adsorption affinity for all nanoparticles. The asphaltene adsorption behavior is correlated to the asphaltene structural characteristics. The comparison of the asphaltene oxidation with or without nanoparticles determines the nanoparticles catalytic effects. The exothermic reactions in the pure asphaltene occur at temperatures close to 600 °C, whereas the asphaltene adsorbed on SiO_2, Al_2O_3, and TiO_2 nanoparticles indicates exothermic peaks at 400, 450, and 500 °C respectively. The estimated activation energies of the asphaltene type 2 oxidation on the nanoparticles indicate an increase in the following order: $SiO_2 < TiO_2 < Al_2O_3$. The SiO_2 contributes most to a decrease in the activation energy and the oxidation temperature in comparison with the other nanoparticles used in this study.

REFERENCES

1. Bouhadda Y., Bormann D., Sheu E., Bendedouch D. et al., "Characterization of Algerian Hassi-Messaoud Asphaltene Structure Using Raman Spectrometry and X-ray Diffraction," *Fuel*, **2007**, *86*, 1855-1864.

2. Nassar N. N., "Asphaltene Adsorption onto Alumina Nanoparticles: Kinetics and Thermodynamic Studies," *Energy Fuels*, **2010**, *24*, 4116-4122.

3. Groenzin H. and Mullins O. C., "Molecular Size and Structure of Asphaltenes from Various Sources," *Energy Fuels*, **2000**, *14*, 677-684.

4. Syunyaev R. Z., Balabin R. M., Akhatov I. S., and Safieva J. O., "Adsorption of Petroleum Asphaltenes onto Reservoir Rock Sands Studied by Near-Infrared (NIR) Spectroscopy," *Energy Fuels*, **2009**, *23*, 1230-1236.

5. Alboudwarej H., Pole D., Svrcek W. Y., and Yarranton H. W., "Adsorption of Asphaltenes on Metal," *Ind. Eng. Chem. Res.*, **2005**, *44*, 5585-5592.

6. Marczewski A. W. and Szymula M., "Adsorption of Asphaltenes from Toluene on Mineral Surface," *Colloid Surf. A. Physicochem. Eng. Asp.*, **2002**, *208*, 259-266.

7. Rudrake A., Karan K., and Horton J. H., "A Combined QCM and XPS Investigation of Asphaltene Adsorption on Metal Surfaces," *J. Colloid Interface Sci.*, **2009**, *332*, 22-31.

8. Franco C., Patino E., Benjumea P., Ruiz M. A. et al., "Kinetic and Thermodynamic Equilibrium of Asphaltene Sorption onto Nanoparticles of Nickel Oxide Supported on Nanoparticulated Alumina," *Fuel*, **2013**, *105*, 408-414.

9. Xinhu T. and Dongyang L., "Evaluation of Asphaltene Degradation on Highly Ordered TiO_2 Nanotubular Arrays via Variations in Wettability," *Langmuir*, **2011**, *27*, 1218-1223.

10. Mohammadi M., Akbari M., Fakhroueian Z., Bahramian A. et al., "Inhibition of Asphaltene Precipitation by TiO_2, SiO_2, and ZrO_2 Nanofluids," *Energy Fuels*, **2011**, *25*, 3150-3156.

11. Lopez-Linares F., Carbognani L., Sosa-Stull C., Almao P. P. et al., "Adsorption of Virgin and Visbroken Residue Asphaltenes over Solid Surfaces. 1. Kaolin, Smectite Clay Minerals, and Athabasca Siltstone," *Energy Fuels*, **2009**, *23*, 1901-1908.

12. Nassar N. N., Hassan A., and Almao P. P., "Thermogravimetric Studies on Catalytic Effect of Metal Oxide Nanoparticles on Asphaltene Pyrolysis under Inert Conditions," *J. Therm. Anal. Calorim.*, **2012**, *110*, 1327-1332.

13. Tarboush B. J. A. and Hossein M. M., "Adsorption of Asphaltenes from Heavy Oil onto In Situ Prepared NiO Nanoparticles," *J. Colloid Interface Sci.*, **2012**, *378*, 64-69.

14. Tarboush B. J. A. and Hossein M. M., "Oxidation of Asphaltenes Adsorbed onto NiO Nanoparticles," *Appl. Catal. A.*, **2012**, *445-446*, 166–171.

15. Nassar N. N., Hassan A., and Almao P. P., "Effect of Surface Acidity and Basicity of Alumina on Asphaltene Adsorption and Oxidation," *J. Colloid Interface Sci.*, **2011**, *360*, 233-238.

16. Nassar N. N., Hassan A., and Almao P. P., "Effect of Particle Size on Asphaltene Adsorption and Catalytic Oxidation onto Alumina Nanoparticle," *Energy Fuels*, **2011**, *25*, 3961-3965.

17. Nassar N. N., Hassan A., and Almao P. P.,

"Comparative of Adsorbed Asphaltene onto Transition Metal Oxide Nanoparticle," *Colloid Surf.*, **2011**, *384*, 145-149.

18. Nassar N. N., Hassan A., and Almao P. P., "Metal Oxide Nanoparticles for Asphaltene Adsorption and Oxidation," *Energy Fuels*, **2011**, *25*, 1017-1023.

19. Hosseinpour N., Khodadadi A. A., Bahramian A., and Mortazavi Y., "Asphaltene Adsorption onto Acidic/Basic Metal Oxide Nanoparticles Toward in Situ Upgrading of Reservoir Oils by Nanotechnology," *Langmuir*, **2013**, *29*, 14135–14146.

20. Husein M. M. and Alkhaldi S. J., "In Situ Preparation of Alumina Nanoparticles in Heavy Oil and Their Thermal Cracking Performance," *Energy Fuels*, **2014**, *28*, 6563–6569.

21. Shirokoff J. W., Siddiqui M. N., and Ali M. F., "Characterization of the Structure of Saudi Crude Asphaltenes by X-ray Diffraction," *Energy Fuels*, **1997**, *11*, 561-565.

22. Solaimany Nazar A. R. and Bayandory L., "Investigation of Asphaltene Stability in the Iranian Crude Oils," *Iran. J. Chem. Eng.*, **2005**, *5*, 3-12.

23. Foo K. Y. and Hameed B. H., "Insights into the Modeling of Adsorption Isotherm Systems," *Chem. Eng. J.*, **2010**, *156*, 2-10.

24. Natarajan A., Kuznicki N., Harbottle D., Masliyah J. et al., "Understanding Mechanisms of Asphaltene Adsorption from Organic Solvent on Mica," *Langmuir*, **2014**, *30*, 9370-9377.

25. Dudasova D., Simon S., Hemmingsen P. V., and Sjoblom J., "Study of Asphaltenes Adsorption onto Different Minerals and Clays: Part 1. Experimental Adsorption with UV Depletion Detection," *Colloid Surf.*, **2008**, *317*, 1-9.

26. Nunez L., Fraga F., Nunez M. R., and Villanueva M., "Thermogravimetric Study of the Decomposition Process of the System BADGE," *Polymer*, **2000**, *41*, 4635-4641.

27. Coats A. W. and Redfern J. P., "Kinetic Parameters from Thermogravimetric Data," *Nature*, **1964**, *201*, 68-69.

Permissions

List of Contributors

Wang Wei
School of Civil Engineering and Architecture, Southwest Petroleum University, Chengdu 610500, China

Zhang Jianzhong and Xiao Guohua
Drilling Technology Research Institute of Jidong Oil Field Company, Hebei, 063000, China

Liu Jun
School of Civil Engineering and Architecture, Southwest Petroleum University, Chengdu 610500, China
Modern Design and Simulation Lab for Oil and Gas Equipment, Southwest Petroleum University, Chengdu 610500, China

Ming Li, Shuang Deng, Yujia Yang and Xiaoyang Guo
State Key Laboratory of Oil and Gas Reservoir Geology and Exploitation (Southwest Petroleum University), Southwest Petroleum University, Chengdu, 610500, China

Jianzhou Jin
CNPC Drilling Research Institute, Beijing 102206, PR China

Wanchun Zhao, Dan Zhao and Dongfeng Jiang
Department of Petroleum Engineering, Northeast Petroleum University, China

Tingting Wang
Department of Electrical Engineering and Information, Northeast Petroleum University, China

Maryam Mirshahani, Mohammad Kassaie and Arsalan Zeinalzadeh
Geochemistry Group, Faculty of Upstream Petroleum Industry, Research Institute of Petroleum Industry (RIPI), Tehran, Iran

Ping Jiang, Lei Zhang, Jijiang Ge, Guicai Zhang, Yang Wang and Haihua Pei
College of Petroleum Engineering, China University of Petroleum (East China), Qingdao 266580, China

Yufeng Yuan
Engineering Technology Research Institute of Jiangsu Oilfield, China Petroleum & Chemical Corporation, Yangzhou 225009, China

Hiep Phan and Tan Nguyen
Department of Petroleum Engineering, New Mexico Institute of Mining and Technology, New Mexico, United States

Eissa Al-Safran
Kuwait University, Kuwait

Olav-Magnar Nes and Arild Saasen
Det Norske Oljeselskap ASA, Norway

Socrates Acevedo
Universidad Central de Venezuela, Facultad de Ciencias, Escuela de Química, Caracas 1053, Venezuela

Henry Labrador and Luis Puerta
Universidad de Carabobo, FACYT. Departamento de Química, Lab. Petróleo, Hidrocarburos y Derivados(PHD), Valencia Edo. Carabobo

Brice Bouyssiere
CNRS / UNIV Pau & Pays de l'Adour, Institut des Sciences Analytiques et de Physico-Chimie pour l'Environnement et les Matériaux, LCABIE, UMR 5254, 64053 Pau, France

Hervé Carrier
CNRS/TOTAL/UNIV PAU & PAYS ADOUR, LFCR-IPRA UMR 5150, 64000, PAU, FRANCE, Av. de Université-BP1155-64013. Pau, Cedex France

Yi Xiangyi
Chengdu University of Technology; Chengdu, 610059, Chengdu, China

Ma Yongxin
CNOOC -Ltd Zhanjiang. Zhanjiang, 524057, Zhanjiang, China

Li Mingjun
Chengdu University of Technology; Chengdu, 610059, Chengdu, China
CNOOC -Ltd Zhanjiang. Zhanjiang, 524057, Zhanjiang, China

Ali Reza Solaimany Nazar, Navvab Salehi, Yavar Karimi and Masoud Beheshti
Chemical Engineering Department, University of Isfahan, Isfahan, Iran

Roha Kasra Kermanshahi
Department of Biology, Alzahra University, Tehran, Iran

Javad Honarmand
Petroleum Geology Department, Research Institute of Petroleum Industry (RIPI), Tehran, Iran

Abdolhossein Amini
School of Geology, College of Science, University of Tehran, Iran

Mohammadreza Zare Reisabadi and Seyyed Saeed Ghorashi
Faculty of Research and Development in Upstream Petroleum Industry, Research Institute of Petroleum Industry (RIPI), Tehran, Iran

Vanessa Cristina Santanna, Tereza Neuma Castro Dantas, Allan Martins Neves, Juliana Rocha Dantas Lima and Camila Nóbrega Pessoa
Department of Petroleum Engineering, Federal University of Rio Grande do Norte, Natal, Brazil

Mahdi Rastegarnia
Department of Petrophysics, Pars Petro Zagros Engineering and Services Company, Tehran, Iran

Mehdi Talebpour
Department of Petroleum Engineering, Islamic Azad University, Science and Research Branch, Tehran, Iran

Ali Sanati and Seyed Hassan Hajiabadi
Faculty of Petrochemical and Petroleum Engineering, Hakim Sabzevari University, Sabzevar, Iran

Hossein Zangeneh
Ahwaz Research Center, Research Institute of Petroleum Industry (RIPI), Ahvaz, Iran

Mohammad Amin Safarzadeh
Tehran Energy Consultants Company (TEC), Tehran, Iran

Ali Bakhtyari, Yasin Fayazi, Feridun Esmaeilzadeh and Jamshid Fathikaljahi
Department of Chemical and Petroleum Engineering, School of Engineering, Shiraz University, Shiraz, Iran

Fatemeh Amin and Ali Reza Solaimany Nazar
Chemical Engineering Department, University of Isfahan, Isfahan, Iran

Index

www.ingramcontent.com/pod-product-compliance
Lightning Source LLC
Chambersburg PA
CBHW050437200326
41458CB00014B/4972